Perspectives in Space Surveillance

MIT Lincoln Laboratory Series

Perspectives in Space Surveillance, edited by Ramaswamy Sridharan and Antonio F. Pensa

Perspectives on Defense Systems Analysis: The What, the Why, and the Who, but Mostly the How of Broad Defense Systems Analysis, William P. Delaney

Ultrawideband Phased Array Antenna Technology for Sensing and Communication Systems, Alan J. Fenn and Peter T. Hurst

Decision Making Under Uncertainty: Theory and Application, Mykel J. Kochenderfer

Applied State Estimation and Association, Chaw-Bing Chang and Keh-Ping Dunn

MIT Lincoln Laboratory is a federally funded research and development center that applies advanced technology to problems of national security. The books in the *MIT Lincoln Laboratory Series* cover a broad range of technology areas in which Lincoln Laboratory has made leading contributions. The books listed above and future volumes in the series renew the knowledge-sharing tradition established by the seminal *MIT Radiation Laboratory Series* published between 1947 and 1953.

Perspectives in Space Surveillance

edited by Ramaswamy Sridharan and Antonio F. Pensa

The MIT Press
Cambridge, Massachusetts
London, England

This book was set in Adbode Garamond Pro by Toppan Best-set Premedia Limited.

Library of Congress Cataloging-in-Publication Data

Names: Sridharan, Ramaswamy, editor. | Pensa, Antonio, editor.
Title: Perspectives in space surveillance / edited by Ramaswamy Sridharan and Antonio Pensa.
Description: Cambridge, MA : MIT Press, [2017] | Series: MIT Lincoln laboratory series | Includes bibliographical references and index.
Identifiers: LCCN 2016030739 | ISBN 9780262035873 (hardcover : alk. paper)
ISBN 9780262549950 (paperback)
Subjects: LCSH: Space vehicles--Tracking--Research--United States. | Space debris--Tracking--Research--United States. | Space surveillance--Research--United States--History--20th century. | Lincoln Laboratory--Research.
Classification: LCC TL4030 .P454 2017 | DDC 623/.69--dc23 LC record available at https://lccn.loc.gov/2016030739

Robert Bergemann
1922–2010

In the early 1970s—when the space surveillance program at MIT Lincoln Laboratory began to take shape—Robert Bergemann was the associate leader of the technical group that included the editors and many of the chapter authors of this book. We today look back at that time and think of Bob as the eclectic genius and patient mentor who made all of this possible, through both his unique technical insights and his willingness to entrust and delegate so much responsibility to relative youngsters just beginning their careers. Bob was well aware of the pressing technical challenges facing the country at that time, and he was constantly inventing creative solutions—most of them in pencil on graph paper with figures and calculations. Two of those creative solutions, the use of high-powered coherent radars and large-field-of-view electro-optical systems for operational space surveillance, formed the foundation on which the forty-plus-year MIT Lincoln Laboratory space surveillance effort was built. We dedicate *Perspectives in Space Surveillance* to the memory of Bob Bergemann and all that he contributed to making that effort possible.

Contents

Preface ix

About the Authors xv

Acknowledgments xix

1 **Historical Perspective** 1

Antonio F. Pensa

2 **Technology Development for Space Surveillance with Narrow-Beam Radars** 15

Ramaswamy Sridharan and Israel Kupiec

3 **Overview of Wideband Radar Imaging Technology at MIT Lincoln Laboratory** 75

Craig Solodyna

4 **Ground-Based Electro-Optical Technology Development** 119

Eugene Rork

5 **Technology Development for Space-Based Electro-Optical Deep-Space Surveillance** 179

Jayant Sharma

6 **Technology Developments in Catalog Discovery** 203

Ramaswamy Sridharan and George Zollinger

7 **Characterization of Resident Space Objects** 237

Ramaswamy Sridharan and Richard Lambour

8 Conjunctions and Collisions of Resident Space Objects   **295**
Richard I. Abbot

9 Resulting Technology Applications   **329**
Joseph Scott Stuart, E. M. Gaposchkin, and Robert Bergemann

Afterword   **363**
Index   **365**

Preface

Examples abound in history of revolutionary steps in technology that have altered the nature and speed of human development and have provided enormous economic and social benefits. A relatively recent example was the launch of an artificial satellite, Sputnik 1, in 1957. The immediate reaction in the United States was fear of losing our technology leadership advantage in the world. To encourage and support the pursuit of science and technology, funding for the National Science Foundation was increased significantly, and the National Aeronautics and Space Administration (NASA) was formed. Since then, the launch and deployment of communications, reconnaissance, weather, navigation, scientific, and other satellites have had a phenomenal impact on life across the globe, to the extent that today most industrial and developing economies are dependent on functions provided by these satellites. To understand the impact of that technology over the past six decades, it is important to understand the driving forces behind creation of the space era.

In the early 1950s, the United States and the Soviet Union were locked in a struggle to amass large numbers of nuclear weapons and, more important, the means of delivering these weapons. The intercontinental ballistic missile (ICBM) was the presumed best answer to the delivery question. The race was on to build and deploy large numbers of nuclear-tipped ICBMs. U.S. assessments of the number of deployed Soviet ICBMs varied from few to many times what the United States had. The real answer could only be determined by observation of missile launch areas deep within the Soviet Union. Flights of the high-altitude U-2 aircraft provided some initial intelligence-gathering capability over such areas, but the Soviets quickly reacted by developing and deploying high-altitude interceptor missiles. The shoot-down of Gary Powers's U-2 over the Soviet Union in 1960 marked a low point in the U.S. Cold War defensive effort and provided much of the motivation for moving intelligence collection to space. The Soviet Union had the same intelligence needs and was also actively pursuing a similar space-based intelligence-gathering capability. Both nations would race to adapt the ICBM into a space launch vehicle and to develop devices capable of orbiting the Earth, observing the ground, and reporting back to Earth. The successful launch of Sputnik 1 in 1957 clearly showed that the Soviets had won the first step of the race by adapting an ICBM to successfully place a man-made satellite into orbit around the globe. The United States would react, but an important part of that reaction would be the need to monitor and assess the growing Soviet space capability.

Congress tasked the Air Force with the responsibility for detecting and tracking man-made resident space objects (RSOs) and for monitoring their status. The initial set of sensors deployed to support the Air Force mission consisted of existing radars and a few wide-field-of-view, film-based cameras known as the "Baker-Nunn system." Over the years, as the satellite population and orbital altitudes grew, a distinction between near-Earth and deep-space orbits emerged, a distinction based on the limitations of the original radars pressed into service. The radar single-pulse detection capability was then limited to a range of about 2,000 nautical miles (about 3,700 kilometers) for a typical-size satellite. The following definition of orbital characteristics was later adopted and is still in use: all resident space objects with an orbital period of less than 225 minutes or, equivalently, a semimajor axis of less than about 12,000 kilometers are defined as being in near-Earth space; all other RSOs, as being in deep space.[1]

The distinction between near Earth and deep space, though arbitrary, is critically important to the content of this book. Up until the early 1970s, the United States had been deploying many critical missions into geosynchronous orbits in deep space, whereas the Soviet Union—mainly because of its extreme northern latitudes—had shown little interest in this orbital region. The U.S. detection, tracking, and characterization capability in near Earth was adequate; the Baker-Nunn system, with its manually intensive film development and hand processing, though also adequate for detecting and tracking the few commercial satellites in deep space, would prove totally inadequate to support the near-real-time critical national security needs that emerged when the Soviet Union began to launch satellites into deep space.

The types of sensor technologies and surveillance strategies fundamentally differ for the two orbital regions. Ground-based radars, the principal source for detection and tracking of near-Earth resident space objects, needed to achieve a thousandfold increase in sensitivity to detect and track RSOs at geosynchronous ranges. Radar sensitivity was and still is directly proportional to the amount of power transmitted and the size of the radar antenna reflector. Most of the radars in use were already transmitting megawatts of power and using large reflectors. The ultimate solution for achieving the

---

1. Most near-Earth resident space objects are less than 3,000 kilometers above the Earth, although two near-Earth LAGEOS (Laser Geodynamics Satellites) orbit at altitudes of about 5,700 kilometers; and most deep-space RSOs are more than about 15,000 kilometers above the Earth.

sensitivity gain needed to detect and track geosynchronous satellites would evolve out of the work of several MIT radio astronomers attempting to map the surface of the moon in preparation for the lunar landing. What those radio astronomers discovered and demonstrated was that, if the phase coherence of the transmitted pulse could be maintained, then the sensitivity would increase directly proportional to the number of pulses that could be added together, whereas, if phase coherence was not maintained, the achieved gain would only be proportional to the square root of the number of pulses added.

In 1972, when the Soviet Union launched what was believed to be an important military satellite into a deep-space orbit, MIT Lincoln Laboratory—as a federally funded research and development center (FFRDC) with extensive experience in high-powered radar and systems development—was called upon to apply the coherent integration work to the detection and tracking of the Soviet satellite. The ultimate success of that effort opened an entire area of deep-space surveillance technology whose development at the laboratory this book describes, as it does the laboratory's development both of technologies later applied in near-Earth surveillance and of radar imaging, a key technology for the monitoring and characterization of RSOs.

In 1972, Lincoln Laboratory was uniquely positioned to support the government in addressing the new area of deep-space surveillance. As a laboratory-based FFRDC, it was one of the Department of Defense's sources for rapid development of solutions to emerging problems. Equally important was the fact that, as a result of its previous work for the department, the laboratory was operating two of the world's most powerful radars—the Millstone Hill L-band and Haystack X-band radars—which played a key role in developing and implementing solutions to space surveillance problems.

Today the Air Force Space Command provides a system consisting of ground- and space-based sensors for detection, tracking, and characterization of all resident space objects larger than about 10 centimeters in size; and, in addition, it operates a facility for synthesizing and processing all space surveillance sensor data and distributing the results. The Air Force Space Command continues to deploy new sensors, new operational strategies, and new data-processing and -dissemination technologies as both the population of RSOs and the number of space-faring nations continue to increase.

The chapters of this book describe many of the specific space surveillance technologies and processes that the laboratory developed and applied from the mid-1970s to about the year 2000. Although much has occurred in this area since 2000, this is left for the current generation of researchers to describe at some future time.

Organization of the Book

The major purpose of this book is to document the considerable space surveillance technologies and techniques developed at MIT Lincoln Laboratory in the first thirty years (1970–2000) of the laboratory's efforts, with emphasis on the earlier years, when the seminal work was done.

Chapter 1 describes the state of play in the field of space surveillance, and the challenge that caused Lincoln Laboratory to devote considerable resources to this work. Chapter 2 describes the adaptation of a narrow-beam, narrowband radar, originally devoted to ionospheric research, to achieve deep-space surveillance in response to the challenge.

Two other technologies followed quickly. The development and use of wideband radar technology for characterization and monitoring of resident space objects are described in chapter 3; the development of ground-based electro-optical technology for deep-space surveillance is described in chapter 4. Much later, the electro-optical technology was adapted to work in space, as described in chapter 5.

Search techniques for newly launched deep-space resident space objects are a fundamental requirement for building and maintaining an RSO catalog. Chapter 6 describes how radars—the only reliable means of near-real-time discovery of RSOs—have been used for that purpose and gives specific examples. The chapter also describes how data from a radar on Eglin Air Force Base has been used to address a related discovery problem, the proliferation of space debris from fragmentation events, in which resident space objects, whether by collision or explosion, each generate from a few to a few hundred debris objects.

Characterization of RSOs and monitoring of their status can be achieved through radar imaging, described in chapter 3, optical data analysis, described in chapters 4 and 5, and analysis of signature data from narrowband radars, described in chapter 7, which also describes an interesting application of radar and optical data fusion to characterize spherical sodium-potassium drops in space.

All space-faring countries recognize that space is getting increasingly crowded with satellites and debris, resulting in a small but growing collision hazard. Crowding in geosynchronous orbit space is of particular concern because space-debris objects produced by fragmentation events accumulate in orbit, there being no forces to cause them to decay into the atmosphere. Chapter 8 describes how the in situ failure of a large communications satellite (Telstar 401) in a densely populated region of the geostationary belt and consequent feared collision hazard led to the development of a system to accurately characterize the hazard and provide timely avoidance maneuver information to satellite owners/operators.

Finally, chapter 9 describes two unanticipated spin-off applications of space surveillance technologies—the application of optical surveillance technology, originally developed for detection and monitoring of deep-space RSOs, for detection and monitoring of near-Earth asteroids; and the application of optical data-processing technology for analysis of serendipitously collected infrared data on resident space objects.

Perspectives in Space Surveillance is the product of a cohesive team at MIT Lincoln Laboratory, whose members have been working together for over forty years in the field. The book's editors—Ramaswamy (Sid) Sridharan and Antonio (Tony) Pensa—and a former colleague, Gerry Banner, are principals on that team and original participants in the laboratory's space surveillance efforts. The authors of its individual chapters are laboratory staff with long experience in the topics they describe. The editors deliberately chose a *"Scientific American"* approach for the book, hence there are relatively few mathematical equations. Finally, the editors deemed it impossible to name the vast array of individuals who participated in the laboratory's space surveillance work over the last forty years—the memory of their contributions lies in the technical content of the various chapters and the references cited there.

About the Authors

Richard I. Abbot is a retired staff member from MIT and MIT/Lincoln Laboratory. His research interests include celestial mechanics, astrodynamics, geodesy, and astrometry. He received BA degrees in astronomy and mathematics from the University of South Florida, and a PhD in astronomy from the University of Texas at Austin. As a graduate student, he focused on dynamical resonances in the solar system, although he also worked on NASA-funded projects for lunar and satellite laser ranging and astrometry of the outer planet satellites for the Voyager fly-bys. After graduate school he became a researcher in the Earth, Atmosphere, and Planetary Sciences department at MIT, where he was involved in developing early geodetic applications with the Global Positioning System. From 1989 to 2015 he worked at Lincoln Laboratory on numerous projects regarding orbit determination for artificial satellites.

E. M. Gaposchkin graduated from Tufts Engineering School in 1957 with a bachelor of science degree in electrical engineering and went on to earn a Diploma in numerical analysis from Cambridge University in 1959 and a PhD in geological sciences from Harvard University in 1967. He served as Head of Satellite Geophysics at the Smithsonian Astrophysical Observatory from 1959 to 1982. From 1983 to 1998 he was a Senior Staff member of Lincoln Laboratory, with research specialties in satellite geodesy, celestial mechanics, space physics, and computer science.

Israel Kupiec grew up in Israel. In 1959 he graduated from the Technion-Israel Institute of Technology with a bachelor of science degree in electronic engineering. He worked for two years as an electronic engineer for TADIRAN, an Israeli electronic conglomerate. In 1961 he moved to the United States and spent one semester in Ohio University Electrical Engineering Graduate School. In 1962 he transferred to Polytechnic Institute of Brooklyn, now Polytechnic University, and earned a master of science degree in electrical engineering in 1963. He continued his graduate studies, part time and full time, in the Electrophysics Department and completed his PhD degree in 1969. He continued his research on modeling the interaction of electromagnetic waves with the wake of a reentry vehicle as a post-doctoral fellow for two years. In 1971 Dr. Kupiec joined MIT Lincoln Laboratory, and he has been there for 45 years. In this long period he participated and led many R&D efforts in the areas of ballistic missile defense, radar design and radar data processing, radar imaging, low observables, and space debris detection and modeling.

Richard Lambour received his BS in physics from Clemson University in 1989 and his PhD in space physics and astronomy in 1994 from Rice University. He worked for two years at the Air Force Research Laboratory on terrestrial space weather forecasting science and algorithm development before joining MIT Lincoln Laboratory in 1996. At the Laboratory he has worked as a data analyst and electro-optic systems engineer for multiple space and ground-based sensor programs covering ultraviolet through long-wave infrared wavelength regimes. He works in the Space Systems Analysis and Test Group.

Antonio F. Pensa is currently Assistant Director Emeritus at MIT Lincoln Laboratory. His current activities include participation in both the U.S. national security space and cyber mission areas. Pensa joined MIT Lincoln Laboratory in November 1969 after receiving a PhD degree in Electrical Engineering from the Pennsylvania State University. Pensa was responsible for the development and implementation of the coherent integration tracking, which led to the realization of a U.S. operational deep space radar capability, he was instrumental in establishing the Space Surveillance Program at Lincoln Laboratory, and subsequently was responsible for the Laboratory programs in Space Control. Pensa is a past member of the Air Force Scientific Advisory Board (SAB), the Defense Science Board (DSB) Task Force on Space Superiority, the DSB/SAB Task Force on National Security Space, and the DSB Task Force on Nuclear Treaty Monitoring and Verification. He is a past member of the Intelligence Science Board (ISB) and the U.S. Strategic Command Advisory Group (SAG) space panel.

Eugene Rork (PhD Physics, Ohio State University 1971) joined Lincoln Laboratory in 1975 to work on electro-optical space surveillance. At the Lincoln Experimental Test Site in NM during its first four years, he helped to develop and optimize sensors to detect artificial satellites from reflected sunlight to make a successful GEODSS proof-of-concept demonstration. He then worked there and at Lincoln Laboratory in Lexington, MA, to extend sensor sensitivity to detect faint space object images, to develop a system to indicate location of clear sky, to develop sensors to detect low altitude satellites in clear daytime sky, and to develop a mathematical model to predict sensor sensitivity in both visible and in infrared spectral bands. Now a retired staff member and subcontractor consultant, he continues to work on new sensor technology for both ground and space-based electro-optical space surveillance.

Jayant Sharma is a staff member in the Applied Space Systems group. He has worked on a variety of space-based sensor projects with an emphasis on data processing and exploitation. His other research interests include astrodynamics with an emphasis on orbit determination. He received BS and PhD degrees in aerospace engineering from the University of Texas at Austin and an SM degree in aeronautical and astronautical engineering from MIT.

Craig Solodyna received his BA degree in physics from the University of California Berkeley, and the MS and PhD degrees in physics from MIT. His thesis work concerned magneto-hydrodynamic wave propagation in the interplanetary medium. Upon graduation from MIT, he was awarded a NASA post-doctoral fellowship for two years doing research on Skylab soft X-ray observations of coronal holes, one of which he discovered, and solar flares. He came to MIT Lincoln Laboratory in 1978 and began working in the new field of wideband radar imaging of foreign satellites using the Haystack X-band long-range imaging radar. For over thirty-eight years he has worked on a variety of programs for the National Air and Space Intelligence Center, the U.S. Air Force, the National Intelligence Council, and other organizations. He is an expert in the fields of wideband radar imaging and the analysis of the Space Object Surveillance and Identification (SOSI) of technical assets of foreign nations.

Ramaswamy (aka Sid) Sridharan graduated from VJT Institute in Bombay with a degree in Electrical Engineering in 1962. After two years in industry, he joined Indian Institute of Technology, Bombay, for an MS in rotating machines in the Electrical Engineering Department, which he finished in 1966. He became a graduate student in electrical engineering at Carnegie-Mellon University in 1967 and acquired a PhD in space sciences in 1972. He joined MIT Lincoln laboratory in 1972 and has worked in surveillance of satellites ever since. Most of his publications pertain to this area and are found as references in the book.

Joseph Scott Stuart joined MIT Lincoln Laboratory in 1993after obtaining a bachelor's degree in Computer Science from the University of Pennsylvania. Scott's early days at Lincoln Laboratory included work on sensors for the Earth Observering-1 satellite. In 1997, he became involved with the initiation of a near-Earth asteroid survey called LINEAR. Entranced by the study of asteroids, Scott entered the graduate program in MIT's department of Earth, Atmospheric, and Planetary Sciences with support from Lincoln Laboratory. For his PhD thesis, Scott used data from the LINEAR survey to estimate the number and properties of the near-Earth asteroids. After completing his PhD in 2003, Scott returned to Lincoln laboratory where he has continued to work on LINEAR and other projects.

George Zollinger is the Leader of Information Integration and Decision Support Group. He joined Lincoln Laboratory in 1996. His areas of expertise include electro-optical systems, software development and space control. He received his MS in computer engineering from Case Western Reserve University. His initial work focused on studying breakup events in space and developing automated mission planning software. He has supported a number of ground and space based space surveillance systems including but not limited to MSX/SBV, SBSS, STSS-D, and ORS-5. He is the program manager of the Optical Processing Architecture at Lincoln (OPAL) effort which enables optical processing of space surveillance for many sensors. He was also the program manager of the DARPA Ibex effort. He is the program manager of the Nyx effort, which enables data fusion of key SSA data in support of indications and warning. His current work involves the development of a laboratory Space BMC3 test bed effort that leverages information from Space, Missile Defense and ISR communities.

Acknowledgments

MIT Lincoln Laboratory's unique ethos of encouraging independence and creativity in its entire staff makes it an exciting and challenging place to work. Without this ethos, the work recorded in these pages would not have happened.

The editors thank all the authors of the individual chapters as well as the many unnamed contributors who made our efforts more productive and effective over the years. We are deeply and fervently grateful to chapter author Craig Solodyna for his tireless help in editing, organizing, and preparing the manuscript. We thank Dorothy Ryan of the laboratory's Communications and Community Outreach Office for her talented and diligent copyediting, which certainly enhanced the quality and readability of the book. And we also thank her for overseeing the redrawing of all the graphics to conform to the laboratory standards.

All the authors would like to thank the staff at MIT Press for all their efforts to bring this text into print.

Finally, and most important, the editors and all the chapter authors thank the U.S. Air Force Space Command (USAFSC) and its predecessor organizations for their enlightened and steadfast support of the technology developments recorded in this book. The more than forty-year association of Lincoln Laboratory with the USAFSC has resulted in significant contributions to national security.

1

Historical Perspective

Antonio F. Pensa

1.1 Introduction

Two critical events that occurred in the history of MIT Lincoln Laboratory fundamentally influenced the establishment of the space surveillance mission. The first event occurred on 4 October 1957, when the Soviet Union launched Sputnik 1 as the first man-made satellite to orbit the Earth. As part of the worldwide response to that historic event, the Millstone Hill radar was used to detect the Sputnik 1 satellite. Along with the receivers at Jodrell Bank in England, Millstone Hill became the first radar to actively detect a man-made object in orbit around the Earth. Although the new technique of closed-loop tracking would become a critically important capability of the radar, it had not yet been completed in 1957: Sputnik 1 was detected by manually steering the antenna. By 1958, however, the Millstone Hill radar was fully operational and could both detect and track Sputnik 2, which had been launched on 3 November 1957.

The second critical event was the Soviet Union's launch of Kosmos 520 on 19 September 1972 at the peak of the Cold War, when, despite massive numbers of nuclear weapons on each side, the United States enjoyed a first-strike capability. A space-based system for real-time monitoring of Soviet missile launch sites, successfully deployed and tested and eventually known as the "Defense Support Program" (DSP), provided the United States with sufficient warning to launch missiles in retaliation for a Soviet strike. Whereas the United States could monitor Soviet launch sites in real time, the Soviets had no corresponding capability: their missile launch sites could thus be targeted and their retaliatory force destroyed before it could be deployed. Indeed, the U.S. capability to destroy Soviet launch sites and render them useless for retaliation was an important aspect of East–West relations at that time. The Soviets understood the criticality of this imbalance and were aggressively working to develop a comparable space-based surveillance capability.

In the early 1970s, the U.S. Spacetrack system was still in its infancy. It consisted mainly of radars for launch support and missile warning, along with two new fence radars, the FPS-85 radar at Eglin Air Force Base in Florida and the Navy Space Surveillance Fence, headquartered at Dahlgren, Virginia. All but the FPS-85 radar shared one fatal limitation: they could detect and track satellite-size objects only out to about 2,000 nautical miles (about 3,700 kilometers). Although the detection and tracking range of the FPS-85 could be extended somewhat farther, as a fixed-phased array, its ability to point was limited.

For coverage beyond 2,000 nautical miles, Spacetrack counted on data from the Baker-Nunn camera system [1], whose globally deployed film-based, wide-field-of-view cameras used long time exposures to reveal satellites as streaks against the local star field. By knowing the exposure time, the start points and endpoints of the streaks could be measured with respect to known stars, thus determining the right ascension, declination, and time of the points. Of course, this process only worked on clear nights and was labor intensive: the film needed to be developed and the points of interest (streaks) measured manually with respect to cataloged stars.

Command and control of the Spacetrack system, especially for new launches, was highly dependent on intelligence. Initial indications of satellite launch activity, such as booster rocket type, launch site, and launch azimuth, were used to alert the system. Based on these indications, various military and civilian launch possibilities were postulated and cataloged as "launch folders," each eventually identified by the booster type and launch azimuth and represented by an estimated initial orbit. Over-the-horizon radar and U.S. launch-detection satellites provided an initial estimate of the launch time and azimuth. Detection times and pointing guidance were then passed to downrange radars with coverage, and subsequent detections and tracks used to refine the initial orbit estimate and to officially catalog the object.

For the launch of Kosmos 520, the Spacetrack system functioned as planned. The satellite was detected and tracked on its first orbital revolution, and an initial orbit was determined and cataloged. The initial orbit achieved was, however, not stable and the apogee drifted at a significant rate, rapidly moving beyond the limited coverage of Spacetrack's ground-based radars. Moreover, the Baker-Nunn camera that could have observed Kosmos 520 near apogee to provide the needed orbital correction was out of service for an extended repair. Spacetrack's signals intelligence (SIGINT) collectors thus lacked the ephemeris data they needed to point their narrow-beam collectors; as a consequence, there was no way of determining whether Kosmos 520 was a launch-detection satellite and, if so, its operational status. This very serious consequence needed to be addressed at many levels.

Lincoln Laboratory's pioneering work on high-powered radar systems for use in the Ballistic Missile Early Warning System (BMEWS) was well known throughout the defense and intelligence communities. The laboratory had recently developed an even more powerful radar known as "Haystack," which had been successfully used to map the surface of the Moon for the recent lunar landing. It was therefore not surprising that this federally funded research and development center (FFRDC), with its extensive reputation for work in high-powered radars and complex system development, was asked to use the Haystack radar to search for Kosmos 520 at apogee. The closed-loop tracking capability of the Millstone Hill radar would be used to search for Kosmos 520 at lower orbital altitudes, where the radar's sensitivity would be best matched to a successful detection. Kosmos 520 was detected, tracked, and cataloged by the Millstone Hill radar; and the rest is history.

Although the above events provided the foundation for what has become a major area of research at Lincoln Laboratory, the significant background surrounding each of those events has never been fully described. Section 1.2 is an attempt to tell that story; section 1.6 will cover the significant period between 1972, when Kosmos 520 was detected, and 1975, when space surveillance research was formally established at Lincoln Laboratory.

The space surveillance, now called "space control," mission area, unlike any of the other large mission areas at the laboratory, was started from the bottom up. There were no directors, division heads, or group leaders involved either in creating the intellectual property of the programs or in presenting them to their eventual sponsors. This was all done by technical staff members and support personnel, although with the tolerance and encouragement of great leaders at all levels of management.

1.2 Background

To understand the background surrounding the important space surveillance events described above, one needs to understand the environment that existed in the country from the mid-1950s through the early 1970s. Although many significant events shaped that environment, the one constant was national concern over the growing Soviet nuclear arsenal. In the early 1950s, Lincoln Laboratory was formed around the development of SAGE (Semi-Automatic Ground Environment) at the North American Aerospace Defense Command (NORAD) and the concern over the Soviet ability to deliver nuclear weapons with long-range bombers.

By the mid-1950s, the concern broadened to the possibility that the Soviet Union would develop intercontinental ballistic missiles that could deliver the nuclear weapons to the continental United States. A radar was desperately needed that could look into the Soviet Union and observe rocket and missile testing. Developing such a radar was considered at the time to be a significant technical challenge, hence the decision was made by the Office of the Secretary of Defense (OSD) to have Lincoln Laboratory build a "preprototype"—the Millstone Hill radar—to demonstrate a method of, and reduce the cost of, meeting that challenge. In 1957, Millstone was nearing "full-up" operational status. Sputnik 1 was detected first by a radio telescope at the Jodrell Bank Observatory in England and almost simultaneously by the new radar at Millstone Hill, both first-ever electronic detections of a man-made object in space.

For the United States, the shock of being beaten into space by the Soviet Union raised real concern that the nation was ill prepared in science, mathematics, and technology to meet the national security challenges that the coming space age would present.

The growth of the resident space object (RSO) population following the launch of Sputnik 1 led to the first formalized effort to catalog satellites. Project Spacetrack, later known as the "National Space Surveillance Control Center" (NSSCC), was located at Hanscom Field in Bedford, Massachusetts. More than 150 individual sites around the world were pressed into service to track satellites and send their observations to the NSSCC. These sites included several range-tracking radars, Baker-Nunn cameras, astronomical observatories, radio telescopes, and the telescope sites operated by the amateur astronomers participating in Operation Moonwatch. Each of the sites was assigned a numerical designation to attach to its data. The Millstone Hill radar, designated "sensor 1," was operated in the evenings by an Air Force crew as a contributing sensor to the Spacetrack system until the Department of Defense capability to maintain the growing space catalog could catch up.

In the period between the Sputnik 1 detection and the Kosmos 520 detection, there were three independent ongoing activities at Lincoln Laboratory that greatly affected the U.S. ability to react successfully to the Soviet launch of Kosmos 520.

1.3   Millstone Radar Propagation Studies

Concern over the growing Soviet nuclear missile threat remained foremost in the national consciousness. Missile defense sites using kinetic energy–based interceptor missiles for terminal defense were under serious consideration for protection of the U.S. retaliatory forces against a Soviet first strike. One of the many technical hurdles facing the development of this capability was the aim-point accuracy required by the interceptors. Could the planned L-band radars generate the required accuracy?

This question was the central focus of the Millstone Radar Propagation Studies program. In support of this effort and under the leadership of Dr. John Evans, the Millstone Hill radar was converted from ultrahigh frequency (UHF) to L-band frequency, and a one-of-a-kind dual-band (UHF/L-band) subreflector was designed and installed on the 84-foot dish. The objective of the program was to establish a more precise understanding of the atmospheric propagation effects in relation to the precision tracking requirements of the missile defense radars. The Navy Transit Satellites were simultaneously tracked by L-band radar and by UHF beacon to quantitatively measure the angle of arrival, amplitude, and phase fluctuations of the received signal.

Several by-products of this program proved to be invaluable to the subsequent development of coherent processing. The Millstone Hill radar became the most exquisitely calibrated radar in the world. Measurements from it could be corrected for a number of factors, such as the differing droop of the radar dish as a function of the elevation pointing, the impact of the solar heating on one side of the antenna tower as a function of both the time of the year and the time of the day, and the effect of the prevailing wind on the dish. Understanding the impact of these factors and the ability to correct for them were unheard of for any radar sensor at the time.

The Millstone Hill radar was a one-of-a-kind closed-loop radar tracking system, with a state-of-the-art (for that time) digital computer integrated into the radar loop. Return signals from the target were digitized through a first-of-a-kind analog-to-digital converter; tracking loops were closed digitally through the computer. Using this capability, a team of engineers developed CAST (Computer Aided Satellite Tracking), a substantial software program that could generate in real time residuals between the actual and the nominal ephemeris.

The other by-product of this program was less tangible but ultimately much more important to the future space surveillance effort at Lincoln Laboratory. The engineers and technicians trained at Millstone and dedicated to the successful operation of its radar system laid the groundwork crucial for modifying the radar to implement coherent integration. They built and completely understood the digital side of this very complex radar system. They were responsible for the design and building of both the UHF and L-band transmitters. They custom-built the receiver and timing system, as well as all of the mechanical equipment, including the antenna. They developed the CAST software, which was to become invaluable in supporting initial satellite-tracking efforts.

What was most significant, however, was the dedication and professionalism of this team, and the willingness of team members to do everything they possibly could to shape this radar system into something well beyond its original design. In the initial days of satellite-tracking operations at Millstone Hill, one could always count on their availability to help out at all hours of the day or night. Their identity was completely wrapped up in the radar system and its success.

1.4   Radar Studies of the Moon

On 25 May 1961, President John F. Kennedy announced to the world that we would land on the Moon and successfully return to Earth by the end of the decade. With this announcement, the race to the Moon between the United States and the Soviet Union was established as yet another point of national focus. The Haystack radar, then a narrowband X-band radar, was being used principally for radio astronomy. In support of the planned lunar landing, NASA initiated and sponsored a program at Haystack that used the X-band radar to map the lunar surface for potential landing sites. The Haystack staff developed a dual-polarization coherent integration process that eventually succeeded in creating a surface relief map of the lunar landing sites. Through this process, Haystack was the first radar to demonstrate that coherent gain in the signal-to-noise ratio was directly proportional to the number of pulses integrated, when phase coherence of the pulses was maintained, as opposed to the square root of the number of pulses integrated, the standard noncoherent gain commonly obtained at that time, when phase coherence was not maintained. That the Haystack radar was fully coherent and fully integrated with a digital computer is what permitted this demonstration to succeed, but, unlike in the Millstone Hill radar, Haystack's radar tracking loop was not closed through the computer.

The work at Haystack played a critical role in the beginning of deep-space space surveillance. In 1971, it was widely believed that satellites in geosynchronous orbit could be detected and tracked only by ground-based optical systems. Ground-based radars were limited by the achievable single-pulse signal-to-noise ratio. In fact, at this time, the established radars in the Spacetrack network could not track satellites with orbital altitudes greater than about 1,600 nautical miles (about 3,000 kilometers). Such satellites could only be tracked with Baker-Nunn cameras.

It was in the early 1970s that, owing to the criticality of space-based nuclear command and control, developers began thinking about the survivability of geosynchronous communications satellites. All their analysis was focused on the detectability of satellites by ground-based optical sensors. Ground-based radars were not considered a threat to satellite survivability since they lacked the sensitivity to detect and track satellite-size objects at geosynchronous ranges at that time. Robert Bergemann, an associate group leader at Lincoln Laboratory responsible for evaluating the performance of radar systems, including the analysis of the Soviet radar defenses, held a different view.

It was in this same time period that Bergemann, whose research was principally involved with the analysis of ballistic missile reentry systems, became aware of the research being carried out at Haystack to map the surface of the Moon in support of NASA and the lunar exploration mission. He postulated that an implementation of coherent integration similar to that used to map the lunar surface could be employed to detect a geosynchronous satellite. Why not, he asked, use the Haystack radar as a surrogate for the large Soviet radars just under construction to detect a satellite in geosynchronous orbit?

The technique Bergemann proposed to use was to be tested on the Lincoln Experimental Satellite 6 (LES-6). The reasoning behind the choice was simple: since LES-6 was a laboratory-controlled asset, its ephemeris was known very precisely, which allowed laboratory personnel to drive the antenna along the satellite track. In this way, they could hold the target in the main beam long enough to accumulate enough pulses to integrate up to a detectable signal-to-noise ratio. This detection was a game changer.

1.5   Satellite Observables Program

The successful use of coherent integration to detect LES-6 led to the establishment of an effort known as the "Satellite Observables Program" (SOP). In 1972, a small team of staff from Lincoln Laboratory began regular satellite detection and "ephemeris" tracking with the Haystack radar. Two nights a week, one or more of the Lincoln team members would spend several hours detecting and collecting data on domestic satellites whose ephemerides were well known.

Data collected in this manner revealed a surprising array of satellite features: stability, wobble, and spin were all measurable in the data. In one now famous observation of the final deployment of a Defense Support Program satellite, the release of the lens cover could be observed in the data, and the actual separation velocity between the cover and the spacecraft could be measured.

The Haystack radar measurements convinced decision makers that residence in the geosynchronous belt could no longer be considered a sanctuary for high-valued assets. In fact, the radar signature of a resident geosynchronous satellite became important, and there ensued an effort at the Air Force Space and Missile System Organization (SAMSO) to determine the contribution of physical components to the radar signature. Lincoln Laboratory and the Haystack radar played a fundamental role in this effort.

The Satellite Observables Program (SOP) began in 1971–1972 and was specifically focused on two tasks: radar signature measurements and radar range measurements. The plan was to measure the radar signatures of known U.S. satellites with the Haystack radar and then to use a radar range to collect data on models of the same satellites. The objective was to determine which components contributed to the radar signature of the target and how best to mitigate those contributions.

This program used a ground-based asset such as a radar to determine physical characteristics of resident space objects, especially those in the geosynchronous belt. Up to this time, space object identification (SOI) consisted mainly of observations from near-Earth radars that were used to distinguish tumbling resident space objects and their time-varying radar signatures from stable RSOs and their steady signatures. Rocket bodies and nose cones (i.e., cylinders and conics) could easily be identified from the specular pattern in their radar reflectivity. These data were used to distinguish the rocket body from the payload during initial detection and prior to assignment of object identification numbers.

The Satellite Observables Program was the first program to demonstrate that significant informational content in the radar signature could be used to characterize a satellite. During the SOP, the spin period of each Defense Support Program satellite was measured with greater precision than what was available to the satellite owners and operators. Minute misalignment of a satellite's spin axis with its Earth-aligned axis could also be measured and proved to be critical to satellite mission planning.

The Satellite Observables Program, which continued at a relatively low level throughout most of the 1970s, also observed satellites of the Defense Satellite Communications System (DSCS); the program's measurements revealed partial deployment of one of the two Earth-pointing antennas on the first DSCS satellite and a significant wobble for the second DSCS satellite. In both cases, the quantitative nature of the radar observations turned out to be valuable in the recovery efforts; indeed, it is not clear that these units could have been recovered without the data provided by the Haystack radar.

1.6 Kosmos 520 and 606

The activity just described is generally believed to mark the beginning of what was first known as the "space surveillance mission" and what is today called the "space control mission" at Lincoln Laboratory. Prior to 1972, the Soviet Union had considerable experience in deploying communications satellites into a highly inclined, elliptical, semisynchronous orbit. With much of its landmass in the higher latitudes, this orbit best served the communications needs of its nuclear command and control and military. The orbit, with apogee twice a day, once in the Eastern and once in the Western Hemisphere, has become known as the "Molniya orbit," so named after the Soviet communications satellites to first use it. The nature of the orbit is such that a satellite there spends most of its orbital period at or near apogee with the resulting excellent coverage of the northern latitudes.

Although what we actually knew about Kosmos 520 prior to its launch on 19 September 1972 is somewhat speculative, it is clear that the U.S. intelligence community was anticipating that the Soviets would attempt to deploy a launch-detection system similar to the Defense Support Program system we had deployed. The persistent (continuous) coverage provided by the DSP system could only be duplicated by using a geosynchronous orbit. Because the Soviets had no experience with that orbit at the time, U.S. intelligence believed that they would deploy their initial launch-detection system into the Molniya orbit and achieve persistence by employing multiple satellites.

In 1972, Kosmos 520 was detected by using the standard procedures for initial detection of the Molniya communications satellites, which took place as the satellites passed through the search fence of the UHF phased-array FPS-85 radar at Eglin Air Force Base at relatively low altitude, with subsequent detection by the Baker-Nunn camera at near apogee.

Whereas the Molniya satellites all had a 718-minute, semisynchronous orbital period, Kosmos 520 experienced a faulty apogee burn and subsequently had an orbital period of 715.3 minutes. This difference resulted in an eastward drift of the satellite of about 1.35 degrees per day.

To ascertain the mission of Kosmos 520, U.S. intelligence sensors, waiting to interpret whatever detectable activity emanated from the satellite, needed an accurate ephemeris, normally provided by the Spacetrack system. But because Kosmos 520 drifted far enough to the east to pass beyond the coverage of the Spacetrack ground-based radars, and because the Baker-Nunn camera in eastern Canada that could have recovered the satellite was down for an extended maintenance period, that ephemeris was not provided, and Kosmos 520 was officially considered "lost" by the system.

With a critically important Soviet target officially "lost" all capability both inside and outside the national satellite-tracking network was called upon to assist in finding Kosmos 520. Satellite-tracking measurements coming out of the Satellite Observables Program had been widely circulated throughout the intelligence and defense communities; Lincoln Laboratory was specifically asked to use the Haystack radar to find Kosmos 520.

The Lincoln team, however, recognized the futility of attempting to search for Kosmos 520 with the Haystack "needle beam" radar. The spatial uncertainty area at apogee—the only place with the requisite spatial stability to support coherent integration—was too large to be covered by the very narrow Haystack beam. Two teams were created. One would use the Haystack radar in the Satellite Observables Program mode to search at apogee, and the other would use the Millstone Hill radar to search at the horizon near perigee. The expectation was that if the target was acquired during a horizon scan, the Millstone Hill radar would be able to switch rapidly into a closed-loop tracking mode and follow it out to some unknown range. The tracking data collected in this manner could be used to calculate a new ephemeris.

Although the Millstone Hill radar is generally credited with reacquiring Kosmos 520, that is not the whole story. A group of observers using an astronomical telescope at an observatory in Cloudcroft, New Mexico, briefly detected what they believed was Kosmos 520. Though not sufficient to establish a new ephemeris, the group's data were sufficient to provide a temporal offset from the last official ephemeris. Using that temporal offset, the Millstone team was able to center the horizon search scan on 16 June 1973, and again on 22 June 1973, and in both cases succeeded in detecting and in closed-loop tracking Kosmos 520. With tracking data from two orbits separated by six days, a new ephemeris could be calculated, one accurate enough to support a Haystack radar ephemeris scan. The radar was used extensively throughout July and August to track the satellite and provide even more precise ephemeris offsets to support the continuous operation of the U.S. intelligence sensors.

Later that fall, the Soviets were preparing to launch the next satellite in the Kosmos series—Kosmos 606. As pleased as the U.S. intelligence community was, especially the Defense Special Missile and Astronautics Center (DEFSMAC), with the support that Lincoln Laboratory provided with reacquiring and tracking Kosmos 520, it was not surprising that the same team was called on for the early acquisition and tracking of Kosmos 606.

The laboratory told the DEFSMAC that the best way to accomplish this would be to have the Millstone Hill radar acquire the satellite on its initial orbital revolution. The plan was simple: the DEFSMAC would provide a prelaunch initial orbit prediction that lacked only the precise launch time for completeness. An arrangement was made for the DEFSMAC to notify staff at the Millstone Hill radar directly to support acquisition at the estimated horizon break at 40 minutes postlaunch.

On 2 November 1973, Kosmos 606 was launched at approximately 0500 EST. Everything worked as planned, and the satellite was acquired by the Millstone Hill radar at 10 degrees elevation and 130 degrees azimuth. These measurements were a few degrees in angle off the nominal prediction provided by the DEFSMAC. The Millstone Hill radar track lasted for about 20 minutes, at which time the target was at a range of about 6,500 nautical miles (about 12,000 kilometers) and the receive signals were below the detectable threshold.

The Millstone Hill radar was not only the first, but also the only sensor to acquire Kosmos 606 on its initial orbit. An interesting side note is that the Kosmos 606 payload, because of the early detection, was assigned a lower catalog number than the rocket body, which remained in low Earth orbit (LEO) and was easily detectable by most of the Spacetrack network sensors. The early acquisition and report allowed for immediate telemetry acquisition by the intelligence community sensors.

Although the Millstone Hill radar continued to acquire and track Kosmos 606 for three more days (3, 4, and 5 November 1973), after the final orbit adjustments were made to increase the orbital period from 710 to 718 minutes, the ephemeris once again began to deteriorate. Millstone reacquired the target on 9 November 1973; this time, rather than a straight scan search and closed-loop track, it used an ephemeris-driven track and employed the Computer Aided Satellite Tracking program to establish the residuals between the nominal and observed actual orbits—both significant milestones in the development of deep-space satellite tracking. Successfully establishing a set of offsets for the Kosmos 606 ephemeris provided the inspiration and motivation to incorporate coherent integration into the CAST program. The resulting software, called "Satellite Acquisition and Tracking Using Coherent Integration Techniques" (SATCIT), has remained the fundamental software for deep space-tracking radars to this day.

Using extreme, for that time, software techniques, a young programmer named Tom Clark managed to accumulate sixty-four coherent return signals in each of three channels, sum, azimuth error, and elevation error, and to complete the required three Fourier transforms in a single interpulse period. This would be an impressive piece of software engineering in almost any environment, but considering that this was 1973 and Clark was using computer technology that was already significantly outdated, it was truly an amazing piece of work.

With SATCIT running on the radar, tracking was no longer limited to ranges that provided a detectable single-pulse signal-to-noise ratio. For several months following the initial success of SATCIT, the focus was on increasing the number of pulses that could be integrated in a single interpulse period. Results quickly progressed from 64 to 128 and then to 256 pulses. All of this was accomplished by clever manipulation of machine language code in what was then already a digital mainframe relic.

Coherent integration of 256 pulses provided sufficient sensitivity to track many satellites out to geosynchronous range. By employing coherent integration, the Millstone Hill radar became the first-ever deep-space autonomous tracking radar—able to track a target, however acquired, by moving the range gate and the antenna (in azimuth and elevation) according to the coherently integrated signals in the range, azimuth, and elevation error channels.

For the next two years, the Millstone Hill radar was operated as an acquisition aid for the Haystack radar, which had proven to have a formidable capability for anomaly resolution. The laboratory was planning for the future launch of communications satellites LES-8 and LES-9; having the Millstone Hill radar for acquisition and Haystack radar data for launch anomaly resolution was the rationale for keeping the Millstone Hill radar in operation during this period.

Maintaining the catalog of geosynchronous and other deep-space satellites was largely the responsibility of the Baker-Nunn camera system in 1973. It was the failure of this system that led to the crisis surrounding the launch of Kosmos 520. Although members of the Lincoln Laboratory Kosmos 520 acquisition team felt strongly that the Millstone radar should be part of the official Air Force Spacetrack component to take full advantage of the considerable Millstone capability in tracking future events, the Air Force had to be convinced that Millstone's operational processes were up to the standards they expected from an operational sensor.

A formal proposal to establish the Millstone Hill radar as a contributing sensor in the Air Force's Spacetrack system was presented to Major General Lee Paschal, commander of the 14th Aerospace Force. Paschal agreed to the proposal but only under the condition that the radar could pass a rigorous test of all its standard operational processes. Two formal one-week test periods were set up. A small group of Air Force officers traveled to Millstone from Colorado Springs. Each morning, they monitored and timed the effort to bring the radar "up" from cold storage. Once it was up and fully calibrated, they would present the operator with a list of satellites to be tracked that day. At the end of the day, all of the radar's tracking data would be transmitted to the Space Control Center (SCC) in the Cheyenne Mountain Complex. There the data would be processed and compared to what were believed to be the correct data. The operational day would end with a telephone call from the SCC to the Millstone Hill radar control room acknowledging receipt of the data.

The young Air Force officers who came to Millstone for the initial tests formed career-long bonds with Lincoln Laboratory and its technical staff. Many of them went on to achieve significant rank; in the ensuing years, when much of the U.S. space surveillance technology and policy was established, they were the biggest proponents of the technology and positions advanced by Lincoln Laboratory.

At the conclusion of the two test periods, the Air Force declared the Millstone Hill radar fully capable and worthy of serving as a contributing sensor to the U.S. space surveillance network. The initial contract between Lincoln Laboratory and the Air Force Aerospace Defense Command was for 40 hours per week of operations. Subsequently, the Millstone Hill radar went on to become the most prolific of all of the contributing sensors. Millstone calibration procedures were eventually established as the foundation for calibration of all the space surveillance radars, and Millstone's coherent integration technology formed the basis for deep-space surveillance at both the ALTAIR UHF radar on Kwajalein Atoll in the Marshall Islands and the FPS-79 radar in Turkey. Together, these three radars formed the deep-space radar network and provided all of the radar tracking of the growing geosynchronous satellite population.

More than just a source of data on deep-space satellites, the Millstone Hill radar became the national center for expertise on these satellites and how to process data from them to generate an equivalent to the LEO catalog. Almost overnight, the radar and its associated staff were recognized as the "world experts in deep space." Hardly any launches from the Soviet Union that were expected to go into deep space took place without one of the Millstone principal staff serving as "quarterback" for the U.S. space surveillance response.

Many of the technologies and processes used throughout the U.S. space surveillance system were developed at Lincoln Laboratory and transitioned to government and industry for application in facilities other than those operated by the laboratory. The transition was first accomplished case by case, but in the early 1980s, it became apparent that a more formalized process was needed. The Space Surveillance Workshop, now known as the "Space Control Conference," held yearly at Lincoln Laboratory was established as a means of sharing capabilities developed by the laboratory with the space surveillance community. The conference quickly evolved into the premier forum for technical presentations from a large number of government-related and private organizations on space surveillance, space situational awareness, and space control; to this day, it remains one of the most important gatherings of the space control research and development community.

Reference

1. Henize, K. G. (January 1957). The Baker-Nunn Satellite-Tracking Camera. *Sky and Telescope* 16:108.

2

Technology Development for Space Surveillance with Narrow-Beam Radars

Ramaswamy Sridharan and Israel Kupiec

In the late 1960s and early 1970s, satellite tracking was separated into two distinct orbit regimes:

1. Below 3,000 kilometers: Satellites and other man-made resident space objects (RSOs) larger than 1 square meter in radar cross section (RCS) were tracked by high-powered microwave radars out to about 3,000 kilometers, the maximum range at which the received signal-to-noise ratio (SNR) permitted reliable target detection.

2. Above 3,000 kilometers: Satellites in orbit beyond 3,000 kilometers were generally tracked optically by using reflected sunlight with a film-based Baker-Nunn camera system.

In the early 1970s, MIT Lincoln Laboratory demonstrated the first-ever use of a high-powered microwave radar to actively track a man-made satellite at ranges beyond the single-pulse signal-to-noise-ratio detection limit. The laboratory went on to develop a substantial part of the initial radar and electro-optical technology for surveillance of RSOs in the deep-space regime [1]. Section 2.1 addresses the development of deep-space surveillance technology for narrow-field-of-view radars (the contemporaneous development of electro-optical technology is described in chapter 4).

Substantial numbers of launches into space on the part of many nations combined with inadvertent or occasionally deliberate fragmentation of RSOs have resulted in a proliferation in space of long-lived debris objects and the consequent increasing probability of collisions between them [2]. The problem is particularly acute in low Earth orbit (LEO) and is nearing a point of cascading collisions among large numbers of small debris objects (smaller than 10 centimeters in characteristic size) [2]. These objects are undetectable by the radars normally used in LEO surveillance. An innovative application of the highly sensitive, narrow-beam Haystack radar has enabled the National Aeronautics and Space Administration (NASA) to develop a model of debris distribution in space. Section 2.2 describes the data-collection effort supporting the development of this model. Scientists in the Orbital Debris Program Office at NASA's Johnson Space Center analyzed the data from the Haystack radar; figures presenting the results of their analyses are drawn from publications by the office in the decades before 2000.

2.1   Deep-Space Surveillance with Narrow-Field-of-View Radars

The technology developments that enabled detection and tracking of deep-space resident space objects with high-power, narrow-beam (small-field-of-view), narrowband radars were largely algorithmic and were applied to existing radars whose hardware systems continually evolved (the search for and acquisition of deep-space RSOs using such radars is addressed in chapter 5 as part of the discussion of catalog discovery). Made possible by the presence of a high-speed digital computer in the radar tracking loop, these algorithmic developments included the following:

1. Use of the fast Fourier transform (FFT) in real time for the coherent integration and combination of sequentially received pulses in amplitude and phase.

2. SATCIT (Satellite Tracking Using Coherent Integration Techniques), the radar control software program of the first prototype system for tracking deep-space resident space objects at long ranges, capable of combining multiple returned pulses using the fast Fourier transform to achieve substantial gain in the signal-to-noise ratio. SATCIT was also extended to noncoherently combining multiple returned pulses from attitude-unstable RSOs, again using the medium of the FFT. All this processing was simultaneously carried out on several overlapping and sequential range gates as well as on the associated angle-error gates (in azimuth and elevation).

3. Advanced algorithmic techniques for the detection and tracking of deep-space RSOs and for the near-real-time metric and radiometric calibration of radars in deep-space tracking mode.

4. Algorithmic optimal combination of the signal data received in both principal and orthogonal polarizations to enhance the signal-to-noise ratio in detection and tracking and, as a by-product, improve the accuracy of the processed result.

5. Algorithmic enhancement of search capability with along-orbit and across-orbit search, local search about a given pointing, time-biased search, orbit element search, and so on.

6. Algorithmic real-time processing and display of radar signature data to support near-real-time assessment of the status of RSOs, particularly active satellites.

7. Postprocessing algorithms for archived pulse-by-pulse signal data to enable the understanding and comparison of radar signature features on RSOs.

Items 1, 2, and 4 were key new developments at Lincoln Laboratory. All the others were specific algorithmic applications of available techniques (generally found in most books on radars)[1] in the deep-space tracking context.

In the opinion of Lincoln Laboratory analysts, the national command and control system for space surveillance—then at Cheyenne Mountain Complex in Colorado Springs—was not adequately prepared in the early 1970s to support the needs of sensors tracking satellites in deep space.[2] This perceived shortcoming led to the development of a comprehensive, interconnected system of processes (the Comprehensive Millstone Hill Radar System) consisting of a suite of near-real-time and non-real-time software tools that enabled the laboratory to demonstrate, in prototype, a functioning enterprise for deep-space surveillance, akin to the system for low-altitude surveillance in use at Cheyenne Mountain.

Most important, the overall effort at Lincoln Laboratory helped to build a team of radar engineers, radar data analysts, and software developers with a thorough understanding of deep-space surveillance problems; they had both the freedom and the ability to develop and implement innovative solutions to these problems. It is important to note that all the algorithmic developments mentioned earlier were transferred to other radars such as the Army/Lincoln Laboratory radars on the Kwajalein Atoll in the Marshall Islands, and the Air Force/General Electric radar at the Pirinçlik Air Base, Turkey. This transfer also applied (in somewhat abbreviated form) to the Air Force FPS-85 radar on Eglin Air Force Base in Florida and to the Air Force/Raytheon Globus-2 radar in Norway. (Similar technology transfer in electro-optical deep-space surveillance is described in chapter 4.)

---

1. See, for example, M.I. Skolnik, *Radar Handbook* (New York: McGraw-Hill, 1970).
2. This conclusion is based on extensive personal interactions between laboratory analysts and analysts at the Cheyenne Mountain Complex.

2.1.1   Development of the Millstone Hill Radar as a Deep-Space Detection and Tracking Sensor

The Millstone Hill radar was built by the laboratory in the mid-1950s as a "preprototype" for the intelligence community's missile tracking radar that was eventually deployed in Turkey to monitor and collect data on Soviet missile tests. The radar pushed the boundaries of technology in many pertinent areas, including high-powered microwave tubes, low-noise receivers, high-precision timing, and precise antenna control. In 1957, just as the radar's development was nearing completion, the Soviet Union launched Sputnik 1. Lincoln Laboratory engineers used the Millstone Hill radar to achieve the first active track of a man-made object in low Earth orbit. This success ushered in a new round of development of high-powered radars to track and collect position data on what was to become a growing population of resident space objects. Since RSOs needed to be detected at a range of about 2,000 kilometers near the horizon (as opposed to a range of 200 kilometers for aircraft detection), a factor of 10,000—10^4 or 40 dB (decibels)[3]—increase in detection sensitivity was needed.

Beginning in the early 1970s, resident space objects in high-eccentricity and geosynchronous orbits were recognized by the intelligence community as being militarily significant. As is well known, the Soviet Union and the United States deployed satellites in high-eccentricity orbits with perigees in the Southern Hemisphere, apogees in the Northern Hemisphere, and orbital periods of about 12 hours. These satellites could observe and communicate with large areas of the Northern Hemisphere over a large part of each orbit; their orbital period enabled them to synchronize with the Earth's rotation and thus travel over the same location at the same time every day. Geosynchronous satellites, on the other hand, were nearly stationary over the equator, with a continuous view of approximately two-thirds of the hemisphere in both the east–west and north–south directions. Satellites in these orbits could not be surveilled in a tactical time frame with just the fair-weather, nighttime capability provided by optical systems. High-powered radars with day and night, all-weather surveillance capability were needed, as was a further increase in sensitivity of 40–50 dB to extend

---

3. Narrow-beam radar transmits electromagnetic energy in a particular direction. The intensity of the energy reaching a target resident space object decreases by the square of the distance to the target. The backscattered energy intensity decreases by a further factor of the square of the distance by the time it returns to the detection receivers in the radar antenna. Hence, when comparing a target distance of 200 kilometers (airplane) to 2,000 kilometers (resident space object), there is a diminution in returned energy intensity of $(2,000/200)^4$, that is, 10^4 or 40 dB. A similar calculation applies when extending the range to 36,000 kilometers.

the detection range from 2,000 to about 36,000 kilometers (geosynchronous range). Because, however, enhancing the single-pulse radar sensitivity by increasing either the transmitted power or the size of the radar reflector was impractical with available technology, new technology was needed as well.

In the mid-1960s, experiments at the laboratory used the newly developed Haystack radar, collocated with the Millstone Hill radar in Westford, Massachusetts, to detect coherent energy reflected from the surface of the Moon and Venus [3, 4]. These experiments demonstrated that, if the location and velocity of a target relative to the radar were known, conventional high-powered, narrow-beam radars (like the Haystack radar or the Millstone Hill radar) could be used to detect that target well beyond the radar's single-pulse detection limit, given the following:

1. The frequency and initial phase of the pulses of transmitted energy could be kept invariant over a long observational period (defining coherent radar).

2. A receiver system could be built both to detect the backscattered signal from the target over a set of transmit pulses and to convert the signal into in-phase and quadrature orthogonal components in digital form.

3. A detection-processing system using the Fourier transform method was available to concentrate the energy from the set of pulses into a narrow window in the frequency domain, thus enhancing the signal-to-noise ratio for detection.

The specific implementation of the fast Fourier transform on the Haystack digital computer that supported lunar surface detection was based on the FFT algorithm developed at Bell Laboratories by James Cooley and John Tukey [5].

2.1.1.1 Demonstration of Detection of a Deep-Space Resident Space Object

In 1970, a group of Lincoln Laboratory experimenters under the direction of Robert Bergemann used lunar surface scattering-measurement techniques [3] to detect a resident space object in deep-space orbit. The Lincoln Experimental Satellite 6 (LES-6) in geosynchronous orbit was chosen as the initial target since its ephemeris was being maintained in house at that time. The position and velocity of the satellite were well known and available to point the Haystack antenna and position the receiver range gate during the data-collection interval. The radar was operated in a pulsed mode, and several series of radar pulses backscattered by the satellite were collected and processed via the fast Fourier transform detection processor.

Apart, of course, from their size, when it comes to detection, deep-space resident space objects and planets differ in two major respects: RSOs move rapidly in orbit as compared to planets, and, unlike most planets, RSOs may also be unstable in attitude—a typical RSO tends to tumble at a rate of every few seconds to every few minutes. The rapid orbital motion of an RSO has to be compensated for by predicting its instantaneous, time-varying position in four dimensions relative to the radar (azimuth, elevation, range, and Doppler) and by moving the radar receiver's electronic window in all four dimensions. It is worth noting here that the long range to the deep-space RSO (and consequent long round-trip delay of a radar pulse) implies that several pulses of electromagnetic energy are usually transmitted before the radar receiver has to be reset to capture the energy backscattered by the RSO from the first pulse. This offset is endemic to RSO detection and tracking in deep space. As an example, the Millstone Hill radar operates at about 30 pulses per second, each pulse being 1 millisecond long, with an interpulse period of about 33.3 milliseconds. The round-trip delay to a geosynchronous resident space object is about 250 milliseconds, or between 7 and 8 pulses.

With perfect compensation of target motion in all four dimensions, an attitude-stable RSO will yield a fast Fourier transform output represented as a single spike in the frequency domain. The peak of the spike will have a signal-to-noise ratio equal to NS, where N is the number of pulses coherently integrated in the FFT process and S is the SNR for a single pulse.

Figure 2.1a, adapted from Guernsey and Slade [6], illustrates the result of the attempted detection of LES-6 on 15 September 1971. Essentially, the transmit frequency and the beginning phase of the sequence of transmitted pulses are precisely controlled to be the same. The backscattered signal from an attitude-stable resident space object is then in phase from pulse to pulse, which enables a coherent integration of the pulses with the resultant signal-to-noise ratio increasing linearly with the number of pulses integrated, whereas the signal's frequency changes depending on the rate of change of the range to the RSO (the Doppler effect), which can be corrected for in processing implemented by the fast Fourier transform algorithm. If the Doppler value is not correctly predicted, then either the linear or the quadratic offset of the returned signal in frequency from pulse to pulse can be corrected in processing.

Figure 2.1b, also adapted from Guernsey and Slade [6], demonstrates the first measurement of the radar signature of a geosynchronous satellite and illustrates the periodicity of the energy in the processed returns attributed to the rotational period of LES-6. Like all cylindrical satellites in orbit, LES-6 was stabilized in attitude by spin around the cylinder's axis, which was nearly parallel to the Earth's rotational north–south axis.

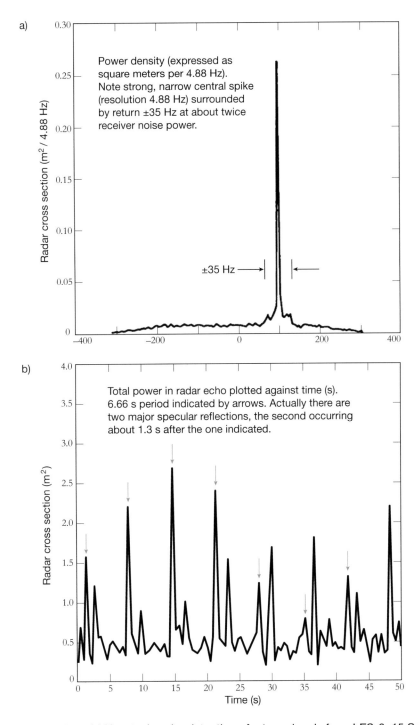

Figure 2.1 (a) Haystack radar detection of return signals from LES-6, 15 September 1971. (b) Processed radar signature from Haystack radar detection of LES-6, 15 September 1971.

2.1.1.2 Deep-Space Satellite Tracking

Although the detection of LES-6 by the Haystack radar demonstrated that coherent gain could be achieved on a man-made resident space object in a deep-space orbit, the method used was to collect return signals from a large number of pulses and then process these offline, an approach that would not support tracking an RSO in real time. In fact, the success of the experiment was predicated on the prior knowledge of the target position and motion with respect to the radar. Satellite "tracking" is really a continuum of acquisition, closed-loop compensation for target motion (known as "closed-loop tracking"), and data collection to establish an accurate orbital element set and to permit analysis of the radar signature to determine satellite characteristics.

Deep-space satellite tracking began in 1975 at the Millstone Hill radar. As a by-product of a 1960s tracking and calibration program, the radar was instrumented to support real-time digital closed-loop tracking. In addition to a 1970 vintage digital computer embedded in the tracking loop, high-speed analog-to-digital converters were available in all the in-phase (azimuth/elevation error) and quadrature orthogonal (principal/orthogonal polarization) channels.[4] The net result was that fast Fourier transform–based coherent processing could—with appropriate computer optimization—be applied in all of these channels, thus implementing real-time closed-loop tracking on multiple coherently integrated pulses.

The detection of LES-6 by the Haystack radar was based on a fully coherent target for which all of the received (signal) energy was concentrated in a single fast Fourier transform bin.[5] Figure 2.2, from an FFT of Millstone Hill radar data, shows that nearly perfect coherent integration is achieved on a geosynchronous target at 39,699 kilometers slant range with an integration of 6×256 pulses (over about 60 seconds) at a pulse-repetition frequency of about 24 Hz.

Many, if not most, of the resident space objects in orbit are uncontrolled in attitude because they are defunct satellites, rocket bodies, or fragmentation debris. Hence the Doppler value of various parts of a typical RSO with respect to the radar line of sight varies from pulse to pulse, which causes the backscattered radar energy to be noncoherent from pulse to pulse. Figures 2.3a and 2.3b illustrate the processing of a

---

4. A conventional radar always transmits energy that is polarized either linearly, with the electromagnetic vector being in one plane along the line connecting the radar to the target, or circularly, with the electromagnetic vector rotating in a plane perpendicular to that line. Energy backscattered by the target has two components, the principal and the orthogonal. In the case of linear polarization, these are in two perpendicular planes, whereas, in the case of circular polarization, the vectors are contrarotating.

5. In fact, due to the spin of LES-6 and its surface structure, the energy was spread over several bins, as seen in figure 2.1a, but most of the energy was either in one or in a very few bins.

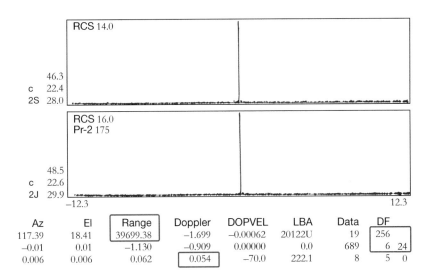

Az	El	Range	Doppler	DOPVEL	LBA	Data	DF	
117.39	18.41	39699.38	−1.699	−0.00062	20122U	19	256	
−0.01	0.01	−1.130	−0.909	0.00000	0.0	689	6	24
0.006	0.006	0.062	0.054	−70.0	222.1	8	5	0

Figure 2.2 Near-perfect coherent integration of 6 × 256 pulses at about 24 Hz pulse-repetition frequency (marked by rectangle) at the Millstone Hill radar on a geosynchronous target (range: 39,699 kilometers; Doppler spread: 0.054 Hz, marked by rectangles). The x-axis is Doppler offset from estimated Doppler frequency, and the y-axis is signal strength in dB. Two channels—a polarimetric and a sum channel—are shown, each covering 150 meters in slant range. The estimated accuracies of the metric data are shown in the bottom line: 6 millidegrees (0.006) in azimuth and elevation, 62 meters (0.062) in range, and 0.054 Hz in Doppler spread, marked by rectangles. Estimated radar cross section of the target is also shown as "RCS."

"noncoherent" target in such a case. Essentially, the attitude of the RSO is changing with respect to the radar in an unpredictable fashion, and thus the backscattered signal is noncoherent (in phase) from pulse to pulse and has to be summed up across all the fast Fourier transform bins in a range gate (the FFT being used as a matter of convenience). Detection of a target is achieved by comparing the total signal in the range gate where the target is expected to the average of the signal in the range gates where neither the target nor the ionosphere[6] is present. Note that in figure 2.3a the target return signal is spread across the entire spectrum, whereas in figure 2.3b it is apparently distributed over 70 percent of the spectrum. Because the spread may vary, however, the entire FFT spectrum is processed for the signal in both cases.

---

6. The ionosphere, which extends from about 90 to about 200 kilometers, reflects radar energy noncoherently and could be mistaken for a target if not blanked out of the radar receiver.

a)

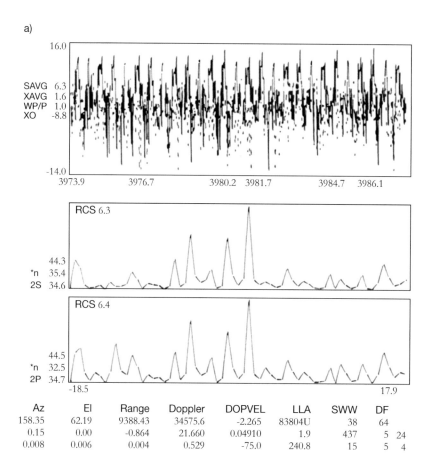

Figure 2.3 (a) Real-time fast Fourier transform output on a tumbling rocket body. The top panel shows the temporal radar signature of the radar cross section of the target during a track, namely, the periodicity caused by the tumble of the object; the middle panel shows the "sum" or range channel in which the target is resident during the noncoherent integration; and the bottom panel shows the result of polarimetric processing. (b) Noncoherent integration on another rocket body at a range of 23,444 kilometers illustrating the spread of the target return signal over only a part of the spectral width of the fast Fourier transform. The top panel shows the temporal radar signature during the track of the target, a clear periodicity caused by the attitude motion or tumble of the target; the middle panel shows the significant energy of the orthogonal polarization channel; and the bottom panel shows the FFT output for the "sum" or principal polarization channel, being the range gate within which the target is confined.

b)

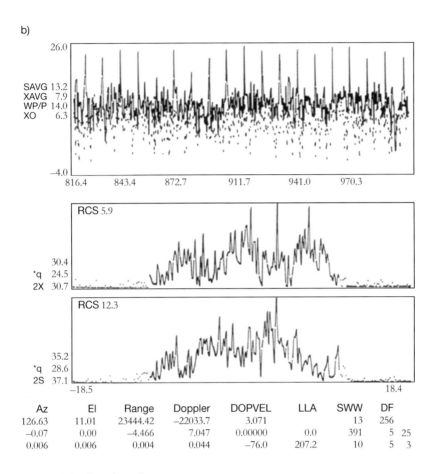

Az	El	Range	Doppler	DOPVEL	LLA	SWW	DF	
126.63	11.01	23444.42	−22033.7	3.071		13	256	
−0.07	0.00	−4.466	7.047	0.00000	0.0	391	5	25
0.006	0.006	0.004	0.044	−76.0	207.2	10	5	3

Figure 2.3   (continued)

2.1.1.2.1   Best-Approximation Algorithm   A key development at the Millstone Hill radar was a best-approximation algorithm for estimating radar receiver signal amplitude (related to the radar cross section of the resident space object) and delay (related to the slant range to the RSO), described by Richard C. Raup and colleagues [7]:

> Low single-pulse SNRs and fluctuating target cross sections are encountered when ground-based radars track high-earth-orbit [deep-space] satellites. This condition requires that the usual filter bank radar receiver architecture be generalized. A generalization can be obtained by estimating

signal amplitude and delay in narrowband radar from the viewpoint of best approximation in a Euclidean space. The best-approximation approach is general enough to address the problem of direct estimates of the cross section and range of the [deep-space] satellite. …

The best-approximation approach leads to a way of thinking about radar receiver design that does more than emulate the analog filter-bank architectures used for many years. With appropriate restrictions, the best-approximation approach generates estimates of target cross section and [target] range identical to those obtained from a filter-bank receiver. Thus the algorithm can be interpreted as a generalization of the matched filter bank. For example, if constant unknown signal amplitude is assumed, then a matched-filter-bank approach for the [deep-space] satellite tracking problem can be constructed with generalized filter elements based on higher-order delay models. The banks can become large because the number of elements in the bank grows exponentially with the order of the delay model.

By using the algorithm proposed in this article, only the outputs of a subset of the generalized filter elements of the bank need to be constructed, and only portions of these filter outputs need to be sampled in time. This partial construction and sampling is essentially what happens whenever the linear subproblem is solved. The results of each sampling are combined with previous samplings by the numerical optimization process to predict which filter elements should be constructed and sampled next. The optimization process refines an estimate of the filter parameters that are best matched to the radar receiver signal, and the time when their outputs will maximize. This procedure is equivalent to finding the best estimate of signal amplitude and delay. (pp. 323–324)

The Millstone Hill radar has successfully used an implementation of the algorithm described in [7] for more than two decades.

2.1.1.2.2 Polarimetric Processing   Another significant development pioneered by one of the analysts at the laboratory in the early 1980s and applied to the Millstone Hill radar was the optimal use of polarization data to enhance the signal-to-noise ratio for detection of deep-space resident space objects [8, 9]. Briefly, the radar transmits circularly polarized radiation (clockwise per consensus) and receives both the principal (counterclockwise) and orthogonal (clockwise) components of the backscattered radiation from the deep-space RSO. The key theoretical development was a set of algorithms for combining the receiver data in both polarizations so as to enhance the SNR for detection and tracking and for radar signature analysis. The polarization ratio for simple scatterers is shown in figure 2.4.

Figure 2.5 graphically summarizes the twelve models developed for simultaneously processing received signals at both polarizations. These models were then implemented in real time in SATCIT, and the best model (defined as the one with the highest signal-to-noise ratio) was used in detection and tracking.

A performance analysis of the model detectors in figure 2.5 is shown in figure 2.6, which demonstrates that a significant gain of up to 3 dB can be achieved by using the sum and the polarimetric receivers rather than the receiver with the maximum target amplitude. Equivalently, a probability of detection of 90 percent is achieved at a 2–3 dB lower signal-to-noise ratio per pulse when using the sum and the polarimetric receivers. For deep-space targets, a 3 dB improvement in SNR is equivalent to halving the number of pulses (and consequently, halving the time) that needs to be integrated for detection or tracking. This improvement in detection and tracking time has been amply demonstrated at the Millstone Hill radar in its operational deep-space tracking role for more than 30 percent of resident space objects in all deep-space orbits [9].

The gain in signal-to-noise ratio due to polarimetric processing also applies to detecting and tracking resident space objects in low Earth orbit. But because most of

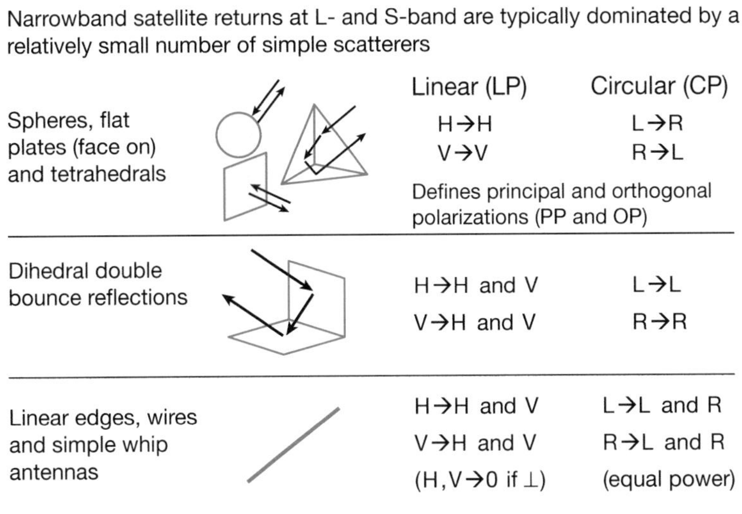

Narrowband satellite returns at L- and S-band are typically dominated by a relatively small number of simple scatterers

		Linear (LP)	Circular (CP)
Spheres, flat plates (face on) and tetrahedrals		H→H V→V	L→R R→L
		Defines principal and orthogonal polarizations (PP and OP)	
Dihedral double bounce reflections		H→H and V V→H and V	L→L R→R
Linear edges, wires and simple whip antennas		H→H and V V→H and V (H,V→0 if ⊥)	L→L and R R→L and R (equal power)

Figure 2.4 Polarization properties of simple scatterers. CP, circular polarization; LP, linear polarization; L, left; R, right; H, horizontal; V, vertical.

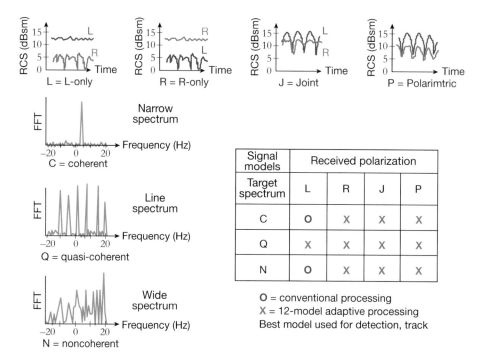

Figure 2.5 Models for adaptive detection processing of polarization data. Twelve algorithms were developed for the four possible polarization combinations (principal, orthogonal, joint, and polarimetric) and three target return signal models (coherent, quasi-coherent with line spectrum, and noncoherent or wide spectrum).

these are detectable on a single pulse, the advantage of polarimetric processing is not as dramatic, except that even RSOs with a low radar cross section on the threshold of detectability will be rendered detectable.

2.1.1.2.3 Tracking of Deep-Space Resident Space Objects   Although it provides accurate metric and radar signature data for space situational awareness, radar tracking of a deep-space RSO, as previously mentioned, has to be enabled in four dimensions simultaneously—slant range, slant range rate or Doppler, and azimuth and elevation angles. The Millstone Hill radar is a monopulse system with angular offset receivers in two orthogonal dimensions. If the RSO is centered in the beam, then the error signal in each dimension is essentially noise; if not, the location of the beam is corrected by using the ratio of the error signal to the primary (or sum) signal and the relative phase to indicate direction of the error. The concept is similar to that for conventional

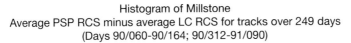

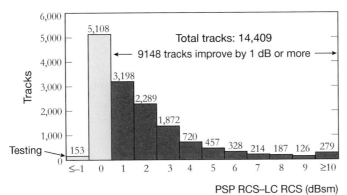

Histogram of Millstone
Average PSP RCS minus average LC RCS for tracks over 249 days
(Days 90/060-90/164; 90/312-91/090)

Figure 2.6   Histogram of performance analysis of polarimetric processing at the Millstone Hill Radar. Note that a 3 dB gain in the signal-to-noise ratio implies half the time for tracking at a constant SNR. LC, left circular polarization; PSP, polarimetric signal processing.

radars, except that, in this case, the processed return signal used is from a sequence of pulses rather than from one pulse. The number of pulses to be integrated is dynamically adjusted to achieve a desired signal-to-noise ratio for tracking. The processing of the error gates for a coherent target sums up the energy in the few Doppler bins containing the target, whereas the processing of the error gates for a noncoherent target sums up the total energy across all Doppler bins in each gate.

The receiver range gates are sampled at twice the pulse-repetition frequency (Nyquist rate), or at a higher rate, over a range span centered on the target. Tracking in range is achieved by constructing 50 percent overlapped range gates. The two adjacent range gates to the sum or primary range gate must contain equal energy for the target to be centered. A coherent target can be centered in range by using the ratio of the adjacent (or error) gate energies in the few Doppler bins containing a target. Centering in range for a noncoherent target is achieved by balancing the total detected energies across all the Doppler bins in the two overlapped range gates.

Range rate or Doppler tracking of a resident space object proceeds as follows. Since the fast Fourier transform bandwidth typically is a small fraction of the receiver bandwidth, large aliasing corrections are made by first varying the pulse-repetition frequency and thus the FFT bandwidth in small increments (a classic Vernier method used in a wide variety of disciplines involving metrology). Thereafter, the offset of the centroid of the narrow target spectrum from the estimated Doppler value represents the further correction that needs to be applied to center the target in the Doppler window. Typically, because of the high transmitter frequency of the radar and the high resolution in frequency of the FFT algorithm, the Doppler measurement for coherent targets is the most accurate of all the metrics, followed by range and then by angles.

Doppler correction for noncoherent or broad-spectrum targets is a little more complicated. Three Doppler windows are constructed as follows:

1. The primary fast Fourier transform window is computed as in the standard FFT in each receiver gate.

2. Two overlapped windows are constructed by rotating the detected signal in each receive pulse by +90 degrees and by -90 degrees and computing the overlapped FFT windows.

3. An error offset is computed by using the difference in the ratio of the energy in the overlapped gates to that in the sum or primary gate, and the Doppler value of the target corrected. A set of such corrections will enable the target to be centered in the receive bandwidth.

It is evident that, given the broad spectrum over which the target energy is spread, the accuracy of Doppler measurement is significantly worse for noncoherent targets than for coherent targets. Nevertheless, it is still a useful metric and accurate enough to aid in orbit estimation.

2.1.2   Comprehensive Millstone Hill Radar System for Deep-Space Surveillance

Early in the enterprise of developing the technology discussed in section 2.1, it became clear to Lincoln Laboratory analysts that the Millstone Hill radar must also serve an extended operational function as a deep-space surveillance sensor for the following reasons:

1. The Department of Defense was severely hampered by the lack of a tactical, all-weather deep-space surveillance sensor. The only surveillance sensors available in the early 1970s were film-based Baker-Nunn cameras [10], which could operate only on clear nights and were not tactical in operation.

2. Technology development in deep-space surveillance would be considerably aided by operating the Millstone Hill radar for extended periods of time to enable testing and continued improvement of algorithms and software in support of future deep-space radar acquisition.

3. Lincoln Laboratory deemed the availability of the radar of significant help in its effort to develop concepts for optical deep-space surveillance (described in chapter 4).

Using an operational sensor for research and development entailed developing a comprehensive system of interrelated software and algorithms, a system that could, through dynamic scheduling of the radar, furnish the Air Force Command Center with metric and radar signature data in response to its tasking requests and that could also participate in the analysis of the catalog of deep-space resident space objects. Because deep-space tracking was a "time hog," locally maintained databases of up-to-date orbital elements of deep-space RSOs were needed to reduce the time to acquire and track them, elements whose accuracy needed to be greater than that routinely available from the Air Force Command Center (between 1975 and about 1990). And, most important, a trained, innovative corps of orbital and radar signature analysts and software developers was essential to develop this system.

As a result of these considerations, the Millstone Hill radar and, by extension, Lincoln Laboratory effectively (though not formally) became an alternate analysis center for deep-space surveillance (albeit in the capacity of prototype development) and a repository for all deep-space surveillance data from all the sensors that could provide such data. To supplement the detection and tracking techniques discussed earlier, important aspects of the algorithm and software development for the Comprehensive Millstone Hill Radar System are described below.

2.1.2.1 Databases

Essential to support the space surveillance enterprise, its detailed databases needed to contain, at a minimum, the following information:

1. Up-to-date orbital elements for all resident space objects that had been cataloged by the Air Force and for all newly launched or discovered objects that had not yet been cataloged. These elements enabled the Millstone Hill radar to reacquire RSOs with a minimum of search time expended.

2. Data on the radar characteristics of the RSOs, including detected radar cross sections (mean and maximum at least) and attitude stability, as evidenced by the spectral width of the target in the receive window in tracking. Later on, data

were added on the optical characteristics, including estimated, normalized visual magnitude. With these data, the integration time and the mode of processing—coherent, semicoherent, noncoherent—needed for detection could be determined ahead of time and the search time for an RSO minimized.

3. Data on the desired frequency of tracks on every RSO and the most recent track by any deep-space sensor, data important in dynamically scheduling the radar resources.

4. A comprehensive set of data on the entire catalog of deep-space RSOs—including all the metric data that had been collected by any sensor—was needed to support analysis of the RSOs. The availability of metric data from other sensors aided in generating the most up-to-date orbital elements and in dynamically allocating the resources of the radar to RSOs that had not been recently tracked.

2.1.2.2   Major Software Subsystems

The primary tracking software system on the Millstone Hill radar, SATCIT was supported by a number of near-real-time and postprocessing software subsystems, most notably:

1. MIDYS (Millstone Dynamic Scheduler), a real-time, dynamic scheduling software that enabled efficient operation of the radar [11]. Its objective was to provide the radar operators with a prioritized set of resident space objects that could be tracked next as SATCIT was finishing a track. The choice of any RSO in this set was based on

    *   tasking on the RSO that reflected its importance to the network and the command and control authority;

    *   detectability of the RSO, based on the radar cross section of, distance to, and stability of the target, essential to achieve a high signal-to-noise ratio for good-quality metric data to be collected;

    *   age of the RSO's element set and how recently it had been tracked by any sensor in the network; and

    *   estimate of how much time would be taken by the search, acquisition, and tracking of the RSO, to help conserve radar resources.

    MIDYS was designed to present to the radar operator a set of RSOs prioritized on the basis of the weighted sum of all relevant factors. The operator was allowed to use additional human knowledge to choose the RSO to track.

MIDYS was very successful in enabling efficient radar operations and served as a template for the development of other such dynamic scheduling software at other sensors.

2. ANODE (Analytic Orbit Determination) software for deep-space RSOs, developed in house for near-real-time operation on the radar computer [12, 13, 14]. Its primary purpose was to enable rapid calculation of orbital element sets based on recent metric data—either from the radar or from other sensors—and thus to shorten radar search time for any RSO. To that end, it used only a limited set of perturbations—first-order effects of the Sun and the Moon and the Earth's gravitational expansion—and was based on classical averaging theory.

3. DYNAMO, a numerical (or special perturbations) high-accuracy orbit estimation software, developed initially at the Smithsonian Astrophysical Observatory and later at Lincoln Laboratory (further described in subsection 2.1.3.2). The major tasks handled by DYNAMO were

   • generating high-accuracy orbits for RSOs to permit more efficient searches by the radar and analysis of the orbits to detect and quantify maneuvers of active RSOs; and

   • maintaining the metric calibration of the radars and optical sensors that constituted the deep-space part of the Space Surveillance System;

4. STARS Satellite Tracking and Reporting System), a non-real-time satellite catalog analysis program to enable calculations of the visibility and detectability of any group of RSOs from one or more ground- or space-based sensors and to facilitate sorting and analysis of the catalog of element sets.[7]

5. Several software tools for the sorting and catalog correlation of uncorrelated target (UCT) metric observations, that is, metric data collected on RSOs whose catalog identity was not known. These were a problem for all sensors and were due either to incorrect or old element sets on existing RSOs or to new RSOs from launches or fragmentation events. Correlation software enabled the analysts at the radar to associate such uncorrelated metric observations with the catalog and compare the element sets of newly discovered objects with the catalog. These tools were essential in developing and maintaining an orderly and accurate catalog of deep-space RSOs.

---

7. Unfortunately, because STARS was only documented with an internal HELP file, no reference is available.

2.1.3   Calibration Systems for Deep-Space Surveillance

Calibrating radars in their deep-space surveillance mode was critical to ensure that the data they collected were accurate and timely. The following subsections will examine both radiometric and metric calibration for deep-space surveillance radar systems.

2.1.3.1   Radiometric Calibration

Although radiometric calibration of radars is a well-understood topic covered extensively in textbooks on radar fundamentals, the nature of deep-space radar surveillance—basically, multipulse processing of receiver data for detection of targets—necessitated enhancements in calibration. Just as in standard radars, the calibration process for deep-space surveillance radars was initiated with a known calibration signal injected at the receivers, followed by a calibrated noise signal similarly injected. Both these signals were pulsed so that SATCIT could process them using its standard operating procedure of fast Fourier transforms. This calibration process was repeated whenever the radar was turned on or changes were made in the radar hardware or software, thus verifying the functioning of the receiver chain and the calibration of the internal system noise.

There were a number of highly attitude-stable, high-RCS satellites available in geostationary orbit and visible to the radar, most notably, the Intelsat communications satellites. The radar searched for and tracked one such satellite to verify that the SATCIT was operating correctly in its deep-space search, acquisition, and tracking modes. The relatively invariant radar cross section and polarization ratio of the attitude-stable, high-RCS target also demonstrated the stability of the radar's radiometric calibration.

Additional radiometric calibration of the radar was achieved by tracking a spherical satellite in orbit. A popular choice was Lincoln Calibration Sphere 1 (LCS 1), in a circular orbit at an altitude of ~2,800 kilometers at ~30 degrees inclination to the equator. Because the radar cross section of the sphere was well known—1 square meter or 0 dBsm (decibels per square meter)—precise radiometric calibration of the radar was possible using its track. (Some of the spheres that could be used for radiometric calibration are listed in table 2.1; readers may also want to look at the discussion on radiometric calibration for the Haystack radar in space-debris mode in section 2.2.)

Table 2.1   Radiometric calibration targets: Calibration spheres

Satellite catalog number	Satellite name	Perigee height (km)	Apogee height (km)	Diameter (m)	Radar cross section (dBsm)
900	Calsphere 01	975.2	1,017.8	0.3556	-10.0
902	Calsphere 02	1,048.7	1,077.4	0.3556	-10.0
1361	LCS 01	2,776.8	2,799.3	1.12903	0.0
1512	Tempsat 1	1,081.3	1,187.5	0.3556	-10.0
1520	Calsphere 4A	1,075.1	1,178.0	0.3556	-10.0
2826	Surcal	757.2	764.5	0.508	-6.93
5398	LCS 04	738.2	835.5	1.12903	0.0

Note: The radar cross-section values are at X Band (about 10 GHz) or higher for all targets.

2.1.3.2   Metric Calibration

The following text is largely adapted from seminal works describing calibration of the Millstone Hill and other high-precision radars in the Lincoln Laboratory stable of satellite-tracking radars [15, 16]. The problems addressed and the technologies developed in pursuing this calibration are, of course, applicable to almost all types of high-precision radars, whether mechanically driven or phased array. Metric calibration is the science of converting the *precision* of the basic observables from a radar system into *accuracy* of the reportable metric data. The basic observables—time delay, delay rate where available, encoder-measured angles, and angle rates where available—must be converted into the reportable metric quantities by using a calibration model.

The precision of the time delay and the delay rate is affected by the measured values of the average transmitted power, the received signal-to-noise ratio, the bandwidth of the transmitted radar pulses, and the droop of the transmitter frequency during a pulse. These characteristics are internal to the radar and have to be used appropriately in correcting and reporting the observables. Angles—typically azimuth and elevation—are measured from the encoders in mechanical systems or converted from the electronic commands in a phased-array system. In mechanically driven radars, the encoder values are affected by various parameters such as the orthogonality of the encoders, the perpendicularity of the rotation axes, and the effects of solar heating and cooling on the antenna support tower and of wind on the antenna. An extensive example of the determination of a calibration model for these parameters is given for the Millstone Hill radar in Evans [16]. The angle observables in a phased array are affected by the basic design parameters such as boresight angle and precision and resolution of phase control of the antenna elements. The first step is to correct the reported observables for these effects to the extent possible before using them in the calibration model. An excellent example of such corrections for the Millstone Hill radar is contained in Gaposchkin [15]. The basic principle is that all identifiable physical factors should be accounted for prior to determining the calibration model.

The second step is to correct the observables for environmental effects, primarily tropospheric and ionospheric effects but also the effects of the plasmasphere (i.e., the extended ionosphere up to geosynchronous altitudes) where significant. Standard tropospheric models based on elevation angle, ground-level humidity, and air density are used to accurately correct the observables for the effects of the lower atmosphere. The effects of the ionosphere are best estimated by using a dual-frequency GPS receiver [17, 18]. Although the effects of the diffuse ions in the plasmasphere (from a few hundred kilometers to the altitude of a geosynchronous satellite) are also best estimated by using a model, at the typical radar frequencies considered, these effects are very small (less than a meter in range). It is likely that, given the present state of knowledge, the accuracy of the metric data is largely limited by the accuracy of the environmental corrections.

The third step is to estimate the constants in the calibration model by using an independent external calibration standard. The standard used by all radars in deep-space surveillance is the set of Laser Geodynamics Satellites (LAGEOS).[8] The characteristics of these satellites are as follows:

- Semimajor axis: about 12,000 kilometers
- Eccentricity: 0.006
- Inclination: 109 degrees for LAGEOS 1, and 52 degrees for LAGEOS 2
- Mean motion: 6.6 revolutions per day
- Shape, mass, radius: spherical, 407.0 kilograms, 30.0 centimeters
- Surface: aluminum with 426 embedded corner reflectors
- Reflectivity coefficient: 1.002
- Radar cross section   0.28 square meter (at L band, approximately 1–1.5 GHz transmitter frequency)

---

8. The LAGEOS set comprises at present two satellites (LAGEOS 1 and LAGEOS 2) funded by NASA and the European Space Agency.

Both LAGEOS 1 and LAGEOS 2 are tracked by a global network of laser ranging stations coordinated by NASA. The calibrated metric data are available from NASA on a website[9] and are generally accurate to better than 10 centimeters after corrections and between 10 and 50 centimeters real time. LAGEOS orbits, with a rated accuracy of 20 centimeters computed by using the laser ranging data, are the most accurately known orbits in the solar system.

The calibration process, shown in figure 2.7, consists of the following steps:

1. Tracking LAGEOS over a wide elevation and azimuth angle during multiple passes and computing the corrected observables (as above);

2. Generating the precision orbit on LAGEOS by using the laser ranging data;

3. Generating the differences (residuals) between the predicted observable using the precision orbit and the actual radar observables;

4. Determining the coefficients in the calibration model by using the residuals;

5. Applying the new calibration model to correct future observables from the radar; and

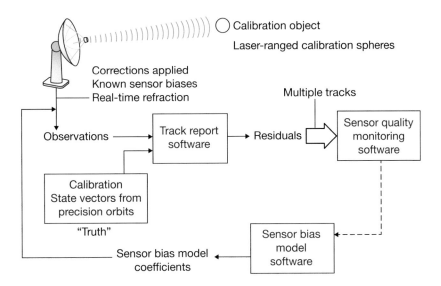

Figure 2.7 Metric calibration process.

9. http://cddis.gsfc.nasa.gov/Data_and_Derived_Products/SLR/SLR_data_and_product_archive.html/.

6. Monitoring the long-term history of the accuracy of the radar data and the coefficients of the calibration model to assess seasonal effects and unmodeled changes in the radar system.

This process

- enables accurate metric data to be reported in near real time with all known corrections applied;
- enables the long-term performance of the radar to be monitored;
- identifies and flags error conditions in the radar; and
- provides a rapid assessment of system modifications.

Calibration is an ongoing endeavor and has to be repeated regularly (typically every day) and immediately after any system modification so as to assure that the system's metric data are accurate.

A similar calibration process can be applied to all space surveillance radars, whether mechanically driven or phased array. In fact, a system built at the laboratory, called the "MIT Radar Calibration System" (MRCS), has been used at the laboratory's radars and at a large phased-array radar, the FPS-85, on the Eglin Air Force Base in Florida. Calibration of optical space surveillance sensors is more easily achieved by using the extensive star catalog provided by the Naval Space Observatory (see chapter 4). However, occasional tracks of LAGEOS are useful in monitoring the calibration of these sensors also, particularly when they are used for acquiring satellite data. Examples of the use of the MIT Radar Calibration System follow.

Figure 2.8 is a display of the calibration in range achieved by using MRCS software at the Millstone Hill radar. Range measurement is the second most accurate metric of the radar after range rate, and thus proper calibration is vital to ensure accurate estimation of orbits using the range data. Because any given radar at Lincoln Laboratory is always under development, there are frequent modifications in the tracking system that affect its calibration. Long-term monitoring ensures that such changes are accounted for in the calibration model for the radar. The various changes in calibration marked in figure 2.8 indicate the detection by the MRCS of tracking system modifications.

A more vivid example of the effects of different modes in the tracking software is shown in figure 2.9. The tracking software has three different modes that can be used depending on the type of target being tracked. Figure 2.9 shows the range calibration difference between these three modes when they are invoked during a single track of one of the laser calibration satellites (LAGEOS 1).

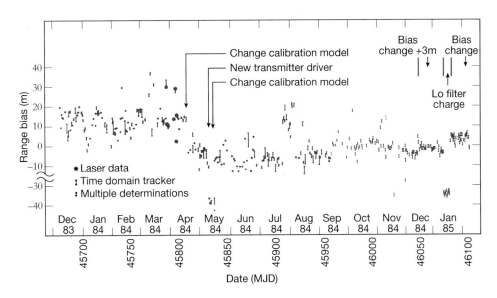

Figure 2.8   Long-term range calibration at the Millstone Hill radar using the MIT Radar Calibration System (MRCS; reproduction of figure 5 from Gaposchkin [15]). MJD, Modified Julian Date.

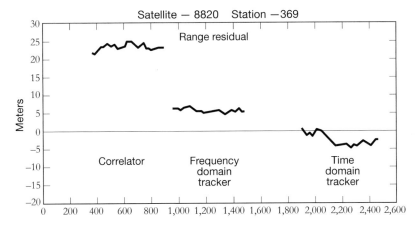

Figure 2.9   Range calibration differences due to differing tracking modes; *x*-axis is time in seconds (reproduction of figure 6 from Gaposchkin [15]).

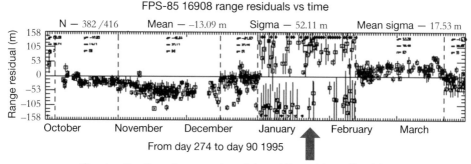

FPS-85 16908 range residuals vs time

N — 382 /416   Mean — −13.09 m   Sigma — 52.11 m   Mean sigma — 17.53 m

From day 274 to day 90 1995

Good calibration allows one to catch and fix problems like this
A 700 ms time bias was introduced when leap second
was applied on 31 December 1995

Figure 2.10   MIT Radar Calibration System (MRCS) monitoring of the range calibration of the FPS-85 phased-array radar.

Figure 2.10 demonstrates the effect of an unmodeled system modification, here an error, for another of the radars used by the Space Surveillance Network (SSN). In this case, an error was committed in accounting for a leap second introduced into the firmware, and thereby a large error resulted in the range calibration. This error was captured by the MIT Radar Calibration System at the Millstone Hill radar and reported by analysts there to the FPS-85 engineers. The long time interval when this error persisted demonstrated the need for a near-real-time MRCS; one such system was subsequently transferred to the FPS-85 radar.

It is essential that the accuracy of orbit determination software used to determine LAGEOS orbits and to compare residuals be greater than that of the data used. Thus, because their geopotential and lunisolar perturbation expansions are truncated to a low order, most analytical orbit determination programs are not useful for monitoring calibration. The best option is to use numerical special perturbations software with the best dynamical models available. One example of such software is DYNAMO, which has been used extensively at Lincoln Laboratory; it was developed at the Smithsonian Astrophysical Observatory by Edward Michael Gaposchkin, who later moved to the laboratory and adapted the software to the type of data available from the Space Surveillance Network. Capable of determining orbits for LAGEOS of 25-centimeter accuracy from available satellite laser ranging (SLR) data, DYNAMO uses

1. Joint Gravity Model 3 (JGM-3; 90 × 90) of the Earth's gravity field,
2. the positions of the Sun, Moon, and planets provided by NASA Jet Propulsion Lab,
3. a simple model of atmospheric drag (constant densities in altitude slabs),
4. a constant acceleration model for thrust due to venting of fuel from satellites, and
5. a specular reflection model for solar radiation pressure;

it accounts for

6. the Earth's albedo pressure and
7. the effects of Earth's body and ocean tides;

and it is able to calculate

8. the values for items 3–6 based on metric data.

The demonstrated success of the Millstone Hill approach to metric calibration resulted in the adoption of similar technology by the U.S. Air Force Space Command for the calibration of both low-altitude and deep-space sensors in the Space Surveillance Network. There are several laser retroreflector spheres with available satellite laser ranging data in orbits much lower than those of LAGEOS for the purpose of detection and tracking by low-altitude radars, which do not have the high sensitivity needed to detect LAGEOS.

Because the laboratory has been at the forefront in developing sensors for deep-space surveillance, including ground-based radars and ground- and space-based optical sensors, an ongoing historical database of the calibration of these sensors has been maintained for nearly two decades at the Lincoln Space Surveillance Control Center (LSSC), located at the laboratory. This database permits long-term analysis and identification of trends in the calibration parameters of the sensors.

2.1.4 Summary of Section 2.1

The techniques for detecting and tracking deep-space resident space objects described in section 2.1 enabled, for the first time in the early 1970s, data collection on deep-space RSOs by ground-based radars. At the Air Force's request, these techniques were first transferred to ALTAIR (and later, TRADEX)[10] at the Kwajalein Atoll in the late

---

10. Originally built as downrange sensors for test missiles launched from Vandenberg Air Force Base, California, ALTAIR (Advanced Research Projects Agency [ARPA] Long-Range Tracking and Instrumentation Radar) is a dual-frequency VHF/UHF radar, whereas TRADEX (Target Resolution and Discrimination Experiment) radar is a dual-frequency L-band/S-band radar.

1970s for their use in deep-space tracking; and, in the early 1980s, to the now-defunct FPS-79 radar at Pirinçlik, Turkey. ALTAIR and the Pirinçlik and Millstone Hill radars constituted a triad of deep-space radars with tactical, all-weather coverage of the entire geosynchronous belt of satellites. Again at the Air Force's request, the techniques used with these radars were transferred in the mid-1980s to the phased-array FPS-85 radar on the Eglin Air Force Base in Florida. These techniques continue to be used on the deep-space radar network with, of course, regular upgrading.

An interesting example of technology transfer in support of experiments was conducted by Lincoln Laboratory analysts during a field experiment at the Arecibo radar in Puerto Rico [19] in 1974. The laboratory's coherent integration techniques were transferred to the Arecibo radar during several visits and used to conduct experiments in detection of small objects in deep-space orbits. One of those experiments resulted in the detection of NASA's Interplanetary Monitoring Platform 6 (IMP-6) at a range of 140,000 kilometers—a distance record for radar detection of a man-made satellite at that time [20].

Although the concept of using data on resident space objects as part of a calibration system was well known, the work described in subsection 2.1.3 advanced the state of the art significantly. To that end, the laboratory proposed that

1. there be one or more calibration satellites whose metric data on and orbital elements were much more accurate than those of normally tracked satellites. Prior to this, calibration was based on processing all the available low-accuracy data on a few satellites from all the Space Surveillance Network radars without regard to the widely varying data accuracies.

2. the orbit determination software be capable of supporting much greater accuracies than the sensor data, whether laser data or radar data. Prior to this, a low-accuracy analytic orbit determination software was used for this purpose.

3. the radars remove all known mechanical, electrical, and environmental errors in the data before they are submitted for calibration against the calibration satellites, and the radars' corrected metric data on the calibration satellites be processed against the accurate orbits of the calibration satellites to estimate bias and standard deviation of the data. The bias can then be used to correct the radar data on all resident space objects, and the standard deviation can be used to weight the data for orbit estimation on the RSOs.

Calibration is not a static process because sensor hardware and software are continually being altered, intentionally or otherwise, and the environmental factors change. Although the methods used earlier often indicated changes in calibration values from day to day, for unknown reasons, it was common to average these changes over a time period for the purpose of using the data. Such a method prevented an understanding of long-term trends in the calibration.

2.2   Man-Made Space-Debris Detection and Processing by the Haystack Long-Range Imaging Radar

Early in the development of deep-space satellite tracking radars, researchers recognized that their inherent high sensitivity could be exploited for detecting small (space-debris) resident space objects. By the mid-1980s, serious concerns over the size and orbital distribution of the space-debris population began to emerge, many motivated by the need to shield the planned space station *Freedom* from collisions with space debris.

Since the mid-1980s, the Haystack Long-Range Imaging Radar (LRIR) and later the Haystack Auxiliary (HAX) radar have collected space-debris data, the first and only statistically significant data on space-debris objects smaller than 10 centimeters in diameter, used by many researchers around the world to model the space-debris population and cited in many articles on this subject. The following subsections describe how Haystack tracking radar expanded its capability to become the main source for detection and characterization of space debris.

2.2.1   Historical Background

Since the launch of Sputnik 1 in 1957, man-made debris has accumulated in space as a by-product of launching satellites. The presence of space debris was recognized from the very beginning of the Space Age, but its existence had little impact on the progress of space exploration. Initially, space debris consisted of deployment hardware: de-spin cables, lens caps, and various other random small objects, in addition to which rocket bodies with remnant fuel were simply abandoned in orbit. Most of this debris was expected to remain in space for a long time; for years, there was no strong reason to worry about it. Throughout the 1960s and 1970s, the population of space-debris objects continued to grow. The largest debris objects—rocket bodies—were detected and tracked by the radars and optical sensors of the early Space Surveillance Network. All the tracked objects were assigned an object number and placed in the resident space object catalog, which was maintained by the Air Force.

In the 1970s, NASA became proactive on the subject of man-made space debris. Heightened concern about the possibility of collision between debris objects and operating satellites stemmed from more than ten years' observations of special events in space, in particular, explosions of upper stages and rocket bodies left in space with remnant fuel. The first such fragmentation event had been observed in 1961. Many similar events followed, leading to concern that the number of space-debris objects might be increasing exponentially, and that the nation lacked adequate models to assess the associated risk to operational space objects. There was an urgent need for both fragmentation-size and orbital debris dispersion models, a need made all the more urgent in light of NASA's plan to deploy large solar-powered satellites, which would be more vulnerable to collision with space debris than smaller satellites. Despite the lack of space-debris data, NASA initiated a serious effort to develop theoretical models in the mid-1970s, at a time when only two large man-made objects were in space, *Skylab* and the first of the Russian *Salyut* space station series, both of which were smaller than the U.S. Space Shuttle.

Established in the 1980s to address the vulnerability of critically important national security satellites to space debris, an Air Force working group conducted important experiments, producing data in support of the development of satellite collision and fragmentation models. In one of these experiments, a scaled-down realistic satellite model was destroyed in the laboratory by means of a hypervelocity collision. The fragments were carefully collected to produce a data set to model the distribution of fragment sizes. The radar cross section of each captured fragment was measured in an anechoic chamber; these measurements were later used to produce a model to convert the RCS of a debris fragment to its size, in this context, its equivalent sphere diameter (ESD). Independently at Kwajalein, NASA measured the RCS of simulated debris fragments by dropping them from balloons and tracking them with the radars of the Kiernan Reentry Measurements System.

At about the same time, NASA was in the initial development phase of the space station *Freedom*. Observations of on-orbit debris impacts, such as shown in figure 2.11, and the *Freedom*'s large cross-sectional area led to the inevitable conclusion that the space station would need to be shielded against debris collision and that the shielding, both primary and secondary, would need to be directly related to the size of the impacting debris objects. In view of the potentially high cost of the shielding and the critical nature of the space station, a statistically significant model for the distribution of space-debris objects was needed, one based on actual data rather than on analytical suppositions.

Lincoln Laboratory proposed the use of the Haystack Long-Range Imaging Radar to sample the space environment for the purpose of building a data-based statistical model; by 1995, it had detected a large number of space-debris objects, producing the first and only statistically significant space-debris data set. The laboratory's use of the

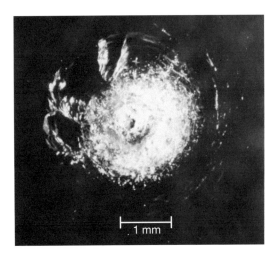

Figure 2.11   Damage to a window of NASA's Space Shuttle (originally space transportation system) STS-7 caused by a collision with a speck of paint.

Haystack LRIR in space-debris-sampling mode is described in the subsections that follow.

2.2.2   Lincoln Laboratory Radars for Space-Debris Detection

In 1984, NASA funded a small program at Lincoln Laboratory to search for and detect space debris by using available telescopes in the Experimental Test System (ETS) in Socorro, New Mexico (described in chapter 4). This challenging effort conducted approximately 30 hours of data collection. A small number of detected space-debris objects appeared to be 1 centimeter in size [21]. The main conclusion from the analysis of these data was that the number of man-made objects between 1 and 10 centimeters in size located in near-Earth orbit (less than 3,000 kilometers altitude) was estimated to be about 99,000—some 11 times the number of RSOs in the North American Aerospace Defense Command (NORAD) catalog. Later Haystack Long-Range Imaging Radar data, however, showed this result to be a gross underestimate (the laboratory's initial estimate failed because it was based on a small data set). This newer finding (1988) generated considerable interest in space debris and in the potential use of high-powered radars to detect the debris.

Since the interest was in detecting resident space objects as small as 1 centimeter in size, it made sense to look to the high-frequency radars as potential detectors. Two options from among the radars that were available to Lincoln Laboratory were considered for that function: the Millimeter-Wave (MMW) radar at Kwajalein,[11] operating at Ka band (35 GHz), and the Haystack Long-Range Imaging Radar (LRIR) in Tyngsborough, Massachusetts, operating at X band (10 GHz). Figure 2.12 shows that at both X and Ka bands, the radar cross section of a 1-centimeter object as a function of its diameter lies in the Mie region, where the RCS is, on average, within ±5 dB of the optical RCS of a sphere, but that it is actually larger at X band. Because the LRIR was also about 19 dB more sensitive than the MMW radar (both having the same beamwidth), the LRIR was obviously a better candidate for space-debris data collection.

An attempt at space-debris data collection was initiated at the Long-Range Imaging radar, which was operated in zenith-pointing stare mode: the antenna was pointed at zenith, and the radar was set to detect objects that crossed the beam. The 1.26-millisecond continuous-wave (CW) waveform was used because it could provide a large range window, from 300 to 1,800 kilometers altitude. About two hours of data were recorded and subsequently processed to produce a single detection. This result

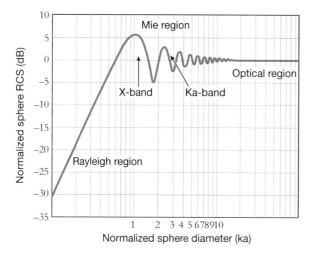

Figure 2.12 Normalized sphere radar cross section (RCS) as a function of the normalized sphere radius.

---

11. The Millimeter-Wave dual-frequency radar is operating at the Kwajalein Atoll as part of the Kwajalein Missile Range in the Marshall Islands. At the time of this discussion (about 1988), the lower band was centered at 35 GHz (Ka band) with 1 GHz bandwidth; its upper band was centered at 95 GHz.

was highly encouraging: the data were collected using the original LRIR system configuration with no modifications for debris data collection. Because, however, the radar could not record the angle channels in all the range gates, it could only count the debris pieces and determine their height and range rate (Doppler), and the process was both cumbersome and slow.

A new, faster signal-processing capability incorporated into the radar system made it easier to collect and handle more data. In one week, about 10–15 hours of data were recorded and processed, yielding excellent results. The rate of space-debris detections was increased to between two and four detections per hour, sufficient to convince NASA that Haystack could provide the required data for modeling space debris. In a subsequent negotiation between NASA and the Air Force (sponsors of the Long-Range Imaging Radar operations), NASA agreed to support the development of a lower-power, higher-frequency Haystack Auxiliary (HAX) radar in exchange for operational time on the LRIR that would be used for debris detection.

2.2.3   Haystack Long-Range Imaging Radar

The LRIR was built with Defense Advanced Research Projects Agency (DARPA) funding to gain a better understanding of the effect of higher range and Doppler resolution on the quality of radar images of man-made satellites. In the early 1970s, Lincoln Laboratory pioneered the use of high-range-resolution radars to image these satellites. At that time, the technology of synthetic aperture radar (SAR) was already in service. SAR systems produce two-dimensional range-Doppler images of terrain and ground targets by coherently processing the received signal while the line of sight from the airborne radars to the ground targets rotates. SAR routinely produced images with a moderate range and Doppler resolution of 5–10 meters. Lincoln Laboratory recognized that, with the use of coherent high-resolution radar, the SAR principle could be applied to produce images of rotating satellites from the ground. The satellite rotation about its center of mass provided the necessary rotation for the production of range-Doppler images. This technique is now known as "inverse synthetic aperture radar" (ISAR; discussed along with imaging radar technology in chapter 3).

The original radar images were produced by using data collected by ALCOR (ARPA-Lincoln C-band Observables Radar), located on Roi-Namur Island in the Kwajalein Atoll. With a 500 MHz bandwidth at C band, ALCOR began collecting high-resolution data on U.S. ballistic missiles in 1969. The 500 MHz bandwidth yielded a 0.5-meter range resolution. The initial success of satellite radar imaging with ALCOR data prompted considerable interest in extending the capabilities of imaging radars. Lincoln Laboratory submitted a proposal to DARPA to build a high-power, wideband X-band radar at the Haystack Observatory. The plan was to share the large antenna and the high-voltage power supply of the Haystack planetary radar to reduce the cost of developing and building the components of the new radar: a high-power wideband transmitter, a low-noise receiver, real-time control and signal-processing software, a data-storage subsystem, and data-processing software. New technology subsystems were introduced—high-power wideband traveling-wave tube amplifiers, a digital waveform generator, a wideband-feed, and a correlation-matched filter.

Declaring that the Long-Range Imaging Radar project had successfully achieved all its objectives, DARPA terminated its funding in 1978. As was expected, the 25-centimeter resolution of the LRIR images yielded finer details and more information about satellites than ALCOR's images had; the radar was also able to image rotating satellites at geosynchronous altitude orbits and deep-space inclined orbits.

The feeling at Lincoln Laboratory was that this new capability could and should be put to use. To that end, staff members developed new image analysis techniques, including stroboscopic imaging and extended coherent processing (see the discussion in chapter 3), techniques that, in effect, served as characterization tools for day-and-night, all-weather satellites. It was clear to laboratory analysts that radar satellite images could be very helpful to the intelligence community, which strongly concurred with this view. DARPA therefore authorized funding through the Air Force Space and Missile System Organization (SAMSO) to operate the LRIR for two more years. The laboratory also obtained support for data collection and operations for the radar from the Air Force Aerospace Defense Command (ADC), predecessor to the USAF Space Command and the Foreign Technology Division. At the end of the two-year period, both DARPA and the ADC agreed to continue funding LRIR operations. Thus the Long-Range Imaging Radar, shown in figure 2.13, became, in itself, a laboratory for radar imaging whose objectives were, and still are, to obtain detailed knowledge about active satellites.

Table 2.2 lists the operating parameters of the Long-Range Imaging Radar in 1988, when space-debris data collection was first considered. Because the single-pulse sensitivity of the radar when it uses the 1.25-millisecond waveform is 62 dB, it could detect a sphere 1 centimeter in diameter at an altitude of 1,585 kilometers at the center of the beam with a signal-to-noise ratio of 13 dB. And because space-debris processing requires multiple single-pulse detections, a lower SNR could assure a high probability of detection that the radar could detect 1-centimeter debris objects at higher altitudes.

Figure 2.13   Cut-through of the radome of the Haystack Long-Range Imaging Radar shows one of the boxes being lifted and the subreflector.

Table 2.2   Operating parameters of the Haystack radar

Center frequency	10 GHz
Bandwidth	1 GHz
Peak power	400 kW nominal
Maximum duty cycle	35%
Antenna diameter	36.6 m
Antenna beamwidth (14 dB taper)	0.060 deg
Transmit polarization	Right-hand circular
Receive polarization	Right-hand and left-hand circular
Antenna gain	67.2 dB
System noise temperature	246° K
Waveforms and bandwidth	0.256–3 ms
Continuous wave (CW)	
Narrow band (LFM, linear frequency modulation)	0.256 ms and 10 MHz
Wide band (LFM)	0.256 ms and 1 GHz
Sensitivity (SNR at 1,000 km on an object with 0 dBsm RCS)	53 dB

2.2.4   Upgrade of the Processing and Control Subsystem

In 1988, when the discussions about space-debris data collection with NASA were about to begin, the Long-Range Imaging Radar was already undergoing its first major upgrade since it had been built. Although the upgrade focused on the processing and control subsystem (PACS), its scope was extended to include a new digital waveform generator, low-noise amplifiers, and improved analog intermediate-frequency sections, as well as modern computers and storage equipment to enhance the radar's performance. Indeed, increasing the speed and capacity of the recording system was high on the list of performance improvements. The goal was to have the ability to record all four channels of the radar—both polarization channels and both angle channels—in all the range gates. The upgrade was a pathbreaking effort; the LRIR's upgraded architecture was later adopted by the rest of the laboratory's radars and applied to other Department of Defense radars. The processing and control subsystem upgrade made the LRIR more digital and enabled more complex real-time processing, and it was these new capabilities that enabled the radar to take on the function of space-debris data collection The ability to record both angle channels made it possible to measure the angles and the angle rates of detected space-debris pieces as they crossed the beam in all range gates and, with these measurements, to make a rough initial orbit determination. The ability to record both polarization channels was necessary for the size estimation of the randomly shaped debris pieces.

Figure 2.14 presents a simplified block diagram of the processing and control subsystem. The radar front end, the radar frequency and the intermediate frequency, consists of analog hardware. The signal is digitized after the second intermediate frequency, and the remainder of the processing is then digital. The data are typically digitized at a high sample rate of 1 MHz; the samples are properly decimated for a variety of waveforms. Initially, the space-debris data collection used the two waveforms shown in figure 2.14.

A key innovation of the upgraded processing and control subsystem was the introduction of a digital in-phase and quadrature (I and Q) filter. The analog system measurement of I and Q signals required precise balancing of two signal paths. In practice, the balancing was never perfect, and the imbalance generated spurious signals near zero Doppler. In zenith-pointing stare mode, the detections cluster around zero Doppler; these spurious signals interfere with radar return signals. Because it uses a single signal path, however, the digital I and Q filter is not burdened by spurious signals. As a whole, the upgraded PACS added to the tracking radar the capability of real-time detection and recording.

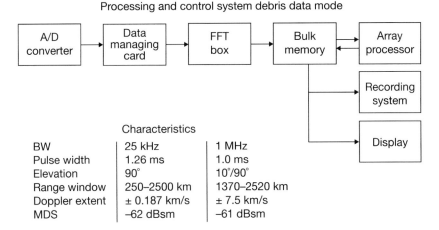

Figure 2.14   Processing and control subsystem (PACS) of the Haystack Long-Range Imaging Radar (LRIR). A/D, analog-to-digital; MDS, minimum detectable signal.

2.2.5   Space-Debris Operations: Data Collection, Processing, and Analysis

The process of space-debris data collection, from real-time detection by the radar to final analysis and cataloging of detected pieces of space debris, is divided into two parts. The Long-Range Imaging Radar function is to detect, process, and validate the data, which are then copied to tapes and sent to the Orbital Debris Analysis System (ODAS) at NASA's Johnson Space Center. The ODAS refines the analysis of the data on each detected object, determines its size and angular rates, and derives the object's orbital elements so that it can be included in NASA's modeled space-debris catalog. Figure 2.15 delineates the processing chain and shows the division line between the radar site and the Orbital Debris Analysis System.

The introduction of the space-debris data-collection function to the Long-Range Imaging Radar required additional calibration routines, additional operator training, and new data-validation software. These requirements were all met.

As a satellite-tracking radar, the LRIR does not have to search for the satellites. Instead, it acquires satellites with known orbits, that is, satellites that are routinely tracked by the Space Surveillance Network. The acquisition process does, however, require a limited search around the nominal orbit to establish a track. On rare occasions, when the orbit is stale, the acquisition fails, and the network is tasked to track the object and refresh its orbital parameters.

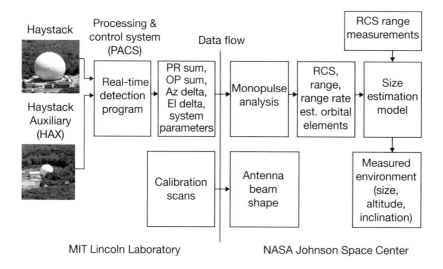

Figure 2.15   Space-debris data-processing chain. Az, azimuth; El, elevation; OP, orthogonal polarization; PR, principal polarization.

When the LRIR is in space-debris data-collection mode, the challenge is to detect uncataloged space objects, which are mostly odd shaped and small (low radar cross section). The radar antenna is stationary and points in a desired direction, and the LRIR is required to detect and discover unknown objects that cross its beam. These tasks must be accomplished during the relatively short time the objects stay in the beam. The beam-crossing path is most likely off the beam axis. For the data to be useful, at least two detections are required. In addition, the off-axis angular position of the object must be measured precisely.

In the tracking operation, the range window is narrow. In space-debris data-collection mode, a continuous-wave pulse with a wide range window and low-range resolution is used. Figure 2.16 shows how the radar can measure the range and range rate of the detected object. This example applies to low-elevation antenna pointing. The range window is divided into eleven overlapping range gates. Each gate consists of 1,024 range samples and corresponds to 153 kilometers. The tracking system hypothesizes that a target is present in one of the gates. To determine whether indeed there is a target in that range gate, the system tests the data by passing them through a large bank (1,024) of frequency filters. This operation amounts to a Fourier transform and is implemented by means of a digital Fourier transform, known as a "fast Fourier transform" (FFT). The presence of a target is proven if the signal crosses the threshold

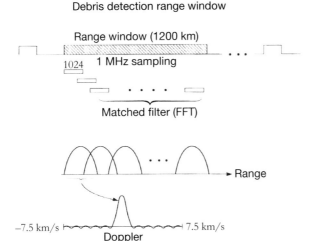

Figure 2.16   Setup of the large range window for detection and ranging.

in one of the filters (bins); this bin will correspond to the Doppler shift of the moving target. When the signal-to-noise ratio is high enough, the signal spills into neighboring range gates. The pattern along the range gates traces the received pulse, whose peak measures the range to the target. The peak Doppler bin measures the Doppler shift and the range rate.

Figure 2.17 shows an actual detection event. The object is at an altitude of 1,415 kilometers, and its size is about 1 centimeter (equivalent sphere diameter). Its range (altitude) and range rate were calculated by interpolating the pulses to determine the peaks precisely.

Here it is worthwhile to briefly discuss the measurements of the radar pointing angles and their rates (described in greater detail in the calibration subsection below). Figure 2.18 shows a trace of an object crossing the beam. Each small circle represents an angle measurement of a single pulse. These actual data were obtained during a fly-through test. The object, an actual resident space object (SSC [Space Surveillance Center] no. 19649), is large with a large radar cross section. The big circle is the one-way receive beam. As can be seen, the object passed through the beam away from its center. The azimuth and elevation of the target are measured at the time of validity (validation), which is the middle of the sequence of detections. The values are the beam pointing plus the offset at the time of validity. The angular rates are determined from the slope of the fitted line.

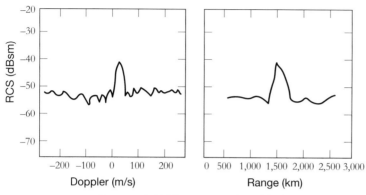

Figure 2.17   Example of a debris-piece detection.

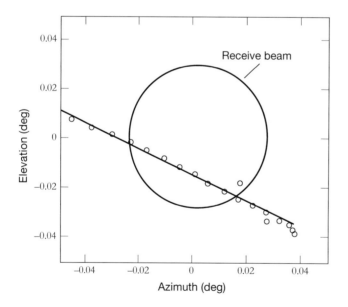

Figure 2.18   Trace of a space-debris object crossing the radar beam.

2.2.6   Space-Debris Detection and Measurement

Very little was known about the shape and size of space-debris objects when the Long-Range Imaging Radar began space-debris measurement operations in 1990. Specific target models for calculating optimal detection thresholds did not exist. Two parameters were needed in order to determine the optimal threshold level: false-alarm rate and signal-to-noise ratio. The false-alarm rate was calculated for each waveform from the number of range gates and Doppler bins. Initially, the LRIR estimated the noise level by continuously averaging the radar return signal, ignoring the possibility that, for some pulses, a target signal below the threshold could be present. Based on the number of detections per hour, two to four, that were observed by the old LRIR, that assumption was acceptable. The estimated noise level and the radar signals were scaled to radar cross-section values and then used to calculate the signal-to-noise ratio.

Although all five standard Swerling models could have been used as target models [22], the threshold level for the detection process was calculated using the Swerling 0 model, which requires the lowest threshold for a given false-alarm rate, and which represents targets with a stable or slowly varying radar cross section. Its use made sense: because small objects with dimensions smaller than the LRIR wavelength (3 centimeters) do not experience strong constructive and destructive interference between scattering centers, their RCS varies slowly as their orientations change. On the other hand, large objects, whose RCS varies with changes in orientation, have a large average RCS and higher signal-to-noise ratio and, hence, are easier to detect.

Concerned about losing track of debris pieces that fall slightly short of crossing the threshold, NASA was not happy with this approach.[12] It asked, instead, that the threshold be lowered and the false-alarm rate be increased to record marginal detections with low probability of detection. To do this, radar operators were allowed to vary the threshold level. But a criterion was needed to control the data-recording volume. It was agreed that operators would allow about thirty events an hour, about ten times the expected detection rate. The LRIR system was programmed to calculate the event rate and display it on the control panel to guide the operators.

The recording hardware implementation is shown in figure 2.19. A ring buffer consisting of six relatively large buffers holds the digitized data. The data are stored in the buffers sequentially; when the sixth buffer is full, the data stream begins to

---

12. Eugene G. Stansbery, Director, Orbital Debris Program, NASA/JSC, personal communication.

Detection and recording

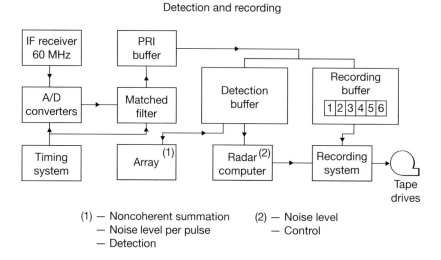

(1) — Noncoherent summation   (2) — Noise level
    — Noise level per pulse   — Control
    — Detection

Figure 2.19   Simplified block diagram of the detection and recording subsystem.

overwrite the first buffer. A detection event triggers data recording. For each detection event, the system records three buffers—the buffer in which the event occurs and its two neighbors. This arrangement provides sufficient data for the Orbital Debris Analysis System to recalculate the noise level and validate the detections.

The LRIR system worked successfully in this manner for a number of years after 1990, but it had a problem with noise-level estimation. Although its running average noise-level estimate handled typical drifts well, the system was affected by other unpredictable random events, such as atmospheric disturbances and prime power variations. To the operators, these random events appear as sudden changes of noise level or detection rate, triggering a restart of the noise-level estimation and causing loss of observation time. For that reason, a constant-false-alarm-rate algorithm was implemented about 1995. This algorithm derives the noise level from a sufficiently large group of noise samples and tracks the variations in the level. It then varies the threshold level automatically to maintain the desired constant false-alarm rate without interrupting the operation.

2.2.7   Radar Coverage beyond Zenith-Pointing Stare

The ideal location for a space-debris-detection radar is on the equator, where the radar can detect debris objects with an orbital inclination down to 0 degrees. Such a radar could accomplish this function by using only a zenith-pointing stare mode. This mode of operation has a number of advantages. With a given sensitivity, the radar can reach higher altitudes, and it experiences the smallest atmospheric losses. The Doppler spread of the detected objects is narrow and provides an excellent way to discriminate between man-made objects and natural meteorites. The latitude of the Long-Range Imaging Radar, however, is 42.6 degrees north and, in zenith-pointing stare mode, the LRIR was able to detect only debris objects with an orbital inclination greater than that latitude. The latitude limit precludes fulfillment of NASA's desire to detect and catalog all space-debris objects in orbit around Earth.

With its high sensitivity, however, the LRIR was able to partially compensate for the latitude limit by pointing the radar beam at lower elevation angles and due south, as illustrated in figure 2.20. By pointing at 7 degrees elevation, it could reach down to 27 degrees inclination, collect data above 400 kilometers altitude, and detect space-debris objects 1 centimeter in size. Although the radar was capable of reaching farther at lower inclination and detecting objects 2 centimeters in size, it could do so only at altitudes above that of the International Space Station and hence was not used.

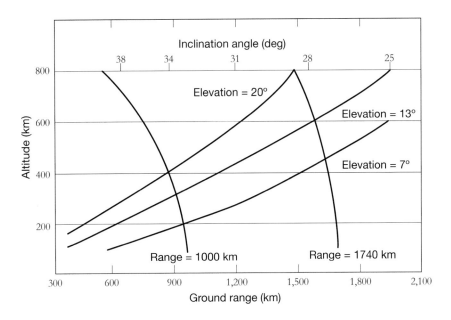

Figure 2.20   Lower elevation charts.

It is interesting to note here that after the LRIR collected sufficient data with its antenna pointed at 10 degrees elevation due south, NASA was surprised to find that the number of detections was higher than what its model had predicted. It took some time for NASA scientists to realize that Lincoln Laboratory had discovered a new mechanism of space-debris generation. It turned out that the inclination 10 degrees due south covers a volume of space in which solid rocket motors (SRMs) perform a few functions, such as injecting satellite payloads into their designated orbits, adjusting an orbit, and terminating an orbit. An SRM burn is not 100 percent efficient. It was suggested that SRMs eject globules of molten slag of various sizes. NASA knew about these but was surprised to discover, from the LRIR data, that the size of the ejected globules was in the range of a few centimeters. The presence of such large pieces was confirmed later by static tests that NASA conducted at the Marshall Space Flight Center in Alabama. A solid rocket motor was fired while being kept in place by mechanical means. The ejected material was captured, collected, examined, and confirmed the above explanation [23].

In later years, NASA analysts used LRIR space-debris data collected at various inclinations to sort out the vast number of detected space-debris objects. Their ability to correlate bands of debris objects with their originating breakups was a major accomplishment [24].

2.2.8   Radar Calibration for Space-Debris Data Collection

In October 1990, when the Long-Range Imaging Radar embraced its new function of space-debris data collection, it did so with no change to its calibration procedures and routines. In the spring of 1991, however, it became apparent to both Lincoln Laboratory analysts and NASA researchers that the data were not sufficiently accurate for estimating the orbits of debris objects. It was therefore decided to dedicate the July and August 1991 missions to improve the calibration procedures and accuracy of the radar: the method of measuring the off-axis position of the target was changed, and other components of the radar calibration system were tuned and tightened. The following subsections, drawing on Lincoln Laboratory Project Report PSI-176 [25], describe the improved radar calibration procedures and routines, as well as the relevant parts of the system hardware and software used

2.2.8.1   Calibration Procedures

The new calibration procedures introduced more frequent monitoring of the stability of the radar, greater use of built-in injected signals, better measurement of off-axis radar pointing, and improved radar cross-section calibration. Five sets of tests were used:

- Radar pointing-angles tests—to monitor elevation and traverse monopulse slopes using orbiting calibration spheres.
- Injection signals tests—to monitor the amplitudes and phases of the four receiver channels; principal polarization, opposite polarization, azimuth, and elevation.
- Channel-phasing tests—to adjust azimuth and elevation channel phases based on azimuth-cut and elevation-cut scans of calibration spheres.
- Sensitivity tests—to monitor radar cross-section calibration, radar sensitivity, and range accuracy using tracking data of satellite spheres.
- Noise tests—to monitor noise levels in all channels using (passive) background measurements in space-debris data-collection mode.

The Haystack Long-Range Imaging Radar antenna is shared with radio astronomers. The LRIR becomes active when the radar box replaces the radio astronomy box on the mount. The calibration process begins before the boxes are exchanged. While still on the ground, the radar box is fully connected to the system to allow the transmitter test, usually conducted at night with full power using a splash plate to reflect the power toward zenith. During the week before the radar box is lifted, the radar undergoes both transmitter and receiver tests with full power to verify its readiness to conduct satellite tracking and space-debris detection. These calibration procedures were found to be adequate and have not changed since the July 1991 mission. A new requirement to conduct radar calibration every 12 hours was added. On occasion, a long space-debris data-collection session required radar calibrations before and after the session. All calibration data were recorded in the radar's database. Each new set of calibration data was compared to the previous one to monitor the stability of the system. If there was an unacceptable difference, the debris data collection was stopped until the problem was resolved.

2.2.9   Off-Axis Angle Measurements

The Long-Range Imaging Radar determines the pointing of the beam by means of a monopulse antenna feed, called "monopulse" because the antenna can measure angle errors with a single pulse. The radar attempts to place the target at the center of the beam. Noise and other pointing imperfections cause small pointing errors. The radar measures these errors and then uses a tracking filter to apply corrections that bring the target closer to the center of the beam. Although this approach works quite well for tracking radars, it proved inadequate for space-debris data collection. Figure 2.21 illustrates how the monopulse concept works well for small errors. When the LRIR began debris data collection, it used the monopulse for large errors (off-sets), and the resulting error measurements were inadequate for initial debris orbit estimation.

In most tracking radars, the antenna feed consists of five horns: one horn whose phase center is placed precisely at the focus of the antenna and four offset horns that generate four offset beams. Although the Long-Range Imaging Radar has a unique modern single-horn design, it is easier to describe the monopulse concept with a five-horn design. The on-focus horn generates the main on-axis beam (also known as the "sum beam") through which the radar transmits and receives the main signal, S in figure 2.21. The offsetting of the other four horns produces beams that are squint-ed in proportion to the horn offset. The beams are offset a half a beamwidth away from the sum beam center. In each direction, elevation and traverse, the signal of the one beam is subtracted from the signal of the other to form a difference signal, D in figure 2.21. The difference signal is normalized with the sum beam to produce a "difference beam," depicted in figure 2.21 as an odd response curve, calibrated by

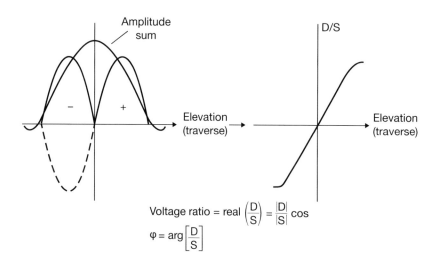

Figure 2.21   Monopulse angle estimate (beam center).

tracking satellite spheres and scanning the beam from one side of the sphere to the other. During an elevation scan, the traverse scan is held constant; during a traverse scan, the elevation scan is held constant. Each pulse of the scan has a known angular offset from the sphere being tracked and a measured value of the corresponding difference voltage. These two quantities allow the processing routine to assign an unambiguous angular offset to the corresponding normalized difference voltage. For each pulse, the radar measures the difference signal in both channels (elevation and traverse) and then uses these difference signals to read off the curve the corresponding angle errors. The angle ϕ in figure 2.21 determines the sense of the offset (i.e., up/down and left/right). Figure 2.21 also shows that the normalized difference response is linear close to the center of the beam and then becomes nonlinear as the offset approaches half a beamwidth.

In order to determine the orbit of the detected space-debris object, an estimate of its range, azimuth, and elevation, as well as an estimate of the rate of change of these parameters, is required. The LRIR is well calibrated to estimate range, azimuth, and elevation (RAE) and the range rate from a single pulse, which it measures using the Doppler shift. The angular rates estimated from multiple pulses during tracking are fine when the angle offsets are small. In zenith-pointing stare mode the radar does not track the detected object but instead records the measurements for use after the end of the debris data-collection session. As was explained above, the angular rates are determined by fitting a line to single-pulse, angle-offset measurements. Initially, analysts treated the monopulse measurements as if they were on a plane using rectilinear coordinates. This approach proved to be inadequate. Fly-through tests showed that the estimated parameters of orbits derived in this manner had unacceptably large errors; eccentricity and inclination were affected in particular.

The Lincoln Laboratory team decided to use spiral scanning of a calibration sphere satellite throughout the beam, hoping that it would be a better way to calibrate and process the monopulse data. Figure 2.22 shows raw spiral data superimposed on the equal-voltage contours of the monopulse model. For each data point, monopulse voltage and offset from the sphere orbit are known and are used in modeling the monopulse data. Figure 2.22 represents the traverse monopulse voltage across the beam and points beyond the 3 dB beam.

Figure 2.23a depicts a three-dimensional plot of unscreened and binned spiral data in which the scan was wider than the beamwidth. Figure 2.23b depicts screened data; all data outside the 3 dB beam have been discarded. It is easy to discern that the top surface is curved and that a plane model would not be as good an approximation of the monopulse data. Instead, this surface is modeled by third-order two-dimensional

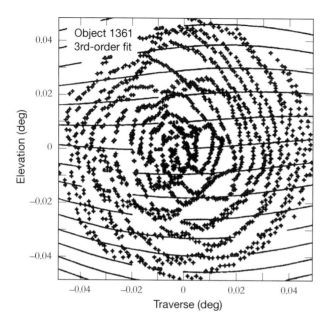

Figure 2.22   Spiral data superimposed on contour lines derived from spiral measurements of sphere 1 square meter in radar cross section.

polynomials. It was found that a third-order polynomial was sufficient to represent the monopulse voltages within the 6 dB two-way beam.

Figure 2.24 shows the constant monopulse voltage contours for traverse and elevation, and also shows that the nonlinear monopulse translates to a nonlinear coordinate system. Near the center of the beam, the spacing of the voltage contours is constant, whereas, away from the center, the contours bunch together and the spacing varies, reflecting the monopulse's increased nonlinearity near the edge of the beam. The beam offset angles can be determined by calculating the specific contours for the measured monopulse voltages for each pulse. The intersection of these contours is the off-axis location of the object for that pulse. In fly-through tests, this method produced more accurate orbit estimates.

Because its processing time was too long, the spiral method was not implemented in the real-time program. Instead, validated spiral scan data were sent to the Orbital Debris Analysis System together with the corresponding space-debris data for post-mission analysis and orbit determination. The validity of the data was checked at the radar site by using the monopulse tracking scheme.

a)

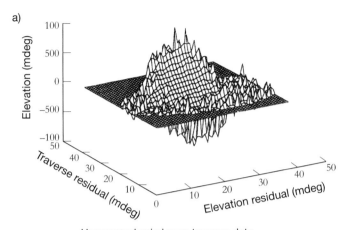

Unscreened spiral scan traverse data

b)

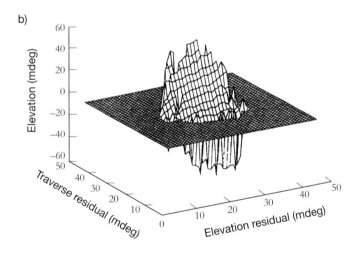

Screened spiral scan traverse data

Figure 2.23   (a) Three-dimensional plot of raw unscreened and binned spiral elevation data. (b) Three-dimensional plot of raw screened and binned spiral elevation data.

a)

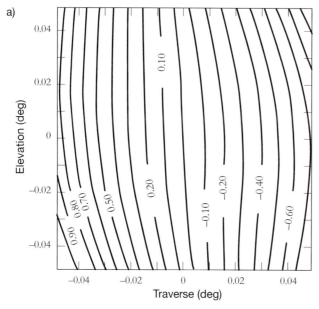

Third-order traverse monopulse contours

b)

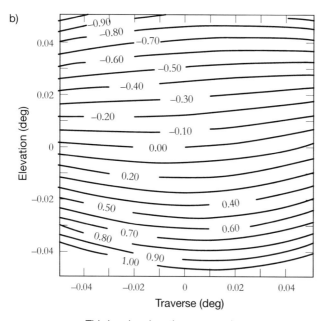

Third-order elevation monopulse contours

Figure 2.24 (a) Third-order traverse monopulse contours. (b) Third-order elevation monopulse contours.

2.2.10   Radar Cross-Section Estimation and Monitoring

In the hypervelocity collision test of a satellite model by an Air Force working group, mentioned in subsection 2.2.1, thirty-nine fragments were collected, examined, and assigned a size, the average of their dimensions being measured along three orthogonal axes. The radar cross section of each fragment was measured in an indoor RCS measurement range. The total radar cross section, sum of both principal and opposite polarizations, was measured for multiple orientations adequately sampling 4π steradians in the 2–18 GHz frequency band. The two polarizations were used because the collision fragments, unlike a sphere, scattered back power in both. Since the fragments' shapes were complex, the assigned RCS was the average total RCS.

A few fragments from the collision test were dropped from a balloon and tracked by the radars of the Kwajalein Missile Range. The radar cross-section data from Kwajalein were added to the indoor range data. The measured RCS was normalized by the square of the wavelength and is plotted in figure 2.25 in red color. The abscissa is the equivalent sphere diameter divided by the wavelength. The curve in black color is a nonlinear fit to the data. The NASA Size Estimation Model (SEM) [23] used the fitted curve to convert the radar's measured RCS to a size estimate. The blue

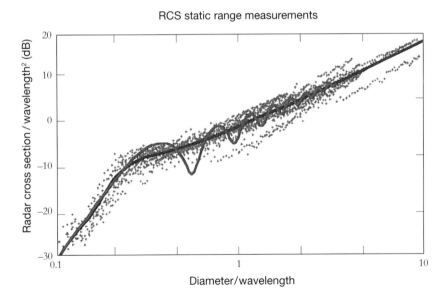

Figure 2.25   Estimation-conversion curve for size of space-debris objects.

curve is the calculated sphere RCS normalized in the same way as the data. The radar cross section of the sphere matches the data almost perfectly in the optical and Rayleigh's regions.

The SEM converts the measured RCS of the space-debris objects in both principal and opposite polarizations to the equivalent diameter of the debris. The radar routinely calibrates the principal polarization channel as part of the sensitivity test conducted when a calibration sphere is tracked. The software scales the received signal to match the known radar cross section of the sphere in dBsm units. The environment of the Long-Range Imaging Radar is not as pristine as that of the indoor range. The LRIR engineer has to deal with weather conditions and instabilities of the radar system, whose stability is monitored by injection signals. When the weather conditions change, the operator initiates calibration tests. During a rainstorm, the radar loses sensitivity: although data collection can continue, the radar will detect fewer objects. However, the radar must be recalibrated to ensure that its object size estimate based on radar cross section stays accurate. If there is no calibration sphere in the radar field of view, data collection stops.

In the indoor range, a standard dipole is used for calibrating the opposite polarization. Unfortunately, there is no reliable dipole in space. Therefore the opposite polarization channel is calibrated on the ground before the radar box is lifted. A standard horn is placed 17 feet away from the feed inside the radome. The horn position is adjusted to produce a null in both angle channels. In this position, the phase center of the standard horn is at boresight. The horn transmits a linearly polarized signal, which produces two equal left and right circular polarized signals. If the radar's opposite and principal polarization channels were identical, the signal levels at the output of the radar receiver would be equal, but because the channels are not identical, the opposite/principal polarization ratio is recorded. A typical value is about 1.37 dB. For the duration of the mission, the calibration system uses the measured polarization ratio to calibrate the opposite polarization channel. The new calibration procedures were adopted in August 1991. Figure 2.26a shows the radar cross-section errors of the principal polarization channel during the space-debris data collection in October 1991.

For the 1991 October space-debris data-collection mission, three measurement points were taken during a rainstorm when the radar experienced 2 dB reduction of sensitivity. The root mean square of the data without the three outlying points is 0.5 dB, a result typical for the Long-Range Imaging Radar. Figure 2.26b shows the radar cross-section errors of the principal polarization channel during the May 1992 mission, when the LRIR achieved better results in clear weather.

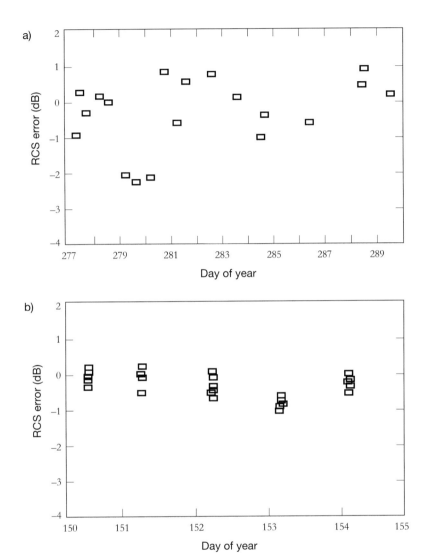

Figure 2.26   (a) Principal polarization (PP) channel radar cross-section (RCS) errors measured on space-debris data-collection mission, October 1991. (b) Principal polarization (PP) channel radar cross-section (RCS) errors measured on space-debris data-collection mission, May 1992.

2.2.11   Impact and Significance of the Haystack Long-Range Imaging Radar Space-Debris Data

Since the inception of space-debris data collection in 1990, the LRIR measurements have provided a means to count and size space-debris objects. With this information, the evolution of the debris population over time has been monitored by the Orbital Debris Program Office at NASA's Johnson Space Center. The data also provide the orbital parameters of each debris object. The debris data are stored in the in static catalog of the center, which also maintains a predicted catalog of debris element sets available to users for their needs. In modeling the space-debris population, NASA analysts calculate the flux of the debris objects, the number of objects crossing a unit area from all directions per unit of time. It is a quantity of interest because debris objects can collide with a satellite from any direction. For Haystack Long-Range Imaging Radar (LRIR) and Haystack Auxiliary (HAX) radar, the flux is equal to the number of detected objects per unit of time divided by the area wrapping the beam.

Details of the findings from LRIR and HAX space-debris data collection can be found in many NASA publications [26]. Figures 2.27–2.29 aptly illustrate the importance of the space-debris data that the Long-Range Imagaing Radar has collected since October 1990.

Figure 2.27 displays the cumulative detection rates for both all detected debris objects and for objects in the Space Surveillance Network catalog. The correlation of the debris data to their source is immediately evident. Since the detection rate is proportional to the population, the chart indicates that the detected debris population is about 100 times the size of the cataloged population. The Haystack Long-Range Imaging Radar detected objects down to 0.5 centimeter in size; the SSN catalog included only about 9,000 objects, implying that population of space-debris objects down to 0.5 centimeter was about 900,000, a number in reasonable agreement with NASA's estimate of about 500,000 objects down to 1 centimeter. It is worth noting that the predicted detection rate in figure 2.27 is based on a 1990 model. The current models are more accurate because they reflect the LRIR and HAX data collected since 1990.

Figure 2.28 shows the estimated density of space debris as a function of altitude derived from a few hundred hours of data measurements. The figure's estimate has survived, with only small changes, for twenty years. The changes are due primarily to space debris produced by a Chinese antisatellite missile test in 2007 [27]. Figure 2.28 shows all the features of the latest estimates based on thousands of hours of

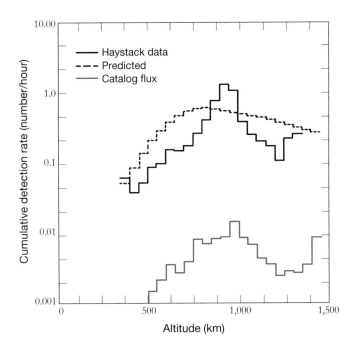

Figure 2.27 Plot of cumulative detection rate of space-debris objects (number per hour) against altitude.

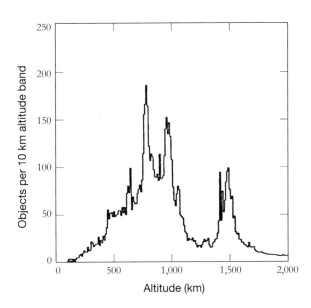

Figure 2.28 Space-debris density as a function of object altitude.

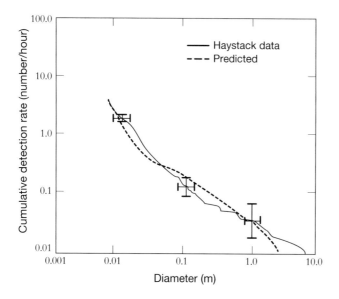

Figure 2.29   Cumulative space-debris detection rate as a function of object size.

observations, most notably, the peaks of the space-debris bands from 800 to 1,500 kilometers altitude.

Figure 2.29, the third chart derived from LRIR data, portrays the cumulative detection rate as a function of debris object size: debris density in space, as measured by the detection rate in the LRIR beam, grows exponentially as the size of the debris objects gets smaller. The number of 1-centimeter debris objects is about 400 times greater than the number of 5-meter objects.

2.2.12   Summary of the Space-Debris Data-Collection Effort

With the implementation of the space station *Freedom* program in 1989 and consequent concerns about collision of space-debris objects with the station and the threat this posed to humans on board, NASA had an urgent need for space-debris data. The Orbital Debris Program Office was charged with assessing the risk to *Freedom* from debris objects 1–10 centimeters in size. At that time, the office lacked the data to establish the validity of existing models. The data NASA had gathered from payloads that returned from space, such as those of the Space Shuttles and the Long Duration Exposure Facility, was limited to space-debris objects smaller than 1 centimeter. The Long-Range Imaging Radar program gave the Orbital Debris Program Office the ability to test and confirm its space-debris models with confidence. If NASA were to launch a large program like *Freedom* today, it could calculate and predict the flux of debris objects and assess the risk it posed with confidence.

At present, the space-debris problem is recognized as serious by the scientific community; there is great concern about the future effects of space debris. As more countries launch satellites, the space-debris population will quickly grow in size. Events like the collision of the Russian Kosmos 2251 and the U.S. Iridium 33 communications satellite in February 2009 will continue to occur, as will events such as the destruction of the Chinese Fengyun-1C weather satellite in January 2007, which doubled the number of objects in the Space Surveillance Network catalog [27]. Indeed, because of the increase in the number of space-debris objects in near Earth, space is becoming a "minefield." Clearing that minefield will be a challenge that may rival the Manhattan Project in effort and cost.

Finally, as described in section 2.2, Lincoln Laboratory demonstrated that high sensitivity was more important than a large field of view in detecting space-debris objects, a finding that would be used to build models of debris distribution in altitude and size. Indeed, the models developed by NASA were dependent to a large extent on the detection data from the narrow-beam (0.05-degree) Haystack radar. These models were used in designing the International Space Station so that the human beings and critical life systems on board could be protected from any catastrophic collision between the station and space debris.[13]

References

1. Grometstein, A. A. 2011. Space Situational Awareness. In *MIT Lincoln Laboratory: Technology in Support of National Security*, 163–193. Lexington, MA: MIT Lincoln Laboratory.
2. Kessler, D. J., and B. G. Cour-Palais. 1978. Collision Frequency of Artificial Satellites: The Creation of a Debris Belt. *Journal of Geophysical Research* 83 (no. A6): 2637–2646.
3. Evans, J. V., and G. H. Pettengill. 1963. The Scattering Behavior of the Moon at Wavelengths of 3.6, 68, and 784 Centimeters. *Journal of Geophysical Research* 68:423.

---

13. Conversation with Eugene G. Stansbery in 2011, when this chapter was being written.

4. Evans, J. V., J. I. Levine, R. P. Ingalls, R. A. Brockelman, G. H. Pettengill, L. P. Rainville, et al. "Reflection Properties of Venus at 3.8 Cm," MIT Lincoln Laboratory Technical Report 456, 6 September 1968.

5. Cooley, J. W., and J. W. Tukey. 1965. An Algorithm for the Machine Calculation of Complex Fourier Series. *Mathematics of Computation* 19:297–301.

6. G.L. Guernsey and C.B. Slade, "Observations of a Synchronous Satellite with the Haystack Radar," MIT Lincoln Laboratory Technical Note 1972-41, December 1972.

7. Raup, R. C., R. A. Ford, G. R. Krumpholz, M. G. Czerwinski, and T. E. Clark. 1990. The Best Approximation of Radar Signal Amplitude and Delay. *Lincoln Laboratory Journal* 3 (2): 311–327.

8. Krumpholz, G. K. "L-Band Satellite Polarimetry at Millstone," in *Proceedings of the 1989 Space Surveillance Workshop*, pp. 191–198, Lexington, MA: MIT Lincoln Laboratory, 1989.

9. Krumpholz, G. R., and T. E. Clark. "Polarimetric Signal Processing at Millstone and the FPS-85," in *Proceedings of the 1992 Space Surveillance Workshop*, pp. 265–272, Lexington, MA: MIT Lincoln Laboratory, 1992.

10. Henize, K. G. (January 1957). The Baker-Nunn Satellite-Tracking Camera. *Sky and Telescope* 16:108.

11. Banner, G. P., and W. Burnham. (July 1981). MIDYS: The Millstone Dynamic Scheduler. *MITLincoln Laboratory Project Report-STK* 111.

12. R. Sridharan and W.P. Seniw, "An Intermediate Averaged Theory for High Altitude Orbits," MIT Lincoln Laboratory Technical Note 1979-25, 27 June 1979.

13. R. Sridharan and W.P. Seniw, "ANODE: An Analytic Orbit Determination System," MIT Lincoln Laboratory Technical Note 1980-1, vol. 1, 24 June 1980.

14. Lane, M. "An Analytical Treatment of Resonance Effects in Satellite Orbits," MIT Lincoln Laboratory Technical Report 841, 19 August 1988.

15. Gaposchkin, E. M. "Metric Calibration of the Millstone Hill L-Band Radar," MIT Lincoln Laboratory Technical Report 721, 19 August 1985.

16. Evans, J. V., A. Freed, and L. G. Kraft. "Millstone Hill Radar Propagation Study: Calibration," MIT Lincoln Laboratory Technical Report 508, 5 October 1973.

17. Coster, A. J., E. M. Gaposchkin, L. E. Thornton, G. R. Krumpholz, and T. A. Cott. "Ionospheric Effects in Satellite Tracking," in *Proceedings of 7th International Ionospheric Effects Symposium*, pp. 80–87, MIT-LIN-MS-10221, 4–6 May 1993.

18. Coster, A. J., E. M. Gaposchkin, and L. E. Thornton. "Real-Time Ionospheric Monitoring System Using the GPS," Lincoln Laboratory Technical Report 954, 24 August 1992.

19. Arecibo Observatory, National Astronomy and Ionosphere Center (NAIC), https://www.naic.edu/ or *Wikipedia*, "Arecibo Observatory," last modified 16 May 2016, https://en.wikipedia.org/wiki/Arecibo_Observatory.

20. Spence, L., A. Pensa, and T. Clark. 1974. Radar Observation of the IMP-6 Spacecraft at Very Long Range. *Proceedings of the IEEE* 62 (12): 1717–1718.

21. Taff, L. G., D. E. Beatty, A. J. Yakutis, and P. M. S. Randall. 1985. Low Altitude, One Centimeter Space Debris Search at Lincoln Laboratory's (M.I.T.) Experimental Test System. *Advances in Space Research* 5 (2): 35–45.

22. Skolnik, M. I. 1970. *Radar Handbook*. 1st ed., chap. 2. New York: McGraw-Hill.

23. Siebold, K. H., M. J. Matney, G. W. Ojakangas, and B. J. Anderson. "Risk Analysis of 1–2 Cm Debris Population for Solid Rocket Motors and Mitigation Possibilities for Geo-Transfer Orbits," in *Proceedings of the First European Conference on Space Debris*, pp. 349–351, Darmstadt, Germany, 5–7 April 1993, Darmstadt: European Space Operations Center, 1993.

24. Settecerri, T. J., E. G. Stansbery, and T. J. Hebert. "Radar Measurements of the Orbital Environment Haystack and HAX Radars, October 1990–October 1998," NASA Report JSC-28744, October 1999.

25. Zuerndorfer, B. W., R. I. Abbot, E. M. Gaposchkin, I. Kupiec, R. W. Miller, and T. J. Morgan. "Calibration of the Long Range Imaging Radar," MIT Lincoln Laboratory Project Report PSI-176, 31 December 1992

26. C.L. Stokley, J.L. Foster, E.G. Stansbery, J.R. Benbrook, and Q. Juarez, "Haystack and HAX Radar Measurements of the Orbital Debris Environment—2003," Johnson Space Center Report 62815, November 2003.

27. Nancy Atkinson. "Chinese Space Debris Collides with Russian Satellite," *Universe Today: Space and Astronomy News* (video), 8 March, http://www.universetoday.com/100608/chinese-space debris-collides-with-Russian-satellite/, 2013.

3

Overview of Wideband Radar Imaging Technology at MIT Lincoln Laboratory

Craig Solodyna

3.1 Introduction

The science and art of wideband radar imaging of satellites and ballistic missiles have matured over the past forty-three years at MIT Lincoln Laboratory. They have stimulated the development of powerful signal-processing techniques that have revolutionized the field of wideband image generation and led wideband image analysts to a deeper understanding of ballistic missiles, orbiting satellites, and aircraft. MIT Lincoln Laboratory could not have developed these techniques without the highly capable radar systems installed at the Millstone Hill complex in Westford, Massachusetts, and at the U.S. Army's facility at the Kwajalein Atoll in the Marshall Islands. Over the years, the laboratory has helped to turn these systems into some of the most sensitive long-range wideband radars in the world. Before moving on to the laboratory's seminal work in signal processing, this chapter will briefly describe these enabling radars.

The world's first ground-based wideband radar was ALCOR (ARPA-Lincoln C-Band Observables Radar; figure 3.1) at the Reagan Test Site (RTS) on the Kwajalein Atoll (figure 3.2) in the Central Pacific Ocean, developed in 1970 to support wideband radar discrimination research. Wideband radar research at Lincoln Laboratory began with a program named "Wideband Observables," whose initial objectives were to verify the nature of wideband return signals from realistic targets and to investigate technology challenges that might arise in developing full-scale wideband radar systems [1, 2, 3].

ALCOR operates at C band (5,672 MHz) and uses a wideband chirp waveform with a bandwidth (BW) of 512 MHz to achieve a range resolution of 53 centimeters, given by the expression 1.81c/2BW, where c is the speed of light in vacuum, BW is the bandwidth of the radar, and the factor 1.81 is associated with the Hamming spectral

Figure 3.1 (Left) 68-foot-diameter ALCOR radome (shown near sunset). (Right) 40-foot-diameter ALCOR antenna dish and pedestal within radome. ALCOR was the world's first wideband imaging radar.

Figure 3.2 Aerial picture of the Reagan Test Site. (Left to right) ALCOR, TRADEX radar, MMW radar, and ALTAIR. ALCOR and MMW radar (under radomes) are wideband imaging radars. TRADEX radar and ALTAIR are narrowband tracking, discrimination, and instrumentation radars.

weighting function, nominally used to reduce the effect of sidelobes in the received radar return signal. ALCOR's wideband waveform is a 10-microsecond pulse that is linearly swept over the 512 MHz frequency range. A high signal-to-noise ratio (SNR) of 23 dB per pulse on a 1-square-meter target at a range of 1,000 kilometers is achieved with a high-power transmitter (3-megawatt peak and 6-kilowatt average) and a 40-foot-diameter antenna. The ALCOR beamwidth is 5.2 milliradians or 0.3 degree. Together with a high-performance antenna mount, this beamwidth enables ALCOR to produce precision target trajectories and provide high-quality designation data to the three other high-performance Kwajalein radars: ALTAIR (ARPA Long-Range Tracking and Instrumentation Radar), TRADEX (Tracking and Discrimination Experiment) radar, and MMW (Millimeter-Wave) radar.

In 1972, the TRADEX ultrahigh-frequency (UHF) radar was replaced by an S-band system (built by RCA under the direction of Lincoln Laboratory), which also had L-band capability. Research on frequency-jump-burst waveforms carried out by the laboratory in the 1960s enabled this technology to be implemented in the S-band (2,950.8 MHz) system. Of the new S-band waveforms, one originally had a wideband signal bandwidth of 250 MHz, achieved by transmitting a string (i.e., burst) of 3-microsecond pulses; each with a different center frequency such that the bandwidth spanned by the burst was 250 MHz. The spacing and number of pulses in a burst were variable, and the maximum repetition rate was 100 bursts per second. Coherent integration of a burst yielded a range resolution of 1.086 meters. Processing of the frequency-jump-burst waveform at S band was implemented postmission. Although the TRADEX radar is commonly known as a narrowband tracking and discrimination radar (which it is at L band), it also was the world's second wideband imaging radar by virtue of its unique S-band frequency-jump-burst waveforms.

The world's third ground-based wideband radar used primarily for space surveillance was the Haystack Long-Range Imaging Radar (LRIR) located in Westford, Massachusetts, operating at X band (10 GHz) with a bandwidth of 1,024 MHz to achieve a range resolution of 26.5 centimeters. The development of the LRIR was sponsored by the Advanced Research Projects Agency (ARPA);[1] upon completion of the radar in 1978, its operations were supported by the U.S. Air Force. As shown in figure 3.3, the Long-Range Imaging Radar has a large-diameter (120-foot) Cassegrain-type antenna. Plug-in radio-frequency (RF) boxes at the vertex of the parabolic

---

1. The Advanced Research Projects Agency (ARPA) was the precursor to today's Defense Advanced Research Projects Agency (DARPA).

Figure 3.3   (Left) Cutaway view of the 120-foot Long-Range Imaging Radar (LRIR) antenna within its 150-foot radome. (Right) LRIR feed horn and transmitter/receiver radio-frequency box in its test dock beside the antenna. The cutaway view shows this box being lifted into the focus of the antenna.

dish support various communications, radio astronomy, and radar functions. The interchangeable boxes, which measure $8 \times 8 \times 12$ feet, are large enough to accommodate the high-power (400-kilowatt peak; 200-kilowatt average) transmitter and associated hardware, as needed. The radar's antenna surface tolerance allows efficient operation up to 50 GHz, thus readily supporting operation at X band (10 GHz). Capable of detecting, tracking, and imaging rotating satellites out to geosynchronous altitudes, the Long-Range Imaging Radar is currently undergoing a metamorphosis; renamed the "Haystack Ultrawideband Satellite Imaging Radar" (HUSIR), it will operate at both X band (10 GHz), with a bandwidth of 1,024 MHz, and at W band (95 GHz), with a bandwidth of 8 GHz. At W band, the range resolution will be 3 centimeters.

The world's fourth ground-based wideband imaging radar was the Millimeter-Wave (MMW) radar (figure 3.4) at the Reagan Test Site, built to extend the general imaging and tracking capabilities of ALCOR and the Long-Range Imaging Radar and to develop millimeter-wavelength radar signatures and wideband imaging of ballistic missiles and satellites. The MMW radar also has a Cassegrain-type antenna consisting of a 45-foot-diameter parabolic dish with a 3.56-foot-diameter hyperbolic reflector. The radar became operational at Ka band (35 GHz) in 1983 and at W band (95.48 GHz) in 1984; both systems initially had a bandwidth of 1,000 MHz, increasing to 2,000 MHz in 1989, although the W-band system was taken offline in 2003. By 2013, the bandwidth of the MMW radar had been extended to 4,000 MHz, resulting in a range resolution of 6.63 centimeters, with a transmitted

Figure 3.4 (Left) Millimeter-Wave (MMW) radar 40-foot parabolic dish at the Reagan Test Site. (Right) View of the MMW radome.

pulsewidth of 50 microseconds at a maximum pulse-repetition frequency of 2,000 pulses per second. The initial peak power was 60 kilowatts at Ka band and 1.6 kilowatts at W band. The maximum peak power at Ka band is currently 100 kilowatts, which is provided by two 50-kilowatt traveling-wave tubes. Radar transmission losses were decreased over the years by the introduction of a beam waveguide system (rather than a conventional waveguide system) and the introduction of a quasi-optical combiner into the feed system in 1993. These changes increased the radar's sensitivity by 10 dB [4, 5]. The combination of metric accuracy, wide bandwidth, and high Doppler resolution makes the Millimeter-Wave radar an excellent sensor for a real-time discrimination test bed. The Kwajalein Discrimination System (KDS) was implemented at the MMW radar from 1988 through 1992 to demonstrate the feasibility of several real-time wideband-discrimination algorithms, many of which have been incorporated into the radar's real-time system.

The world's fifth ground-based wideband radar used primarily for space surveillance was the Haystack Auxiliary (HAX) radar (figure 3.5), also located in Westford, Massachusetts, which operates at Ku band (16.7 GHz), with a bandwidth of 2 GHz. HAX was developed to meet the competing requirements of U.S. Space Command, which was interested in tracking foreign satellites, and NASA, which was interested in collecting data on space debris. The Haystack Auxiliary radar operates with its own antenna, transmitter, radio-frequency hardware, and receiver, but it shares control, signal-processing, and data-processing systems with the Haystack Long-Range

Figure 3.5   (Left) Haystack Long-Range Imaging Radar (LRIR) and Haystack Auxiliary (HAX) radar in Westford, Massachusetts. (Right) 40-foot HAX dish operates in the Ku band (16.7 GHz) and has a bandwidth of 2 GHz.

Imaging Radar; it began wideband imaging and space-debris detection and tracking in 1993.

The general subject of wideband imaging is known throughout the world; indeed, thousands of technical papers have been published on the topic in the foreign open-source literature.[2] To illustrate wideband imaging in Germany, TIRA (Tracking and Imaging Radar) in Wachtberg-Werthoven, Germany (figure 3.6) is a Ku-band (16.7 GHz) wideband imaging radar that normally operates at a bandwidth of 800 MHz,[3] but it can also operate at a bandwidth of 2 GHz [6]. In this chapter, either TIRA wideband images or Lincoln Laboratory's computer-simulated wideband images will be shown.

The following sections discuss topics fundamental to wideband image generation and wideband image analysis: range profiles, range-time-intensity plots, Doppler-time-intensity plots, range-Doppler imaging, autofocusing of wideband radar images, extended coherent processing, bandwidth extrapolation, and ultrawideband coherent processing. Most of the discussions are summarized from a number of unclassified Lincoln Laboratory internal documents and reorganized for presentation here.

---

2. A literature search performed in March 2016 using the Compendex (Comprehensive Engineering Index), the Inspec database, and the NTIS (National Technical Information Service) database, for example, identified 349 papers from Germany, 50 papers from Russia, 295 papers from Japan, and 942 technical papers from China on wideband imaging.
3. Both wideband images and wideband image analysis techniques have been documented by German scientists in open-source publications and on the TIRA website [7].

Figure 3.6   (Left) Tracking and Imaging Radar (TIRA) facility. (Right) Cutaway view of the 34-meter-diameter TIRA antenna. (Both images provided courtesy of the Fraunhofer Institute in Germany.)

3.2   Range Profiles

The ability of wideband radar to resolve targets in range can be used to characterize and identify targets on the basis of the radar cross-sectional distribution of individual scattering centers. Range profiling has some advantages over range-Doppler imaging: it can be used even when range-Doppler imaging fails because of insufficient variation in aspect angle or lack of Doppler motion (e.g., a geostationary, attitude-stable satellite), or because of undersampling caused by too low a radar pulse-repetition frequency.

A range profile is defined as the range distribution of scatterers along the radar line of sight (RLOS), where the vertical axis reflects the strength (in dBsm) of the scatterer, and the horizontal axis reflects the range position (measured in terms of the range gate). The highest-fidelity range profile is obtained for the radar having the highest bandwidth. Figure 3.7 shows simulated range profiles for a missile. Usually a time sequence of range profiles is displayed, with one profile slightly displaced above another in the "range-time-intensity" (RTI) plot, where the vertical axis is now taken to be time. Each individual entry in the RTI plot is the range profile of the object measured at a particular instant of time.

3.3   Range-Time-Intensity Plots

Figure 3.8 shows a computer-simulated range-time-intensity (RTI) plot of a rotating dumbbell. The radar line of sight is incident from the left. A simple plot of radar cross

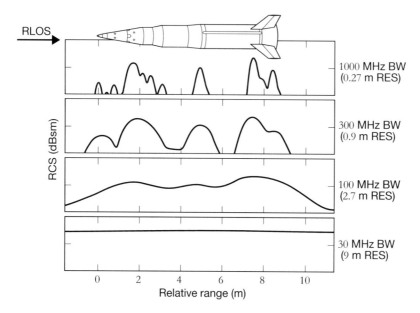

Figure 3.7 Computer-simulated wideband range profiles for a missile obtained with different radar bandwidths. The highest fidelity is obtained for the radar having the highest bandwidth. BW, bandwidth; RLOS, radar line of sight; RES, resolution in range.

section against time is shown in the first image. Time increases along the vertical y axis. Three RTI plots are shown for different bandwidth radars ranging from 300 to 4,000 MHz. The apparent length is determined by the true length multiplied by the cosine of the aspect angle. The range resolution is determined as previously stated and is shown in figure 3.8.

Figure 3.9 (left) shows a photograph of an aircraft; figure 3.9 (right) shows a range-time-intensity plot of the same aircraft taken by a wideband radar over approximately 30 seconds, during which time the aircraft was executing a standard turn. The vertical axis is time (in seconds) and the horizontal axis is range (in meters). The range profiles of the aircraft are displaced one slightly above the other in the RTI plot. The specular response at approximately 15 seconds corresponds to the time when the radar line of light is perpendicular to the side of the aircraft. The range-time-intensity plot gives a lower-bound estimate of the aircraft's true length (22 meters).

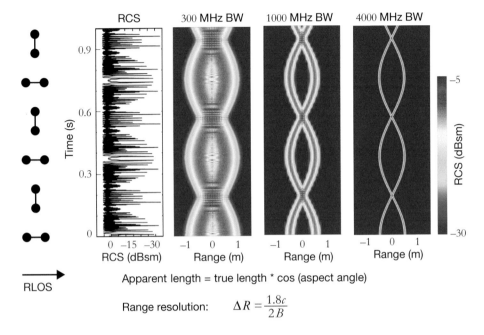

Figure 3.8 Range-time-intensity plots for a rotating dumbbell for three radars with different bandwidths.

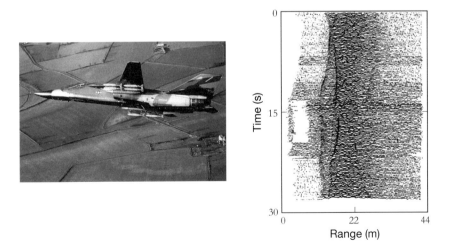

Figure 3.9 (Left) An aircraft in flight. (Right) Range-time-intensity (RTI) plot of the same aircraft taken with a wideband radar.

3.4   Doppler-Time-Intensity Plots

These show the Doppler frequency of an object along the horizontal axis as a function of the time along the vertical axis. Figure 3.10 shows a computer-simulated Doppler-time-intensity (DTI) plot of a rotating dumbbell for three different integration times T_{int}. The Doppler frequency resolution is $\Delta f = 1.8/T_{int}$. Hence the best resolution is obtained for the longest integration time.

3.5   Range-Doppler Imaging

A wideband radar image consists of a Fourier transform of a selected sequence of properly phase-referenced wideband radar return signals reflected from a target. The output from the Fourier transform is often referred to as a "range-Doppler image." It is assumed that the radar has a signal-to-noise ratio sufficient to reliably track the target and a bandwidth sufficient to resolve individual scatterers in range. The major Doppler component of a satellite is caused by the translational motion of the satellite

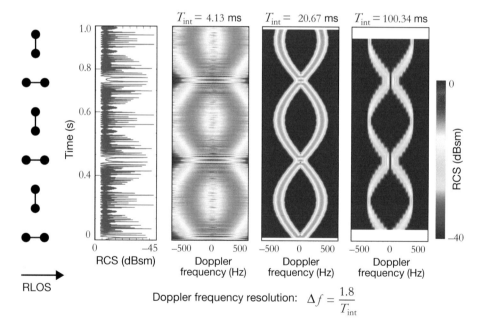

Figure 3.10   Doppler-time-intensity plot for a rotating dumbbell for three different integration times.

in orbit. This translational component must be removed from each pulse before Fourier processing. To accomplish this, a smooth trajectory is calculated by fitting a polynomial function describing the satellite's orbit to the metric data. The resulting fit is used to eliminate tracking jitter and to apply a phase correction to each pulse in order to remove the Doppler translational component. When this has been done, the Doppler components that remain are those caused by the rotation of individual scatterers about their common center of mass. This rotation must be determined to convert Doppler frequency (in hertz) to cross range (in meters). The result of this conversion is the radar image of most interest—the range/cross-range image. In the case of either Earth-referenced stable or inertially stable satellites, determining rotational motion is straightforward, whereas, in the case of highly complex motion involving a combination of spin, tumble, and precession, it is quite difficult.

Figure 3.11 summarizes the salient concepts in range-Doppler radar imaging. Range resolution Δx_{range} is independent of the range to the target and is solely determined by the bandwidth of the radar. For the Long-Range Imaging Radar, which has a bandwidth of 1,024 MHz, the range resolution Δx_{range} is 26.5 centimeters. For the Haystack Auxiliary radar, which has a bandwidth of 2,048 MHz, the range resolution

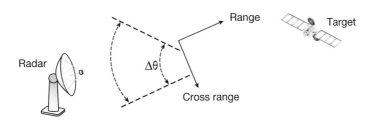

Range resolution is provided by the radar bandwidth

$$\Delta X_{range} = \frac{1.81\ C}{2BW}$$

LRIR: BW = 1024 MHz, ΔX_{range} = 26.5 cm

HAX: BW = 2048 MHz, ΔX_{range} = 13.25 cm

Cross range resolution is provided by the sampling of the target aspect angle $\Delta\theta$ caused by the relative motion between the radar and the target

$$\Delta X_{cross\ range} = \frac{1.81\lambda}{2\Delta\theta}$$

LRIR: λ = 3 cm, $\Delta X_{cross\ range}$ = 26.5 cm ($\Delta\theta$ = 5.87°)

HAX: λ = 1.79 cm, $\Delta X_{cross\ range}$ = 13.25 cm ($\Delta\theta$ = 7.03°)

Figure 3.11 Illustration of the basic concepts in wideband radar imaging.

Δx_{range} is 13.25 centimeters. Cross-range resolution is provided by sampling the target aspect angle $\Delta\theta$ caused by the relative motion between the radar and the target according to the formula $\Delta x_{\text{cross-range}} = 1.81\lambda/2\Delta\theta$, where λ is the wavelength of the radar, $\Delta\theta$ is the aspect angle interval over which the image is made, and 1.81 is the Hamming weighting factor used to reduce sidelobes. It is a common practice at Lincoln Laboratory to make range-Doppler images that have the same resolution in both cross-range and range directions. For the Long-Range Imaging Radar, which has a wavelength (λ) of 3 centimeters, a cross-range resolution $\Delta x_{\text{cross-range}}$ of 26.5 centimeters is obtained if the total change in aspect angle $\Delta\theta$ in one image is 5.87 degrees. For the Haystack Auxiliary radar, which has a wavelength (λ) of 1.79 centimeters, a cross-range resolution $\Delta x_{\text{cross-range}}$ of 13.25 centimeters is obtained if the total aspect angle change $\Delta\theta$ is 7.03 degrees.

The basic concept of range-Doppler imaging is illustrated by a rotating dumbbell, as shown in figure 3.12. For such a target, the velocity components seen by the radar are a function of the distances of individual scatterers about the center of mass and the rotational motion of those scatterers. The dumbbell has two scatterers, and a Fourier analysis of return signals from this target would contain two Doppler components. The Doppler components might be located in the same radar-range-resolution cell, depending on the time at which the target is observed. Computer-simulated images of the dumbbell are shown on the right.

For a wideband radar with good range resolution, most scatterers on a target would be resolved in range. Fourier processing then further resolves the scatterers in Doppler. Because there is a correlation between them, the scatterers' location can be used to determine the observed Doppler component if the attitude motion of the target is either known or can be estimated.

The processing interval in an image is usually chosen to be short enough that scatterers do not move through range and Doppler resolution cells, in which case, only linear approximations are used in the imaging algorithms, and the range-Doppler image is commonly referred to as a "linear image." This concept is illustrated in figure 3.13. One must limit the time interval over which a sequence of pulses forming the image is processed. A scatterer remains within a single range-resolution cell over the time interval T_R, provided that the range walk, defined as $|(dR/dt)|T_R$, is less than the range resolution of the radar. A scatterer remains within a single Doppler resolution cell over a time interval T_D, provided that the Doppler walk, defined as $|(d^2R/dt^2)|T_D$, is less than the Doppler resolution of the radar, defined as $1.81\lambda/2T_{\text{INT}}$, where T_{INT} is the coherent processing time interval. If the processing interval is long enough that these linear approximations are violated and scatterers move through

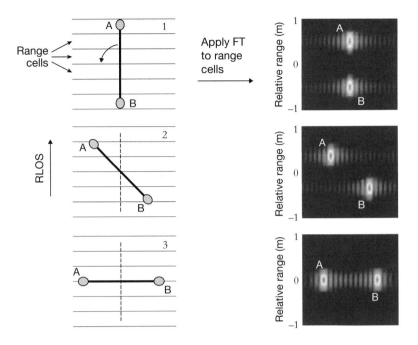

Figure 3.12   Computer-simulated wideband radar images of a dumbbell target. A dumbbell target (left) consists of two identical point scatterers (A and B) shown in red (right). Due to the target's rotational motion about its center of mass, the two scatterers are separated by five range gates in frame 1 and four range gates in frame 2. In frame 3, both point scatterers appear in the same range gate. FT, Fourier transform.

range and Doppler resolution cells, a special nonlinear algorithm for extended coherent processing (ECP; discussed later in the chapter) must be used to calculate the image. If the ECP algorithm is not used, the images will be smeared both in range and in cross range.

Range-Doppler images are useful in identifying the external shape of a target. Figure 3.14 illustrates a sequence of computer-simulated images of an idealized rocket body consisting of a number of specular surfaces and diffraction edges. As the range/cross-range images show, a strong return signal (i.e., specular) occurs when the radar line of sight is perpendicular to a reflecting surface. Specular return signals from the cylinder are seen in images B and B′. Specular return signals from the conic are seen in images D and D′. Image A shows a slipping point return acquired when the RLOS is perpendicular to the curved nose of the rocket body. Rather than a broad

To prevent smearing caused by scatterers migrating through range and Doppler resolution cells, one must limit the time interval over which one processes a collection of pulses

Limitation due to Range Walk

A scatterer remains within a single range resolution cell over a time interval, T_R, as long as

$$\text{Range walk} = |\dot{R}| \, T_R \leq \frac{1.81 \times c}{2 \, BW} = \text{Range resolution (m)}$$

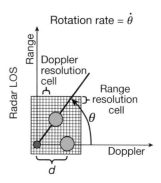

Rotation rate = $\dot{\theta}$

Limitation due to Doppler Walk

A scatterer remains within a single Doppler resolution cell over a time interval, T_D, as long as

$$\text{Doppler walk} = |\ddot{R}| \, T_D \leq \frac{1.81 \times \lambda}{2 \, T_{INT}} = \text{Doppler resolution (m/s)}$$

where T_{INT} is the coherent processing time interval. For maximum processing interval, let $T_D = T_{INT}$

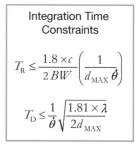

Integration Time Constraints

$$T_R \leq \frac{1.8 \times c}{2 \, BW} \left(\frac{1}{d_{MAX} \, \dot{\theta}} \right)$$

$$T_D \leq \frac{1}{\dot{\theta}} \sqrt{\frac{1.81 \times \lambda}{2 d_{MAX}}}$$

Figure 3.13   Illustration of the concepts of range walk and Doppler walk.

specular return, as would be obtained from a flat surface, a single high radar cross-section return is obtained. Extended-range returns and their associated sidelobes are seen in image E since the radar line of sight is incident on the open rocket nozzle. When radar energy is incident on an open structure (such as an open rocket nozzle), it reflects multiple times on internal surfaces before reflecting back to the radar. These returns are received later in time by the radar and are treated in the wideband imaging software as occurring at a greater, or extended range. (Extended-range returns on actual, rather than simulated, wideband images will be seen when TIRA wideband images of the Space Shuttle are presented later in this chapter.) Figure 3.15 shows that, if each image is then rotated to compensate for the viewing angle (i.e., for the angle between the radar line of sight and the body symmetry axis), a composite image, which is the superposition of all the images, can be formed to show the outline of the rocket body.

Because a radar is able to measure range and Doppler only along the radar line of sight, range/cross-range imaging locates scatterers and specular surfaces on a body by

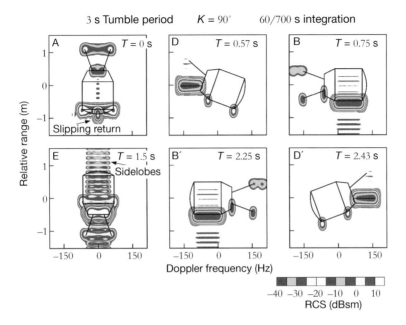

Figure 3.14   Computer simulation of speculars and diffraction edges for a tumbling rocket body.

displaying projections onto the image plane determined by the RLOS and the body's angular velocity. The body shape obtained from range-Doppler imaging will be an outline of the actual body shape (as defined by scatterers and specular surfaces located on the body) projected onto the image plane.

Wideband radar imaging is thus seen to be a five-step process:

1. Determination of a precise orbit fit from radar metric data;
2. Removal of the translational Doppler components and correction of tracking jitter using the precise orbit;
3. Fourier transformation of a selected sequence of wideband pulses to make a range-Doppler image;
4. Determination of the rotational motion of the target; and
5. Conversion of the range-Doppler image to a range/cross-range image by employing the target's rotational motion and the precise orbit.

From examining a sequence of range/cross-range images, an initial guess of the shape and size of a detected satellite is made. A wireframe model is mathematically

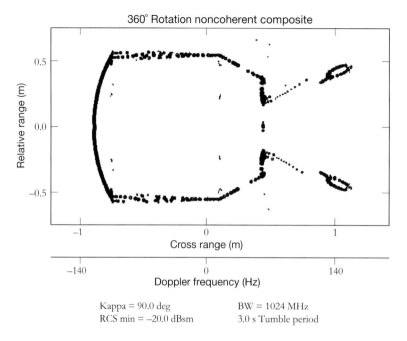

Figure 3.15   Computer simulation of the composite image of the tumbling rocket body seen in figure 3.14.

calculated and "flown" along the satellite trajectory and then superimposed on the sequence of wideband images. The creation of composite images often helps in properly visualizing the satellite.

Figure 3.16 shows a TIRA wideband image of the Japanese satellite Midori II, taken when the satellite was stable, along with an artist's drawing. The resolution of the TIRA wideband radar image is 33.9 centimeters because the bandwidth of the TIRA Ku-Band (16.7 GHZ) radar is 800 MHz. Because of an aspect angle difference between the artist's drawing and the TIRA wideband image, the solar panel is greatly foreshortened in the drawing, although, in general, there is good agreement between scattering features seen on the TIRA wideband image and physical features on the Midori II satellite.

Because the motion of the satellite is known, the TIRA wideband image is properly scaled in cross range. A wireframe model could be overlaid on a sequence of wideband images and propagated forward (and backward) in time along the entire trajectory of the satellite. When the size and shape of the satellite are not known a

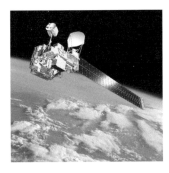
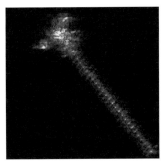

Figure 3.16 Japanese satellite Midori II. (Left) Artist's drawing (provided courtesy of NASA). (Right) TIRA wideband image (provided courtesy of the Fraunhofer Institute in Germany).

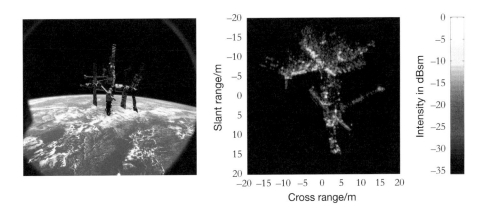

Figure 3.17 Russian space station *Mir*. (Left) Photograph taken by an astronaut on board the U.S. Space Shuttle *Discovery* (provided courtesy of NASA). (Right) TIRA wideband image (provided courtesy of the Fraunhofer Institute in Germany).

priori, this technique of overlaying a wireframe model onto a sequence of wideband images is often used to identify new satellite features. Additional satellite features can be added to a preliminary estimate of the wireframe model when scattering features appear on wideband images at different aspect angles with respect to the radar line of sight.

Figure 3.17 (left) shows a photograph of the Russian space station *Mir*; figure 3.17 (right) shows a TIRA wideband image of *Mir*, which captures its complex shape in remarkable detail.

Figure 3.18 shows a sequence of four TIRA wideband images of the Space Shuttle, which clearly show the distinctive shape of the shuttle with its payload doors open. The brightest return signal in all four wideband images is a specular return occurring at the rear of the payload bay. The returns, which appear to be emanating out-range from the specular, are the extended-range returns discussed earlier. The extended returns are particularly strong due to the large number of multiple reflections caused by the open payload bay.

Figure 3.19 shows a TIRA wideband image of the ill-fated Russian spacecraft *Fobos-Grunt* along with an artist's drawing of the spacecraft. *Fobos-Grunt* was launched 8 November 2011 from the Baikonur Cosmodrome, but subsequent rocket burns intended to set the craft on a course to Mars failed. This failure left the spacecraft stranded in a low Earth orbit. Efforts to reactivate the spacecraft were unsuccessful, and it underwent an uncontrolled reentry on 15 January 2012, reportedly over the Pacific Ocean west of Chile. The TIRA wideband image, included to demonstrate the utility of wideband imaging in failure analysis, clearly shows the multiple main

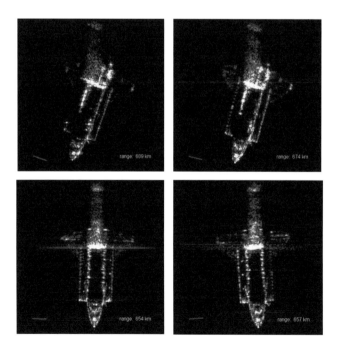

Figure 3.18   TIRA wideband images of U.S. Space Shuttle (provided courtesy of the Fraunhofer Institute in Germany).

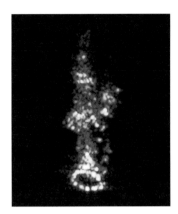

Figure 3.19   Russian spacecraft *Fobos-Grunt.* (Left) Artist's drawing (provided courtesy of the NASA Space Sciences Data Coordinated Archive). (Right) TIRA wideband image (provided courtesy of the Fraunhofer Institute in Germany).

propulsion tanks at the bottom of the image. Complex scattering is seen from the truss structure holding the Mars orbiter and the Phobos lander, and extended-range returns are seen emanating from the top of the satellite.

3.6   Autofocusing of Radar Images

In wideband radar imaging, smeared images occur either because of an incorrect motion solution or because of imperfect phase compensation. An incorrect motion solution can affect the images in one of two ways. First, it can produce incorrect processing intervals. If the processing intervals are too short, then the images will have degraded cross-range resolution. If the intervals are too long, target scatterers will migrate through range and cross-range resolution cells and cause the resulting image to be smeared. This type of image distortion is most pronounced at the ends of the target. Regions of the image close to the center of the target's rotation are better focused. Second, an incorrect motion solution can produce incorrect cross-range scaling, which causes the image to appear both smeared and distorted.

Imperfect phase compensation can result from any of the following causes: trends in the residuals of the orbit fit estimation, imperfect atmospheric refraction compensation, unaccounted-for radar equipment drifts, and nonlinear trends in phase residuals. Modeling the orbit fit of the target for atmospheric effects is never perfect. The phase residuals after compensation exhibit nonlinear trends. In most cases, however, the magnitude of such trends is relatively small, and the resulting images are of good quality. In a few cases, nonlinear residuals are large enough to cause image distortion. The removal of phase errors requires phase measurements of every pulse in the image-processing interval. Subdividing the linear image-processing interval into small subintervals is a convenient way to remove phase errors. Each subinterval yields a low-resolution partial image, which, in turn, yields an independent phase measurement. This method collects sufficient numbers of measurements for empirical error modeling. The error model is then used to interpolate and remove phase errors from the raw data.

The following are three methods for obtaining phase information from an image: (1) identify a persistent point scatterer on the target to serve as a phase reference and measure its phase variations; (2) use the cross-range displacement of the target as seen in partial images formed from subintervals to measure the phase derivative; and (3) use the cross-correlation between consecutive partial images to measure the second derivative of the phase. Although, in theory, all three methods yield equivalent results, in practice, there are significant differences between them, especially for extended targets. In the interests of brevity, only the cross-correlation technique will be discussed here.

Figure 3.20 illustrates the relative displacement between two partial images. The cross-correlation function of the two partial images is a measure of the relative shift Δd. If there is no phase error, then all partial images will be centered, and Δd will be equal to zero. For phase error that varies linearly across the image interval, all the partial images will be shifted by an equal amount. For phase error that varies nonlinearly, the partial images will have differing relative shifts across the image interval.

In the presence of a relative shift, perfect registration can occur only after the shift has been removed. The cross-correlation function between two shifted images will peak at the correct relative displacement. Inversely, the peak of the cross-correlation function will measure the relative shift. It is computationally efficient to perform the cross correlation in the frequency domain. Each image first undergoes a Fourier transform to the frequency domain. The back transformation of the product of the individual transforms then provides the cross-correlation function. To measure the shift, a peak-finding algorithm that performs fine interpolation is used.

The time of validity of a partial image is the middle time of the data interval. The cross-correlation function measures a shift whose time of validity is the mean of the time of validity of the two partial images. Dividing the estimated shift by the difference of the time of validity of the two partial images provides an estimate of the derivative of the image shift with respect to time. This derivative, in turn, is related to

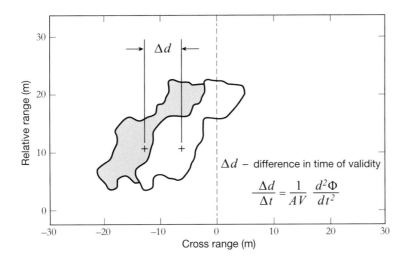

Figure 3.20 Illustration of the relative displacement between two partial images. AV, angular velocity.

the second derivative of the phase error as follows: $d^2\varphi/dt^2 = \text{AV}(\Delta d/\Delta T)$, where ΔT is the difference in the time of validity and AV is the angular velocity. It is inherently difficult to identify a persistent point scatterer in a series of images formed over the trajectory of a target. The use of an image feature as a phase reference when it is not a point scatterer also results in large phase errors. In addition, it is difficult to ensure that a point scatterer will persist throughout the complete processing interval because, as the target rotates, individual scatterers can shadow one another and move in and out of view. Hence partial image cross-range displacement and cross correlation between consecutive partial images have been used to implement autofocusing in radar image generation.

The autofocusing technique corrects phase errors regardless of their source and nature. Image data collected at low elevation always exhibit nonlinear phase residuals. It is not always clear whether the cause of such error is solely atmospheric in nature; more likely, it is a combination of the effects of wave propagation and target motion mechanisms that are not modeled in the orbit estimation process. Autofocusing depends only on the quality of the radar data, and its performance depends on the data-sampling rate and the rate's relation to the temporal variation of the phase errors.

As an illustration, consider the case of autofocusing applied to actual sphere data acquired at low elevation. The pulse-to-pulse phase is uniquely defined, and its

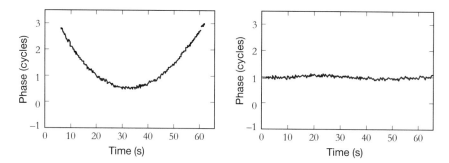

Figure 3.21   (Left) Pulse-to-pulse quadratic phase variation for a sphere; (right) variation after compensation.

variations are independent of object orientation. Figure 3.21 (left) shows the pulse-to-pulse phase variation after orbit compensation for the first 60 seconds of a low-elevation pass of a sphere. It is easy to see the strong quadratic dependence of the phase. A more detailed analysis revealed that higher-order terms were also present.

A wideband image of the sphere made from such data would be smeared. Figure 3.21 (right) shows the pulse-to-pulse phase with perfect compensation. The observed small rapid fluctuations are due to random noise effects. A wideband image of the sphere made from such compensated data would be well focused and sharp, limited only by system noise.

3.7   Satellite Motion Determination

To properly scale a range-Doppler image, the satellite motion must be known. If the motion of the satellite relative to the radar line of sight is known, the range-Doppler image can be properly scaled to form a range/cross-range image, and the locations of the radar returns in the image will match a geometric projection of the body onto the radar image plane. The radar image plane contains the RLOS and is perpendicular to the angular velocity vector. If the satellite motion is not known, the relationship between Doppler frequency and cross-range position will be undetermined, and a reliable cross-range measurement cannot be made. In order to determine the size and shape of a satellite from wideband radar images, the satellite motion must be determined with sufficient accuracy. Knowledge of its motion can also be important in determining either the mission of a foreign satellite or the failure mode of a domestic one.

In many instances, satellite motion can simply be assumed prior to calculating a wideband radar image. Many operational satellites are Earth-referenced stable, inertially stable, or horizon stable. For rapidly spinning satellites that give conic or cylinder specular returns, satellite motion can be determined from specular timing information alone. A motion solution permits the calculation of an entirely new class of wideband images through the use of the extended coherent processing (ECP) program discussed in the next section.

3.8   Extended Coherent Processing

The concept of correlation imaging is familiar in diverse areas of investigation such as synthetic aperture radar (SAR), radar mapping of planets, acoustic imaging, and holography [8]. Figure 3.22 (top) illustrates the sinusoidal motion of a scatterer on a tumbling rocket body and the spherical coordinate system used in analysis. For correlation imaging of rigid satellites, the image space is a three-dimensional (3-D) coordinate system (x, y, z) that rotates with the target. A correlation image is defined as follows:

$$G(x, y, z) = \Sigma \; w_i S_i \exp \left(-j4\pi P_i / \lambda\right)$$

where $R_i = R(x, y, z, t_i)$ is the range to the point (x, y, z) calculated from precise orbital and rotational models at the time t_i; t_i is the time on target of the ith pulse; $S_i = S(R_i, t_i)$ is the complex number consisting of the in-phase and quadrature components of the pulse compressed radar returns sampled at the range R_i; and w_i is a real weight subject to the constraint $\Sigma_i \; w_i = 1$. The correlation image itself is a display of the function $|G(X, Y, Z)|^2$, as seen in figure 3.22 (bottom left).

The functional form of $G(x, y, z)$ is best understood by considering the target to be a collection of scatterers rotating as a rigid body and by noting that the pulse compressed returns from a single scatterer at (x, y, z) have the phases $4\pi R_i / \lambda$ and one amplitude peak each at $R = R_i$. At the image point (x, y, z) coincident with the scatterer, $G(x, y, z)$ samples S_i at its peak, and the factor $\exp(-j4\pi R_i / \lambda)$ cancels the variation of the signal's phase from pulse to pulse. Each term in $G(x, y, z)$ has maximum amplitude and the same phase, so that the sum has its maximum value. At other image points, the sum is reduced by the following two factors: (1) sampling S_i away from its peak; and (2) making phase corrections that are not matched to the signal S_i.

The properties of the correlation image depend mainly on the aspects sampled by the pulses used in calculating the image. An *aspect* denotes an orientation of the radar line of sight relative to the target coordinate system. The weighting factors w_i can be

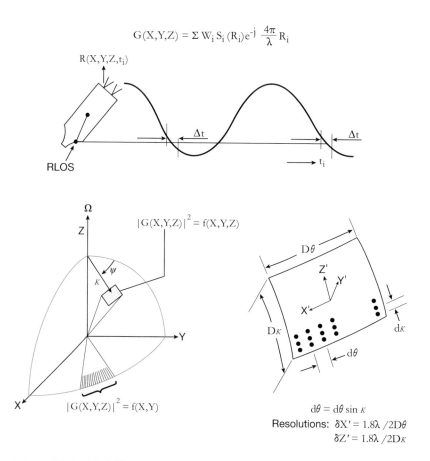

Figure 3.22 (Top) Illustration of the sinusoidal motion of a scatterer on a tumbling rocket body; (bottom left) correlation image.

used to optimize the image in various ways, such as to suppress sidelobes or to suppress ambiguous images. If the data interval Δt used in correlation imaging is so short that the relative range to every scatterer varies approximately linearly with time over Δt, then the correlation image reduces to the familiar range-Doppler image.

The function $G(X, Y, Z)$ can conveniently be calculated in the three-step ECP process. The first step is to subtract the range and phase variations of the target's center of mass (calculated from an orbit fit from each pulse of the raw data). The second step is to calculate a conventional range-Doppler image from short subintervals of data. The third step is to coherently combine these range-Doppler images to approximate $G(X,$

Y, Z). At every image space grid point (x, y, z), the relative range and relative range rate are calculated for the center time of the range-Doppler image and then phase corrected by subtracting from its phase the following quantity:

$$P_{corr} = 4\pi/\lambda[R(x, y, z, t_i) - R(0, 0, 0, t_i)],$$

where the quantity in square brackets is the relative range. The complex number is then added into $G(X, Y, Z)$. Compared with calculating $G(X, Y, Z)$ pulse by pulse, this three-step process has been shown to reduce computer time by an order of magnitude. At Lincoln Laboratory, a computer program also named "Extended Coherent Processing" was developed in the mid-1970s. It allows wideband image analysts to calculate $G(X, Y, Z)$ over any one-, two-, or three-dimensional grid of points that rotates with the target .

As illustrated in figure 3.23, extended coherent processing improves the Doppler (or cross-range) resolution in a wideband image. The six images shown in figure 3.23 (left) represent conventional range-Doppler images calculated from short subintervals of data. ECP uses an image signal–processing model corresponding to a rotating-point motion to remove range walk and Doppler walk. This approach is illustrated in figure 3.23 (top right), which represents the sum of the first three images on the top left. The point scatterer, denoted by an open circle, exhibits both range walk and Doppler walk since coherent integration took place over too wide an angle. ECP enables coherent processing over this wide angle by multiplying all the phases on the image grid by $\exp j4\pi(Y_i - Y_c)/\lambda$, rotating each frame relative to the center frame through its corresponding rotation angle, interpolating the complex signal in the rotated frames,

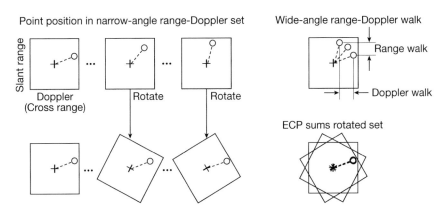

Figure 3.23   Illustration of extended coherent processing (ECP).

and adding each of the frames. Here Y_i is the initial image and Y_c is the current image. The signals coherently build up to peaks at the true peak locations. The extended coherent processing image is shown in figure 3.23 (bottom right).

There are three classes of extended coherent processing images: high-resolution images; stroboscopic images; and three-dimensional images. Each class is associated with a particular type of data set. If one continuous interval of data (larger than can be used in linear range-Doppler imaging) is used, the ECP image is a high-resolution (or wide-angle) image. If the radar data use many equal intervals selected from successive target rotation periods, the extended coherent processing image is either a stroboscopic or a three-dimensional image.

3.8.1   High-Resolution Images

Figure 3.24 (left) shows the effects of the breakdown of the linear image approximation in a high-resolution image when too large a coherent integration interval is used. The data consist of thirty-six simulated point scatterers (each of magnitude 0 dBsm) forming a square grid. A wideband image was made by integrating over 7.2 degrees of aspect angle change at X band (10 GHz). The radar has a bandwidth of 1,024 MHz. Since the integration angle exceeds 5.87 degrees (see figure 3.11), the scatterers, separated in range and cross range from the scatterer located at the (x, y) position

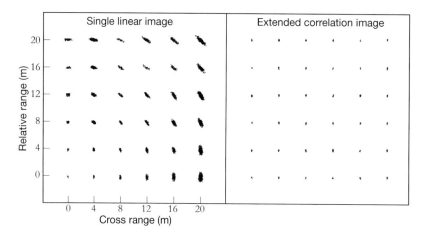

Figure 3.24   High-resolution image created from computer-simulated radar data shows a grid of thirty-six point scatterers. (Left) Wideband image results from conventional processing. (Right) Wideband image results from extended coherent processing (ECP).

(0, 0), exhibit both range walk and Doppler walk over the coherent integration interval. The true location of each scatterer is at the center of its respective image area. When extended coherent processing is used to calculate conventional range-Doppler images from short subintervals of data, rotated, phase corrected, and then added together, a properly focused image of the thirty-six point scatterers is obtained, as shown in figure 3.24 (right). This image has higher cross-range resolution because $\Delta x_{\text{cross-range}} = 1.81\lambda/2\Delta\theta$ and the interval $\Delta\theta$ is larger than intervals in conventional imaging.

3.8.2   Stroboscopic Imaging

Using extended coherent processing, Lincoln Laboratory developed a new imaging technique in which high-quality images of rapidly rotating deep-space satellites could be calculated. This technique, called "stroboscopic imaging," allows several wideband images to be coherently combined from similar aspects to form a single stroboscopic image [8], which has a signal-to-noise ratio higher than the SNR of individual images. The stroboscopic image allows target features obscured by noise in the individual images to be identified and analyzed, although coherent summation can be performed only after the peak amplitude responses between successive rotation periods have been aligned and the phase variations between successive images have been canceled out. The removal of the phase variation between successive images requires that very precise orbital and rotational motion parameter estimates be derived and used to align the phases of the individual wideband images. This process allows their summation to form the resultant stroboscopic image.

The major effort required in the development of stroboscopic imaging was the development of techniques for determining precision orbital and rotational models to permit coherent imaging over multiple rotation periods. For the Long-Range Imaging Radar, a range error of 1 millimeter results in a phase misalignment of 24 degrees. Thus the LRIR requires that the combined orbital and rotational motion models predict the range variations of possible scatterers to millimeter precision over the time period spanned by the coherent radar data. This basic requirement is one for "precision" rather than "accuracy" since a constant range (or phase) bias in the motion models does not degrade the coherent integration. Only the time variations of range must be correctly modeled to this precision. Stroboscopic imaging can be performed on any rapidly rotating geostationary satellite whose rotation axis has a constant (or extremely slowly changing) orientation relative to the Earth. At geostationary ranges, stroboscopic imaging results in both an improved signal-to-noise ratio and a suppression of cross-range ambiguous images.

3.8.3   Three-Dimensional Imaging

When radar data densely sample a solid angle of aspects, as seen in figure 3.25, three-dimensional (3-D) imaging becomes possible [8]. Rapid target rotation can scan the solid angle in the Ψ direction while the changing radar line of sight caused by the orbital motion slowly scans in the κ direction, as in figure 3.25 (left). Here κ is the aspect deviation angle, that is, the angle between the radar line of sight and the target's angular velocity vector. The Hamming weighted 6 dB cross-range resolutions δX, δZ and the cross-range ambiguous intervals $\Delta X'$, $\Delta Z'$ are given by the following:

X' direction	Z' direction
(increasing θ)	(increasing κ)
$\delta X' = 1.81\lambda/2\Delta\theta$	$\delta Z' = 1.81\lambda/2\Delta\kappa$
$\Delta X' = \lambda/2\delta\theta$	$\Delta Z' = \lambda/2\delta\kappa$

where $\Delta\theta = \Delta\psi \sin \kappa$ and $\delta\theta = \delta\psi \sin \kappa$. Here $\delta\theta$ is the aspect angle change occurring between successive pulses and $\delta\kappa$ is the aspect deviation angle change occurring between successive target rotations. When a solid angle of target aspects is densely sampled, as in figure 3.25 (right), the resolutions in the two cross-range directions X' and Z' depend on the extents of aspect change $\Delta\theta$ and $\Delta\kappa$, respectively. As shown above, the discrete sampling of aspects with steps $\delta\theta$ (per pulse) and $\delta\kappa$ (per rotation) cause cross-range ambiguous images in the X' and Z' directions, respectively.

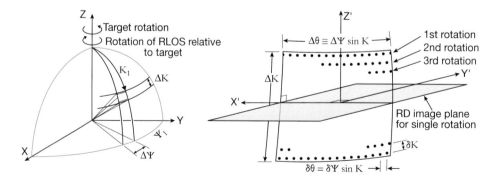

Figure 3.25   (Left) Target aspects sampled for a three-dimensional image. (Right) Detailed enlargement of the aspects sampled on the surface of a unit sphere. The dots represent pulse aspects. RD, radar.

The first requirement in the selection of data for three-dimensional imaging is that $\Delta X'$ and $\Delta Z'$ be greater than the target's X' extent and Z' extent, respectively. If the size of the target requires both these ambiguous intervals to exceed 15 meters, for example, then having a target rotation period of 10 seconds, a radar pulse-repetition frequency of 1,000, and sin κ approximately equal to 1 ensures that $\Delta X'$ is equal to 25 meters. To get $\Delta Z'$ greater than 15 meters requires that $d\kappa/dt$ be less than or equal to 10^{-4} radians per second. For geosynchronous satellites, this requirement is usually satisfied. If $d\kappa/dt$ is not known in advance, long data intervals may be required to ensure getting a data interval with the required κ rate. Long data intervals may also be needed to determine good orbital and rotational motion models. A general conclusion one can draw from this discussion is that three-dimensional imaging requires extensive, carefully planned data collection.

3.9   Bandwidth Extrapolation

Sometimes it is necessary to calculate wideband images that have a resolution greater than conventional radar hardware constraints allow. Although meeting this requirement seems impossible at first glance, it is indeed possible—by using the bandwidth extrapolation (BWE) technique discussed in this section [9].

A wideband radar measures the radar cross section (RCS) of a target with pulses over a band of frequencies, namely, the bandwidth of the radar. The frequency response of the target is converted to a range profile by "compressing" the pulses, a process achieved by taking the Fourier transform of the target's RCS frequency response, as shown in figure 3.26 (top). This leads to a range signature having a resolution, measured 6 dB down from the peak response, of $\Delta r = c/2BW$, where c is the speed of light. The range signature, unfortunately, contains sidelobes that are only 13 dB down from the peak response. This undesirable consequence results from abruptly truncating the spectrum at the radar's inherent bandwidth. The sidelobes can be substantially reduced by weighting the spectrum so that there is a smoother transition to zero at the edges of the spectrum. With Hamming weighting, the peak-to-sidelobe ratio is increased to 30 dB, and the range resolution, at the 6 dB points, is reduced to $1.81c/2BW$, as shown in figure 3.26 (bottom).

Pulse compression can be viewed as a spectral estimation problem, in which the "spectrum" to be estimated is the range signature. Modern spectral estimation techniques are based on parametric modeling of the signal, in this case the frequency response of the target. Typical models include the all-pole (i.e., linear prediction) model, the all-zero model, and the pole-zero model. To apply these techniques to the

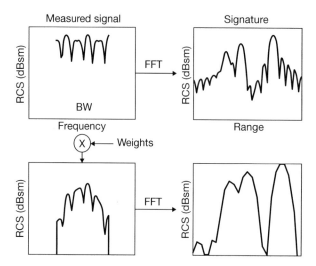

Figure 3.26 Conventional Fourier processing leading to high sidelobes (top) and reduction of sidelobes achieved with Hamming weighting (bottom). FFT, fast Fourier transform.

pulse-compression problem, it is necessary to examine the expected form of a target's radar cross-section frequency response.

At high frequencies, where target dimensions are large compared to a wavelength, the radar cross section of a target, $\sigma = ||V(f)||^2$, can be approximated by

$$V(f,\theta) = \sum_i A_i(f,\theta)\exp\left(\frac{j4\pi fR_i(\theta)}{c}\right),$$

where f is the radar frequency, θ is the nominal aspect angle of the target, $A_i(f, \theta)$ is the amplitude of the ith scattering center, c is the speed of light, and R_i is the range to the ith scattering center. The summation extends over all the apparent scatterers on the target that are located at reflection and diffraction points, as well as at virtual points associated with both multiple reflections and second-order diffraction. The ith scattering center is characterized by its amplitude $A_i(f, \theta)$ and its effective range R_i. In typical radar applications, the maximum bandwidth is approximately 10 percent of the center frequency, and the frequency variation of the scattering amplitude can be approximated by a low-order polynomial (e.g., a Taylor series) about the center frequency.

A frequency response that is well approximated by $V(f, \theta)$ can be modeled by using a linear prediction or all-pole model. The linear prediction model for a uniformly sampled frequency spectrum $v[n] = V(n\delta f, \theta)$ meets the condition

$$v[n] = \begin{cases} -\sum_{i=1}^{p} a[i]v[n-i] \text{ forward} \\ -\sum_{i=1}^{p} a^*[i]v[n+i] \text{ backward,} \end{cases}$$

where δf is the frequency step between data points; $a[i]$ are the model coefficients; and p is the model order (which represents the number of scatterers). This relationship holds when the number of scatterers is known and the model coefficients are exact. This model is also referred to as an "all-pole" or "autoregressive model." Numerous techniques exist for estimating the coefficients of the linear prediction model from the measured data. This chapter will only consider Burg's algorithm [10], which determines the coefficients that minimize the sum of the forward and backward prediction error. The prediction error is defined as the magnitude of the difference of the signal predicted by $v[n]$ and the measured signal. Moreover, Burg's algorithm is computationally efficient and results in a linear prediction filter that is stable, namely, one that will not admit any exponentially growing signals. Initially, these estimation techniques used the spectrum model, based on $v[n]$, to analytically generate the magnitude squared of the range signature, but a range signature generated by the analytic method does not contain phase information and has a large variation in scatterer location; moreover, its accuracy is completely dependent on the accuracy of the model and the estimation procedure.

A hybrid technique proposed by Stephen Bowling that combines parametric spectral estimation with linear prediction theory is the basis of bandwidth extrapolation (BWE) [11]. The BWE technique is applied by first using Burg's algorithm to estimate the linear prediction model coefficients $a[i]$ from the measured data by using an estimate of the model order p and then using the linear prediction model iteratively to extend the data outside the measured spectrum. This extension process is accomplished by using the signal model $v[n]$ and the measured signal data as initial conditions, as illustrated in figure 3.27 [12]. The expanded pulse spectrum is then weighted and compressed via the Fourier transform. A comparison of the conventional, traditional parametric spectral estimation techniques and bandwidth extrapolation is shown in figure 3.28. This procedure has the advantages of maintaining both the phase coherency of the range signature and the measured data.

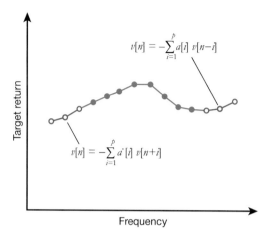

Figure 3.27 Illustration of bandwidth extrapolation (BWE). A linear prediction model of the signal is combined with the measured radar data (*in blue*) to synthesize extrapolated data (*in red*) outside the measured bandwidth. The extrapolation for the higher frequencies is identical to the extrapolation for the lower frequencies, except for conjugated model coefficients.

The two top graphs in figure 3.28 show a measured signal that is one-half the bandwidth of that shown in figure 3.26 and the resulting range signature after conventional Fourier processing. It is evident that the reduction in bandwidth has made separating the scatterers in range more difficult. The middle right panel in figure 3.28 shows the spectral estimation method, which uses a mathematical model of how the radar cross section varies with frequency. The measured signal is used to estimate the parameters in the model, which, in conjunction with them, is used to estimate the range signature of the target, as shown in the middle right graph of the figure. There are three major problems associated with this technique. First, because no phase information is associated with the estimated range signature, any further processing of the signature is precluded. Second, the amplitude estimates are unreliable. Third, these estimates also have a large variance. It is important to realize that the estimated range signature is solely dependent on the accuracy of the model. Bowling's method is illustrated in the bottom two graphs in figure 3.28, where the model parameters are used to "grow" data outside the measured bandwidth. The expanded bandwidth data (shown in red) are then processed in the conventional manner; Hamming spectral weighting is applied and then Fourier transformed to produce the range signature. As seen from the plot on the lower right in figure 3.28, the range resolution has been

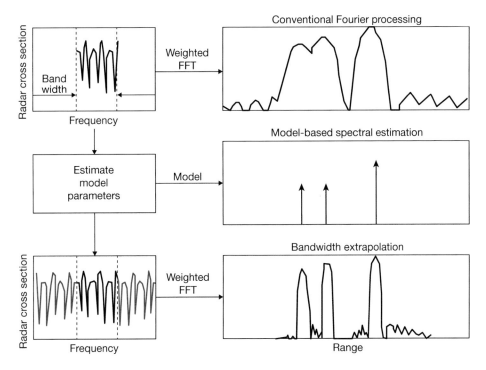

Figure 3.28 Overview of the bandwidth extrapolation process.

improved by a factor of 2, with no corresponding increase in sidelobes. Bandwidth extrapolation addresses all the shortcomings of conventional Fourier processing. Since the original data are kept, the range response is not solely dependent on the accuracy of the model. The use of the Fourier transform reduces the variance of the amplitudes and preserves the phase of the range signature. Sidelobe weighting has direct impact on the performance of model-based spectrum estimation techniques.

Figure 3.29 (top) shows the range signature for a calibration sphere that was measured by the Long-Range Imaging Radar. The measured signal (outer curve) is the radar cross section plotted against frequency over 1 GHz of bandwidth. The bandwidth of the signal is "grown" by a factor of 2.5. The enhanced signal (inner curve) indicates the resulting range signal of this process. It is clear that the resulting resolution has been improved.

Bandwidth extrapolation is based on the assumption that, at high frequencies, the radar energy scattering from a target as a function of frequency can be reasonably

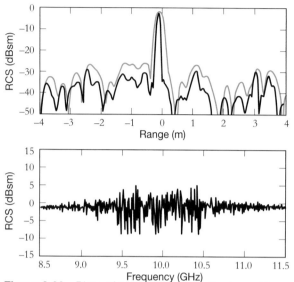

Figure 3.29   Plots of sphere data from the Long-Range Imaging Radar. (Top) Inner curve is the result of bandwidth expansion.

well approximated by a linear prediction model. Namely, the radar return signal at one frequency can be determined as a weighted linear combination of radar return signals at other frequencies. In applying bandwidth extrapolation, the measurements are used to find the weighting coefficients that make the model best fit the data, and then the model is used to predict the target returns outside of the measured bandwidth. At this point, conventional pulse compression proceeds as if the radar had measured data at the wider bandwidth. Numerous techniques exist for estimating the coefficients of the linear prediction model from measured data. None yield significantly better results than Burg's algorithm, and none work better in terms of efficiency and stability. In combination with the BWE process, the Burg algorithm tends to behave well even if the model order is overestimated. Two key features contribute to the success of the bandwidth extrapolation approach: (1) all the measured data are retained; the model is used to extend the data, not to replace them; and (2) conventional pulse compression is used after data expansion. Because the weighting process used in pulse compression reduces the influence of the extrapolated data on the final range signature, any errors in the extrapolated data are mitigated. The reliability of the results depends more on the accuracy of the extrapolated data as the expansion factor increases. Bandwidth extrapolation typically improves the range resolution by a factor of 2 to 3 and often provides striking improvements in the quality of wideband radar images.

3.10   Ultrawideband Coherent Processing

Another powerful signal-processing technique that can be applied to multifrequency data further surpasses the limitations of bandwidth extrapolation: ultrawideband (UWB) coherent processing [13, 14].

Although bandwidth extrapolation improves resolution, the BWE algorithm is based on signal-processing models that characterize a complex target as a collection of point scatterers, each having a frequency-independent scattering amplitude. Bandwidth extrapolation algorithms are often sufficient for typical wideband signal processing in which the waveforms have a small fractional bandwidth (typically 10 percent), compared to the radar center frequency. Over ultrawide frequency bands in which the radar bandwidth is comparable to the radar center frequency, however, the scattering amplitude of scatterers on a complex target can vary significantly with frequency. Spheres, edges, and surface joins are examples of realistic scatterers that exhibit significant amplitude variations as a function of frequency. Ultrawideband signal models must be flexible enough to accurately characterize these non-pointlike scatterers. Many scatterers exhibit scattering behavior proportional to f^α. For example, the radar cross sections of flat plates vary as f^2, those of conic sections vary as f^1, and those of doubly curved surfaces (e.g., a sphere) vary as f^0. The radar cross section of a curved edge varies as f^{-1}, whereas the RCS of a cone vertex varies as f^{-2}. One goal of ultrawideband processing is to detect these frequency-dependent terms in the measured radar cross-section data and to exploit them for scattering-type identification.

Building a true ultrawideband radar can be quite expensive. A more cost-effective approach is to use conventional wideband radars to sample the target's response over a set of widely spaced subbands, as illustrated in figure 3.30, where a radar's S-band and X-band waveforms are used to collect coherent target measurements over their respective widely spaced subbands. Coherently processing these subbands together makes it possible to accurately estimate a target's ultrawideband radar signature. This concept increases processing bandwidth and improves both range resolution and target characterization.

A number of technical issues must be addressed to perform ultrawideband processing. First, a robust signal-processing method must be developed to compensate for the lack of mutual coherence between the two subbands. Next, an appropriate UWB signal model must be fit to the sparse subband measurements. The fitted signal model

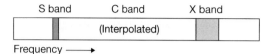

Figure 3.30   Ultrawideband (UWB) processing concept applied to S-band and X-band wideband radar signature data. S-band and X-band measurements are coherently processed together to provide an interpolated estimate of a target's UWB radar signature.

must accurately characterize UWB target scattering and provide for meaningful extrapolations outside the measurement subbands. Figure 3.31 illustrates the approach developed to accomplish ultrawideband processing [13, 14]. An estimate of the target's UWB radar signature is obtained by coherently combining sparse subband measurements. Although only two subbands are illustrated here, it is straightforward to apply this concept to more than two subbands.

The UWB process illustrated in figure 3.31 is divided into the following three steps (details of the processing are found in [13, 14]):

1. Process multiband radar data samples for the in-phase and quadrature channels to make the radar subbands mutually coherent.

2. Optimally fit an ultrawideband all-pole signal model to the mutually coherent subbands. The fitted model is used to interpolate between and extrapolate outside of the measurement bands.

3. Apply standard pulse-compression methods to the enlarged band of spectral data to provide a superresolved range profile of the target.

3.10.1   Demonstration of Ultrawideband Processing in a Static-Range Experiment

Figure 3.32 (left) shows a model of a conical target having a length of 1.6 meters. The conical target is ideal for ultrawideband processing experiments because it has several scatterers that exhibit significant radar cross-section variations as a function of frequency. The spherical nose tip has a radius of 0.22 centimeter. The conical target has one groove located 22 centimeters from the nose tip and one groove located 44 centimeters from the nose tip. The midbody of the conical target has a groove located 1.4 meters from the nose tip, one slip-on ring (not seen), and three grooves. Figure 3.32 (right) shows moment-method radar cross-section results indicating that the grooves exhibit the expected f^3 scattering behavior at the low frequency end of the spectrum, with break points that depend on the size of the groove.

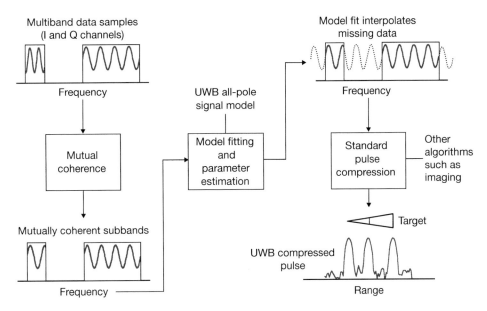

Figure 3.31   Ultrawideband (UWB) process flow to estimate a target's UWB radar signature. Sparse multiband data samples for the in-phase (I) and quadrature (Q) channels are selected. Mutual coherence processing allows two or more independent radar subbands to be used in the model-fitting step that follows. An all-pole signal model is then fitted to the sparse subband data samples and used for interpolation and extrapolation outside of the measurement bands. Standard pulse-compression methods are then applied to the ultrawideband target data.

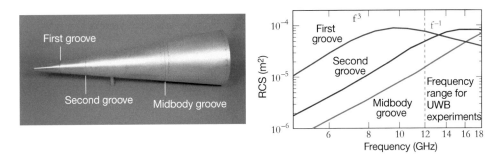

Figure 3.32   (Left) Conical target for ultrawideband (UWB) processing experiments. The target is 1.6 meters long with a spherical nose tip having a radius of 0.22 centimeter and three grooves, two near the front of the model and one at midbody. (Right) Moment-method radar cross-section (RCS) calculation for the three major grooves on the conical target.

The Lincoln Laboratory static-range radar facility was used to collect coherent radar measurements over a wide span of frequencies and viewing aspects of the target. Measurements were taken from 4.64 to 18 GHz in 40 MHz steps. The conical target viewing angles, relative to a nose-on view, ranged from 5 to 95 degrees in 0.25-degree steps. To demonstrate ultrawideband processing, a segment of data collected in the 12–18 GHz frequency band shown in figure 3.33 was used. Figure 3.33a shows an uncompressed radar pulse corresponding to an aspect angle of 20 degrees. To test the ultrawideband processing algorithms, the bandwidth of the uncompressed radar pulse was reduced to two 1 GHz wide subbands as illustrated in figure 3.33b. Figure 3.33c shows the compressed pulses for the two 1 GHz subbands and for the full-band data set. The bandwidth of the two subbands (shown in blue and red) is insufficient to resolve many scatterers on the target, whereas the full-band compressed pulse resolves all the significant scatterers on the conical target. Ultrawideband pole estimates were obtained by applying the sparse subband spectral estimation technique discussed earlier. Figure 3.34, panels a and b show a comparison between the estimated UWB target response and the true UWB radar measurements. The model and the measurements are in excellent agreement.

Since the radar measurements were taken over a wide range of viewing aspects, two-dimensional radar images of the conical target could be generated. Figure 3.35, panels a and b show the lower and upper subband images, respectively. The resolution is insufficient to resolve many of the scatterers on the conical target. Figure 3.35, panels c and d show the true and estimated ultrawideband images, respectively. All four images were generated by applying extended coherent processing to the corresponding compressed pulses over the full range of viewing aspects. Target symmetry was used to process the data as if a range of viewing aspects from −95 to +95 degrees had been sampled. The UWB images provide a clear picture of the conical target and show considerable detail. The sparse subband images closely match the full-band image and provide an accurate estimate of the locations of many scatterers on the conical target. These results suggest that ultrawideband processing of sparse subband measurements can significantly improve range resolution and provide accurate characterizations of targets over ultrawide bandwidths.

Ultrawideband processing has been successfully applied to actual conical targets and associated deployment hardware yielding remarkable images. The practical payoff of UWB technology is that radar measurements need not be taken over the full UWB processing interval; signal processing can be used to a certain extent to compensate for any missing data. Another important benefit of UWB processing is that the frequency dependence (i.e., f^α) of individual scatterers can be more accurately estimated. Such

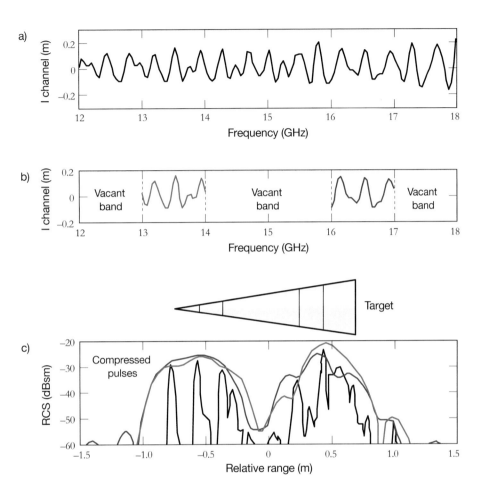

Figure 3.33   (a) Uncompressed radar pulse of the conical test target with viewing aspect 20 degrees from nose-on. (b) Sparse subband measurements used to predict the conical target's response over the full band from 12 to 18 GHz. (c) Compressed pulses for the sparse subbands and the full-band data sets. The full-band compressed pulse (*black*) resolves all of the significant scattering centers on the conical target. I channel, in-phase channel.

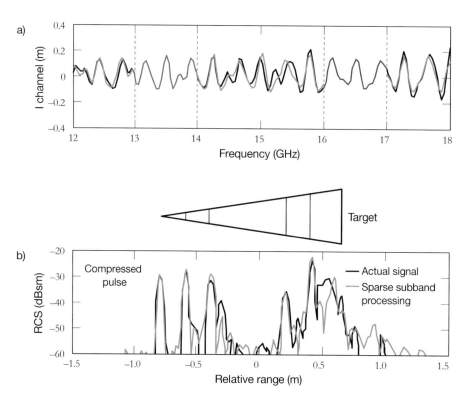

Figure 3.34   Comparisons between the estimated and the true ultrawideband radar measurements. (a) Uncompressed radar pulse for the prediction model (*brown line*) and the actual radar measurements (*black line*). (b) The corresponding compressed pulses that resolve the scatterers on the target.

accuracy helps ultrawideband image analysts better identify the scatterers that compose a complex target.

3.11   Summary

For more than four decades, Lincoln Laboratory has been a world leader in the development of wideband imaging radars and of advanced signal-processing techniques for finer and finer image resolution. The design and operation of true wideband imaging radars (e.g., ALCOR, LRIR, MMW and HAX radars) and of a tracking radar that has wideband capability using certain waveforms (TRADEX radar) have inspired the laboratory to develop a host of sophisticated surveillance techniques (real-time

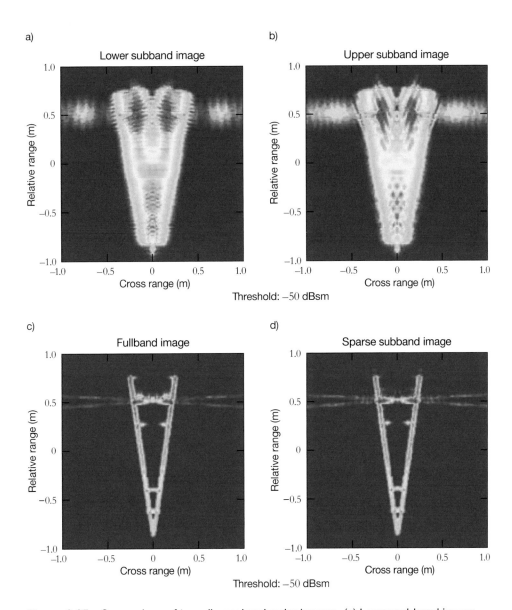

Figure 3.35 Comparison of two-dimensional radar images. (a) Lower subband image. (b) Upper subband image (c) Full-band image uses actual radar measurements over the full 12–18 GHz frequency range. (d) Sparse subband image uses the sparse subband measurements with ultrawideband prediction.

imaging, autofocusing, satellite motion determination, extended coherent processing) and advanced signal-processing techniques (bandwidth extrapolation, ultrawideband processing) to advance the state of the art of wideband radar imaging. They surpass the laboratory's initial development of conventional range-Doppler imaging and are directly applicable to two of the laboratory's key mission areas: space control and ballistic missile defense. The operational radars and advanced signal-processing techniques developed by the laboratory have been successfully applied to solving critical national security problems and have helped keep the United States at the cutting edge of wideband imaging technology.

References

1. Camp, W. W., J. T. Mayhan, and R. M. O'Donnell. 2000. Wideband Radar for Ballistic Missile Defense and Range-Doppler Imaging of Satellites. *Lincoln Laboratory Journal* 12 (2): 267–280.

2. Roth, K. R., M. E. Austin, D. J. Frediani, G. H. Knittel, and A. V. Mrstik. 1989. The Kiernan Reentry Measurements System on Kwajalein Atoll. *Lincoln Laboratory Journal* 2 (2): 247–276.

3. Avent, R. K., J. D. Shelton, and P. Brown. 1996. The ALCOR C-Band Imaging Radar. *IEEE Antennas & Propagation Magazine* 38 (3): 16–27.

4. Abouzahra, M. D., and R. R. Avent. 1994. The 100 kW Millimeter Wave Radar at the Kwajalein Atoll. *IEEE Antennas & Propagation Magazine* 36 (2): 7–19.

5. McHarg, J. C., M. D. Abouzahra, and R. F. Lucey. 1996. 95 GHz Millimeter Wave Radar. *Proceedings of the Society for Photo-Instrumentation Engineers* 2842:494–500.

6. Leushacke, L. "Die Grossradaranlage TIRA,"http://www.100-jahre-radar.fraunhofer.de/vortraege/Leushacke-Die_Grossradaranlage_TIRA.pdf, 30 April 2004.

7. Fraunhofer-Institut für Hochfrequenzphysik und Radartechnik FHR. http://www.fhr.fraunhofer.de.

8. Ausherman, D. A., A. Kozma, J. Walker, H. M. Jones, and E. C. Poggio. 1984. Developments in Radar Imaging. *IEEE Transactions on Aerospace and Electronic Systems* AES-20 (4).

9. Borison, S. L., S. B. Bowling, and K. M. Cuomo. 1992. Super-Resolution Methods for Wideband Radar. *MIT Lincoln Laboratory Journal* 5 (3): 441–461.

10. Burg, J. P. "Maximum Entropy Spectral Analysis," in *Proceedings of the 37th Meeting of the Society of Exploratory Geophysicists*, Oklahoma City, October 1967,

reprinted in *Modern Spectrum Analysis*, pp. 34–41, edited by Donald G. Childeers, IEEE Press, 1978.

11. Bowling, S. B. "Linear Prediction and Maximum Entropy Spectral Analysis for Radar Applications," MIT Lincoln Laboratory Project Report RMP-122, DTIC #AD-A042817, 24 May 1997.

12. Moore, T. G., B. W. Zuerndorfer, and E. C. Burt. 1977. Enhanced Imagery Using Spectral Estimation Based Techniques. *Lincoln Laboratory Journal* 10 (2): 171–186.

13. Cuomo, K. M., J. E. Piou, and J. T. Mayhan. 1997. Ultra-Wideband Coherent Processing. *Lincoln Laboratory Journal* 10 (2): 203–222.

14. Cuomo, K. M., J. E. Piou, and J. T. Mayhan. (June 1999). Ultrawide-Band Coherent Processing. *IEEE Transactions on Antennas and Propagation* 47 (6): 1094–1107.

4

Ground-Based Electro-Optical Technology Development

Eugene Rork

4.1 Introduction

The involvement of MIT Lincoln Laboratory in the electro-optical detection of artificial Earth satellites has recently been documented in a book and in a short technical journal article covering the laboratory's early involvement [1, 2]. In the early 1970s, the U.S. Air Force realized that existing radar and optical space surveillance systems could not fulfill the mission of detecting the growing number of satellites in orbit, particularly in deep-space orbits out to geosynchronous altitudes. Although narrow-beam radars like the Millstone Hill and Haystack radars in Westford, Massachusetts, could detect small satellites in deep-space orbits, they were slow in their searches (a few square degrees per hour). And the Baker-Nunn film-based cameras (each having a field of view of 5 × 30 degrees) were not sensitive enough to detect small deep-space satellites by using reflected sunlight, nor could they provide real-time information [3].

In 1970, low-light-level television imaging technology had advanced to the point where it was thought that a TV camera and a modest-size telescope could make a practical electro-optical sensor, with a far greater sensitivity than the existing Baker-Nunn film cameras for conducting rapid searches of the sky for deep-space satellites. Satellites in high-altitude orbits are in sunlight most of the time and could be detected by reflected sunlight in a clear night sky by a suitably sensitive optical sensor. Although detecting satellites in infrared from thermal self-emission was considered, infrared sky background radiance was much higher than nighttime visible-band radiance, and the infrared cameras were not nearly as well developed as their visible-band counterparts. Detection from visible-band solar reflection seemed much more practical by comparison. Consequently, in 1971, Lincoln Laboratory began the Telescope Detection and Ranging (TDAR) program to evaluate electro-optical sensor technology with the goal of designing sensors to rapidly detect and search for deep-space satellites by using reflected sunlight in the visible band. It was believed that optical search rates would be significantly faster than those of radar (on the order of a few hundred square degrees per hour instead of a few square degrees per hour), although optical sensors were restricted to clear skies and local nighttimes. Sensors with a telescope aperture not exceeding 1 meter in diameter were considered practical from a cost perspective. The laboratory then built a sensor site, the Experimental Test Site (ETS), to test prototype proof-of-concept sensors. The sensor technology was later transferred to a commercial vendor for the new Air Force Ground-Based Electro-Optical Deep-Space Surveillance (GEODSS) program, which built a geographically distributed network of ground-based electro-optical space surveillance sensors.

In the 1980s, the laboratory developed electro-optical focal planes for the telescopes that were silicon-based charge-coupled devices (CCDs). Because CCD focal planes were distinguished by a large number of pixels with low readout noise, and by high quantum efficiency and low dark current, with high uniformity in both, CCD imagers had much higher sensitivity than existing low-light-level vacuum-tube TV imagers. The technology was applied to the laboratory's telescopes as well as to a number of astronomical telescopes with great success, then transferred to a commercial vendor to retrofit the Air Force's GEODSS system. Section 4.2 describes the development of both low-light-level vacuum-tube television and solid-state coupled-charge device sensors for ground-based surveillance of deep-space resident space objects.[1] (Chapter 5 covers the transition of CCD electro-optic technology to a space-based telescope for space surveillance in the 1990s.)

4.2   Development of the Experimental Test Site

Satellites in deep space are detected as unresolved points of light that move differently from pointlike stars. Both satellite and star images are detected in the presence of sky background light and local effects of atmospheric seeing, extinction, and refraction. Figure 4.1 illustrates the phenomenological considerations of satellite detection through the atmosphere by a ground-based sensor in the presence of stars and the Moon.

For the satellite surveillance mission, an electro-optical sensor can measure a detected satellite's angular position (referred to as "metric data") and its brightness and

---

1.   The terms "satellite" and "resident space object" or "RSO" are used interchangeably in this chapter, although, besides satellites, RSOs include rocket bodies and other orbiting objects.

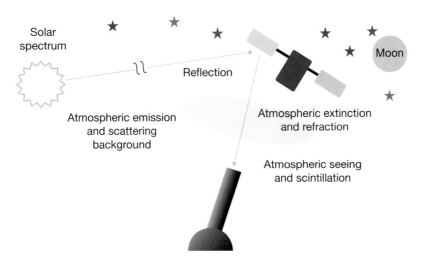

Figure 4.1   Detecting a satellite from solar reflection with a ground-based electro-optical sensor.

color (referred to as "photometric data") as a function of time. However, passive metric measurements made by one sensor do not provide a direct range measurement, which has to be inferred, through orbital parameter fitting, from multiple angular position measurements from that sensor.

In 1970, the camera tube type considered the most promising for the satellite surveillance mission was the electron-bombarded silicon diode–array target camera tube, or Ebsicon [2]. Between fall 1972 and spring 1973, to evaluate the feasibility of satellite detection with an electro-optical sensor, laboratory analysts took an Ebsicon camera to the Lowell Observatory in Flagstaff, Arizona, where they successfully recorded the detections of approximately two dozen satellites in deep space. The analysis of these data, combined with the results of ongoing sensor technology and phenomenology investigations, led the laboratory to build the Experimental Test Site sensor in Socorro, New Mexico, starting in 1974, to prove the concept that an electro-optical sensor could detect deep-space satellites during rapid searches of the sky. First consideration was given to a sensor having a field of view of no less than 1 degree, a budget-limited aperture of 31 inches, an Ebsicon camera at the focal plane, and readout at the commercial TV rate of 30 Hz. The ETS sensor consisted of two identical computer-controlled telescope systems in two domes separated by 59 meters. Each dome contained, on a single equatorial astronomical mount, a 31-inch and a 14-inch telescope, both equipped with specially developed image-intensified Ebsicon

(I-Ebsicon) electro-optical cameras. One telescope system became operational in September 1975 and the second a few months later. Figure 4.2 is an aerial picture of the site showing the two telescope domes. Figure 4.3 shows the 31-inch telescope (with its comounted 6-inch-diameter photoelectric photometer) and the 14-inch telescope and associated cameras inside one of the ETS domes.

From 1975 to 1977, the laboratory developed and demonstrated experimental techniques for deep-space satellite surveillance at the Experimental Test Site; the Air Force then contracted the laboratory to write the specifications that would be sent out to industry for competitive bidding on an operational GEODSS system and then to monitor the system's development. Between 1975 and 1995, to enable the development of deep-space electro-optical surveillance and of the GEODSS system in particular, the laboratory developed the following:

Figure 4.2 Lincoln Laboratory's Experimental Test Site (ETS), located at the Stallion Center on the White Sands Missile Range, New Mexico, was constructed to prove the concept of real-time electro-optical satellite surveillance.

Figure 4.3 Experimental Test Site (ETS) 31-inch and 14-inch telescopes with I-Ebsicon cameras mounted on a computer-controlled equatorial mount in a dome enclosure. A second identical system is located in another dome 59 meters away. A long 6-inch telescope, with a photoelectric photometer for sky brightness and atmospheric extinction measurements, is shown attached to the 31-inch telescope.

- A large-format image-intensified Ebsicon TV camera;
- Hardware and software operating systems to control telescope pointing, signal and data processing, and operator interface;
- Software to correct the telescope mount for pointing and tracking errors (including hysteresis) in the drive system;
- An analog (and later digital) moving-target indication system to find point images that moved differently from stars in video images;
- An automated system to take accurate angular positional measurements of detected satellites;

- An automated photometer system to record satellite brightness and color as a function of time for satellite identification;
- Algorithms to keep track of detected satellites for reacquisition after initial detection (dead reckoning), initial orbit determination, and refined orbit determination (for satellite catalog inclusion);
- Rapid search techniques to find known and unknown satellites and to process data without impacting search time;
- Dynamic algorithms to schedule observation of satellites based on importance, solar phase angle, and sky clarity;
- Sensor-integration and frame-averaging techniques to search for and detect faint satellite images;
- Real-time techniques and systems to indicate satellite feature deployment and orbital maneuver;
- A panoramic sky-monitoring sensor (PANSKY) to indicate location of clear sky;
- Sensors to detect low Earth orbit (LEO) satellites in daytime;
- Medium-wave infrared (MWIR) sensors to detect LEO satellites both in daytime and in nighttime shadow;
- Large-format charge-coupled device (CCD) imagers to replace the GEODSS Ebsicon imagers (1996);
- Large-format frame-transfer CCD technology transfer to help industry make operational GEODSS cameras;

Later sections in this chapter will describe a number of these developments and their significance. It is fair to say that electro-optical satellite surveillance owes its success to the laboratory's most significant and highest-risk accomplishment in that regard—developing an adequate TV surveillance camera. Understanding the phenomenology of both target and sky background was vital to this success.

4.2.1   Phenomenology Affecting the Design of an Electro-Optical Sensor

4.2.1.1   Satellite Brightness from Reflected Sunlight

At the geosynchronous range, the expected brightness of satellites from solar reflection is a key factor in their detectability. The apparent optical brightness of a satellite is complex to model and depends on the following factors: range, physical size, shape, surface material reflectivity, solar illumination angle with respect to the satellite and the observer and aspect angles of the satellite body in its orbit with respect to the Sun and the observer. The spectrum of reflectance and the spectral response of the detector are also factors in satellite brightness and are analyzed in this subsection.

The solar reflectance spectrum from a white or gray painted satellite has the same shape as the solar spectrum. Figure 4.4 shows the solar spectrum for cases of no atmospheric extinction (i.e., zero air mass as seen from space) and for two air masses, as seen from a 30-degree elevation angle from sea level through clear atmosphere [4]. Integration of the zero air mass plot of figure 4.4, from 0.3 to 1.0 microns (μm) in wavelength (which includes the visible band), yields the result that the Sun irradiates 2.88×10^{21} photons per second per square meter of photon flux (or 925 Watts per square meter of power) on a resident space object. A satellite covered with solar cells may have a spectral reflectivity like that plotted in figure 4.5 [5].

Figure 4.6a shows one of twenty-six small (0.7 meter in diameter) communications satellites of the Interim Defense Communication Satellite Program (IDCSP) that were considered hard to detect in the geosynchronous orbit [6]. The IDCSP satellites are covered with blue-colored solar cells having a spectral reflectance similar to that in figure 4.5.

Figure 4.6, panels b and c show larger satellite payloads in the equatorial geosynchronous orbit (in the early 1970s) that have cross-sectional areas ranging from a few

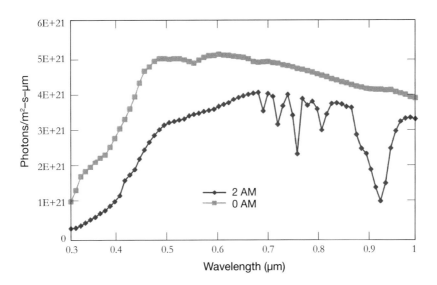

Figure 4.4 Solar spectra for 0 and 2 air masses (AM).

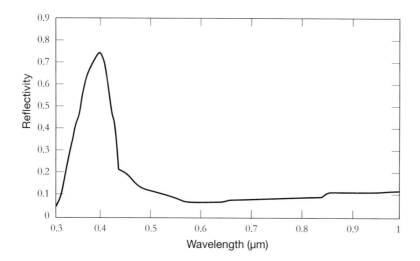

Figure 4.5   Solar cell spectral reflectivity.

to several square meters [6]. These satellites were thought to be easy for an electro-optical sensor with a high search rate to detect.

It is common practice to express the brightness of Sun-illuminated resident space objects in the apparent visual magnitude units used by astronomers for characterizing stars. Detection of a solar-type star (spectral shape like that plotted in figure 4.4) of a given visual magnitude means that a white or gray resident space object of the same brightness can be detected. The apparent visual magnitude of an RSO, designated by m_v, refers to its brightness at any range as seen by the sensor and is used to specify sensor detection capability. The absolute magnitude of an RSO—its apparent magnitude at a specified distance—is used to characterize it, and is designated by M_v in the V band.

The following statements explain how stellar magnitude is defined and calculated and include examples of magnitudes of some familiar space objects. The brightness of the star Vega provides the basis against which all other brightness values are characterized. Vega's apparent visual magnitude is defined as zero. Fainter objects have greater visual magnitudes (greater than zero), whereas brighter objects have lesser visual magnitudes (less than zero). Human observers can see stars as faint as magnitude 6 on a clear night. The difference in apparent visual magnitudes $(m_1 - m_2)$ between two stars, whose fluxes at the focal plane are F_1 and F_2, respectively, is given by

a)

b)   c)

Figure 4.6   (a) One of twenty-six Interim Defense Communication Satellite Program (IDCSP) communications satellites in geosynchronous orbit that were launched between 1966 and 1968. (b) Defense Support Program 1 (DSP-1) satellite, launched in 1970. (c) Defense Satellite Communications System 1 (DSCS-2), launched in 1971.

Table 4.1   Examples of apparent visual magnitude for celestial objects

Name	Apparent visual magnitude (m_v)
Polaris	+2
Vega	0
Pluto	+14.9
Venus	−4
Full Moon	−12.7
Sun	−26.7

$$m_1 - m_2 = -2.5 \log \frac{F_1}{F_2}$$

Table 4.1 shows the brightness in apparent visual magnitudes of some familiar celestial objects. In these cases, the magnitudes are in the visible band, as defined by the astronomer's V-filter spectral transmittance [7].

4.2.1.2   Brightness Estimates for Resident Space Objects with Simple Shapes

The apparent brightness of an RSO depends on factors related to the viewing geometry between Sun and observer as well as the size, shape, and surface reflectivity properties of the RSO. Figure 4.7 shows expected absolute magnitude (M_v) brightness plots calculated for several simply shaped satellite bodies at the equatorial geosynchronous range of 35,788 kilometers as a function of the Sun-satellite-observer solar phase angle (adapted from Krag [8]).

The plots in figure 4.7 were calculated from trigonometric phase functions to estimate the brightness of modeled satellites of simple shapes in order to estimate the expected brightness of actual satellites of more complex shapes. The absolute magnitude of the specular sphere satellite, slightly greater than 14 at all phase angles, as shown by the straight plot line numbered 6 in figure 4.7, is easiest to understand because this shape reflects sunlight equally intensely in all directions.

The apparent brightness of a specular sphere satellite of reflectivity-area product equal to 1 square meter at geosynchronous range is expected to be 13.75 m_v with no atmospheric extinction. If, however, the reflectivity of a solar cell–covered satellite is 10 percent (as could be the case for the satellite in figure 4.6a), the apparent brightness of the 1-square-meter specular sphere satellite would be ten times fainter than

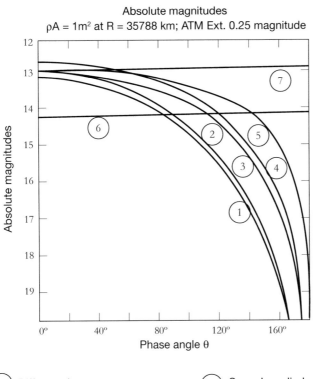

Figure 4.7 Expected absolute magnitude brightness of several simple satellite shapes of reflectivity-area product 1-square-meter cross-sectional area at the equatorial geosynchronous range of 35,788 kilometers (adapted from Krag [8]).

13.75 m_v, or 16.25 m_v. Figure 4.6a represents the smallest approximately spherical satellite launched into geosynchronous orbit in the 1970s. That being the case, to spot the majority of satellites there, an electro-optical sensor would have had to detect satellites in the 16 m_v range.

4.2.1.3   Expected Deep-Space Satellite Angular Motion with Respect to Stars

The angular rates for deep-space resident space objects (with ranges from 6,000 kilometers to beyond 40,000 kilometers) vary between 300 and 6 arc seconds per second. Since the equatorial geosynchronous orbit was considered to be of greatest interest to the Department of Defense from a national security point of view, a sensor design to optimize detection at 15 arc seconds per second (the angular rate for RSOs in geosynchronous orbit) was sought first for the laboratory's Experimental Test Site.

4.2.1.4   Sky Background Brightness and Star Clutter

From a ground-based sensor, resident space objects are detected through the atmosphere as points of light that move differently from the stars (also seen as points of light) in background light emanating from the atmosphere and sky. Such background light includes scattered moonlight, light from chemical activity in the upper atmosphere (airglow), light scattered by interplanetary dust (zodiacal light), light from unresolved stars, and man-made light pollution. In addition, clouds and haze can either accentuate or occlude sky background light. Figure 4.8 shows the number of stars per square degree in the sky that are brighter than a given magnitude as a function of galactic latitude [9].

Figure 4.9, panels a and b show examples of clear sky background spectra at good observatory sites (Kitt Peak in Arizona and Mount Palomar in California) for no Moon and for a full Moon at zenith, respectively. The units of the plots are in photons per second per square meter per micron per square arc second.

4.2.1.5   Atmospheric Seeing, Extinction, and Refraction

It is well known in astronomy that "atmospheric seeing" is the blurring of images caused by index of refraction variations in the atmosphere resulting from thermal gradients. Such variations can occur in time periods as short as 1 millisecond. Seeing is worse at shorter wavelengths because blue light is normally refracted more than red light in a refractive medium. Most effects of changes in the index of refraction occur where the atmosphere is thickest, that is, at low altitudes, but refraction changes high

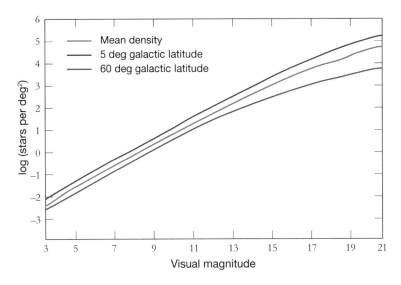

Figure 4.8 Expected number of stars per square degree (on a logarithmic scale) brighter than a given visual magnitude (*x*-axis) from Allen [9]. For example, if a sensor with a field of view of 1 degree operates with a sensitivity of 16 m_v, it must detect satellites against a clutter of 350–6,000 stars, with a mean of 2,000 stars.

in the atmosphere can cause an image to leave a telescope aperture momentarily, resulting in image scintillation or twinkling. Atmospheric seeing can make faint satellite images more difficult to detect by reducing the average fraction of energy from an unresolved point source that would normally fall in a sensor pixel. As expected, high-altitude sites, with less atmosphere to look through, have the best seeing. Typical visible-band seeing values at good high-altitude observatory sites are about 1 arc second at zenith, as compared to 2–3 arc seconds at zenith at low-altitude sites. The atmospheric seeing is correspondingly worse at lower elevations because of the line of sight through more atmosphere.

Atmospheric extinction is a measure of the absorption of light as it passes through the atmosphere. As with atmospheric seeing, high-altitude sites have lower atmospheric extinction than low-altitude sites. Absorption is also wavelength dependent, with blue light having higher extinction than red light because it scatters more than red light in the atmosphere (Rayleigh scattering making the sky blue). Typical extinction values in the V band at good observatory sites range from 0.15 to 0.25 m_v per air mass, depending on atmospheric clarity and water-vapor content. Since extinction is

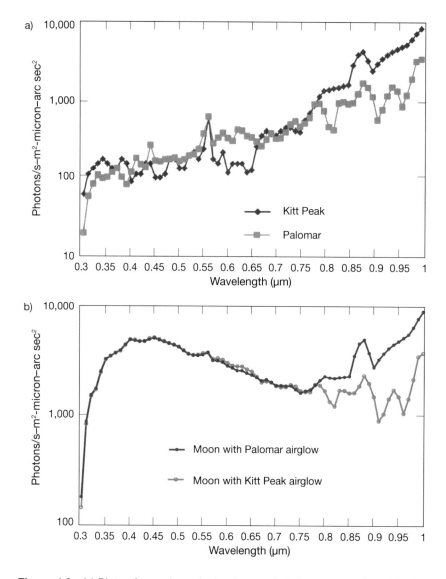

Figure 4.9   (a) Plots of moonless sky background airglow spectra from Kitt Peak [10] and Palomar [11]. (b) Plots of sky background brightness with LOWTRAN 7, full Moon at zenith [12], plus Kitt Peak [10] and Palomar [11] airglow spectra.

a function of the air mass through which the sensor is looking, high sensor elevation angles have lower extinction than low-elevation angles. Calculation of air mass as a function of elevation angle, taking the Earth's curvature into account, is described in Hardie [13].

Atmospheric refraction changes the apparent elevation angle of both satellite and star images in the night sky. Refraction must be corrected not only in telescope pointing for satellite and star acquisition but also for satellite positional measurements, which must be extremely accurate for orbit determination. Astronomers have sufficiently accurate and well-documented atmospheric refraction correction algorithms, which depend on elevation angle and atmospheric temperature and pressure as inputs [14].

4.2.2 Low-Light-Level Television Sensors

An ideal TV imager for satellite surveillance should have the following characteristics:

- High quantum efficiency over the solar spectrum;
- High charge storage and readout capacity;
- Low blooming (increased size of bright point image);
- Low lag (frame-to-frame image persistence);
- Low self-noise;
- Short readout time;
- High spatial resolution;
- Large focal surface;
- Uniform responsivity and resolution;
- Negligible (or correctable) geometric distortion;
- Small vacuum volume;
- Low number of defects; and
- Ruggedness.

Figure 4.10 shows how an electron-bombarded silicon diode–array target (Ebsicon) TV camera works [15]: a photon from a star (or satellite) image goes through a fiber-optic faceplate to the curved surface S-20 photocathode, where its energy dislodges (via a Poisson process) a photoelectron that is accelerated in the camera tube to the silicon diode array target. Upon impact there, it dislodges approximately 2,000 electron-hole pairs. The target diodes are reverse biased with a positive voltage across the resistor (R) such that the holes migrate across the diodes and are filled in with

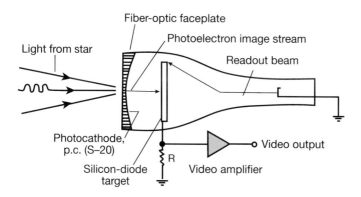

Figure 4.10   Functional diagram of Ebsicon camera tube.

electrons from the raster-scanning electron readout beam. The electrons on the front of the target pass through the resistor to ground. The video signal is taken from the voltage drop across the resistor. The quantum efficiency of the S-20 photocathode as a function of wavelength of solar illumination is shown in figure 4.11.

Effective camera tube self-noise is reduced in the Ebsicon by the same mechanism used in a photomultiplier tube. The many electron-hole pairs created in the camera tube target by each photoelectron from the camera photocathode provide an amplified signal as well as an amplified photon-photoelectron conversion noise, which may then exceed the camera tube self-noise, making sensor sensitivity limited primarily by sky background noise.

4.2.2.1   Development of the Intensified-Image Ebsicon Camera

The Westinghouse WX-32719 Ebsicon tube (see figure 4.12) had a 40-millimeter photocathode and a newly developed target structure known as a "deep-etched metal-capped [DEMC] target." Compared to conventional resistive sea and metal pad targets, the DEMC target improved beam acceptance with reduced blooming [2]. It was therefore selected as the primary camera tube for the Experimental Test Site. With this camera tube, a 31-inch telescope with a 1-degree field of view would have an f-number of 2.9.

On the other hand, a zoomable image intensifier tube, the Varian Model VL-116, with an 80-millimeter-diameter focal surface could shrink an 80-millimeter image down to its 40-millimeter-diameter output phosphor. It could be optically coupled to a 40-millimeter tube to give the combination an 80-millimeter-diameter first focal

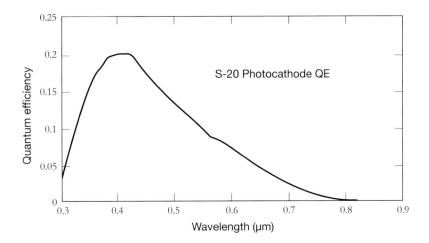

Figure 4.11   Plot of spectral quantum efficiency of an S-20 photocathode similar to those used in Ebsicon camera tubes [16].

Figure 4.12   Westinghouse Wx-32719 Ebsicon camera tube.

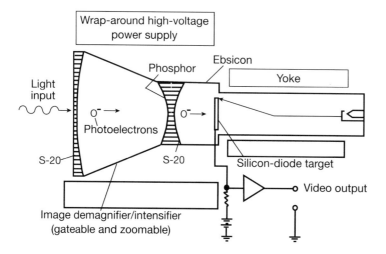

Figure 4.13   Functional diagram of an image intensifier coupled to an Ebsicon camera tube, making an I-Ebsicon camera tube [15].

surface. The resulting f-number for the 31-inchtelescope (with a 1-degree field of view) was a more reasonable 5.8. Figure 4.13 is cross-sectional drawing of this image intensifier coupled to the Ebsicon tube. The designation "I-Ebsicon" applies to the imager formed with these coupled components.

The second image intensification stage of the Ebsicon added a gain factor of about 25, for a total prereadout gain factor of approximately 50,000, thus enhancing the ratio of the signal to the camera readout noise. However, the intensifier tube had the following disadvantages: (1) increased pixel size (decreased resolution); (2) increased shading; (3) increased ionic scintillations (random point flashes); (4) decreased charge storage capacity (decreased dynamic range); (5) increased complexity; and (6) increased size, weight, and cost. Despite these disadvantages, its two major advantages of increased focal plane size and higher signal-to-noise ratio led to its selection.

4.2.2.2   Telescope Procurement

The laboratory specified two telescope systems for the Experiment Test Site. Manufactured by Boller & Chivens, each system consisted of a 31-inch f/5 Ritchey-Chrétien Cassegrain telescope and a 14-inch f/1.7 folded Schmidt telescope on a single equatorial mount having 21-bit incremental encoders (0.62–arc second resolution) on both axes. The 31-inch telescopes each covered a field of view of 1.16 degrees over an 80-millimeter focal plane diameter, whereas the 14-inch telescopes each covered a field of view of 7.5 degrees over the same focal plane diameter. The 14-inch telescopes were added to provide coverage for resident space objects in lower-than-geosynchronous orbits, which were fast moving and brighter because of shorter range. Figure 4.3 shows one telescope system with its I-Ebsicon cameras mounted.

The image quality of the 31-inch f/5 telescopes, covering the solar spectral band, was measured at the telescope factory and found to put 80 percent of incident energy from a distant point source in a 1-arc-second-diameter circle over its 80-millimeter-diameter focal plane, well within the typical atmospheric seeing conditions at the site of 2–3 arc seconds. Consequently, the telescopes would not significantly contribute to image degradation. The image quality of the 14-inch f/1.7 telescopes was measured from star images recorded on photographic glass plates upon installation at the site and images were found to be about 3.5 arc seconds in diameter across its 80-millimeter-diameter focal plane.

4.2.2.3   Operation of the Intensified-Image Ebsicon Sensor

Cameras with the I-Ebsicon imager could operate at the standard TV rate of 30 frames per second with 525 TV lines per frame (including the interlaced scan option of scanning every other line at 60 frames per second), or they could be set up for 875 scan lines per frame, with or without interlacing. For additional sensitivity, the readout beam could be shut off to allow up to 999 1/30-second TV frames to be integrated on the target. Although a large number of frames could saturate the camera tube target, an image could be accumulated by averaging unsaturated shorter exposure frames on an external TV scan converter/storage terminal. The TV rate operation of the I-Ebsicon camera allowed it to use the extensive video signal–processing and recording technology developed for the broadcast industry, and the TV rate scan provided convenient viewing for an operator.

4.2.3   Sensor Model

To predict the operational sensitivity of the I-Ebsicon cameras combined with Experimental Test Site telescopes, laboratory analysts built a mathematical model, whose key aspects are described in the following subsections.

4.2.3.1   Calculation of Satellite Signal Photoelectrons

An electro-optical sensor detects a resident space object from photoelectrons produced by solar-reflection photons from the target object's body. An RSO with an irradiance spectrum $N_{Tgt}(\lambda)$ in photons per second per square meter per micron produces a number of photoelectrons, S, which can be calculated as follows:

$$S = \left(\int_{\lambda_1}^{\lambda_2} d\lambda \, N_{Tgt}(\lambda) q(\lambda) f(\lambda) \right) A_{eff} t_{exp} k_{str} \tau_{atm} , \qquad (4.1)$$

where $N_{Tgt}(\lambda)$ = target RSO photons per square meter per second per micron
$q(\lambda)$ = imager spectral quantum efficiency
λ_1 = shorter wavelength limit of integration
λ_2 = longer wavelength limit of integration
$f(\lambda)$ = filter spectral transmittance (if filter is included)
A_{eff} = effective collecting area of telescope (square meters)
t_{exp} = exposure time (seconds)
k_{str} = straddle factor (fraction of image energy in 1 pixel)
τ_{atm} = atmospheric transmittance (presumed constant in this spectral band)

Equation 4.1 can be used in two ways. If a satellite's spectral irradiance spectrum $N_{Tgt}(\lambda)$ is known, the number of photoelectrons S produced in the sensor from it can be calculated. Also, if S is known from the sensor, and the spectral shape of the satellite's irradiance spectrum $N_{Tgt}(\lambda)$ is known (e.g., for a white or gray satellite whose irradiance spectrum would be the same as that of the Sun), then the function $N_{Tgt}(\lambda)$ is determined from equation 4.1 and S.

To use equation 4.1, however, the terms k_{str} (straddle factor) and τ_{atm} (atmospheric transmittance) must be evaluated. The straddle factor can be estimated by first measuring the average video signal voltage from the video line of the image with the highest signal and then measuring the video signal of any adjacent video lines showing the image. The straddle factor is the ratio of the average video signal voltage in the central line to the sum of video signal voltages of all video lines containing the image.

Atmospheric transmittance (τ_{atm}) can be measured with a photoelectric photometer by first recording photoelectron numbers from calibrated solar-type stars at near-zenith and low-elevation angles and by then determining how much signal is lost at the low-elevation angles as compared to the near-zenith direction. The photometer's spectral quantum efficiency is the same as that of the I-Ebsicon photocathode. An extinction coefficient in m_v per air mass may be computed from the data by the procedure explained by Hardie [13].

4.2.3.2   Calculation of Satellite Brightness in Visual Magnitudes.

For a white or gray satellite, a visual magnitude value can be assigned to S. The first step is to solve equation 4.1, replacing $N_{Tgt}(\lambda)$ with $N_{Sun}(\lambda)$ from figure 4.4 to calculate S_{Sun}. In this case, the Sun is treated as a point source having the same straddle factor value as the target satellite. Given that the brightness of the Sun is $m_v = -26.74$, the satellite's visual magnitude can be calculated directly:

$$\text{White (or gray) target } m_v = -2.5 \log (S/S_{Sun}) - 26.74. \tag{4.2}$$

If the shape of the satellite's reflectance spectrum is different from the shape of the solar spectrum, that is, if the satellite is not white or gray in color, the m_v value calculated from equation 4.2 is that of a modeled white (or gray) satellite that produces the same number of photoelectrons in the sensor as the actual satellite. The calculated m_v in this case is called the "sensor magnitude" of the satellite.

4.2.3.3   Calculation of Sky Background Photoelectrons

A satellite is detected by an electro-optical sensor in the presence of sky background light, such as those shown in figure 4.9a and b. The number of photoelectrons from sky background sources, B, may be calculated as follows:

$$B = \left(\int_{\lambda_1}^{\lambda_2} d\lambda \, N_B(\lambda) q(\lambda) f(\lambda) \right) A_{eff} t_{exp} \alpha^2 , \tag{4.3}$$

where $N_B(\lambda)$ = background photons per square meter per second per square arc second per micron
$q(\lambda)$ = imager spectral quantum efficiency
λ_1 = shorter wavelength limit of integration
λ_2 = longer wavelength limit of integration
$f(\lambda)$ = filter spectral transmittance
A_{eff} = effective collecting area of telescope (square meters)
t_{exp} = exposure time (seconds)
α^2 = pixel angular area (square arc seconds)
Sky background brightness can be measured in the convenient units of m_v per arc square second with a photoelectric photometer. If the photometer (with its known field of view in square arc seconds) is calibrated with solar-type stars, it can measure sky brightness in m_v per square arc second provided that the spectral shape of the sky brightness spectrum is solar. Although, in general, the brightness spectral shape of the

night sky is not solar, its m_v per square arc second value measured by the photometer, with its spectral quantum efficiency similar to that of the camera and with its solar star calibration, produces the same number of photoelectrons in a sensor as does the actual sky. This measurement is precisely what is needed for characterizing the sky background brightness for the prediction of sensor detection sensitivity. Equation 4.4 calculates the number of photoelectrons in the sensor from a measured sky brightness value in m_v per square arc second:

$$B = \left(\int_{\lambda_1}^{\lambda_2} d\lambda N_{\text{Sun}}(\lambda) q(\lambda) f(\lambda) \right) 10^{-0.4(m_v + 26.74)} A_{\text{eff}} t_{\text{exp}} \alpha^2 , \qquad (4.4)$$

where m_v = sky brightness in m_v per square arc second
$N_{\text{Sun}}(\lambda)$ = zero air mass solar spectrum
   All other terms are defined in the same manner as for equation 4.3.

4.2.3.4   The Signal-to-Noise Ratio of a Satellite Image

Astronomers recognized that electro-optical sensors detect a point image in the presence of sky background light from detected photoelectrons with a signal-to-noise ratio given by $S/(B)^{1/2}$, where S is the number of photoelectrons per pixel detected from the satellite image and B is the number of photoelectrons per pixel detected from the sky background [17]. These sensors, however, can have several noise sources that need to be included in the signal-to-noise ratio expression $S/(B)^{1/2}$ in addition to the temporal sky background noise. As explained earlier, they include temporal noise from sensor readout and dark current, and spatial noise from quantum efficiency and dark current variations. A development goal of the laboratory is to make the self-noise of an electro-optical sensor sufficiently low that its sensitivity is limited primarily by $(B)^{1/2}$. The signal-to-noise ratio for a sensor to detect a point source target in the presence of noise may be written as

$$\text{SNR} = S/\sigma, \qquad (4.5)$$

where S = required number of satellite image photoelectrons per pixel for detection
σ = root mean square noise per pixel, which is given by

$$\sigma = [S + (\eta_q S)^2 + B + (\eta_q B)^2 + N_{\text{DC}} t_{\text{exp}} + (N_{\text{DC}} t_{\text{exp}} \eta_{\text{DC}})^2 + (N_{\text{sys}})^2]^{1/2}, \qquad (4.6)$$

where S = shot noise variance (or average number of) of signal photoelectrons per pixel

B = shot noise variance (or average number of) of sky background photoelectrons per pixel

η_q = imager quantum efficiency variation (percent root mean square)

N_{DC} = dark current (electrons per pixel per second)

η_{DC} = variation of dark current (percent root mean square)

N_{sys} = system noise (equivalent electrons per pixel root mean square)

The goal is to calculate S, the required number of photoelectrons per pixel for a particular sensor to detect a satellite image with a specified signal-to-noise ratio. Before this can be done, equation 4.5 for the SNR must be modified to include the image-intensified camera tube prereadout gain.

4.2.3.5   Signal-to-Noise Ratio with Intensified Imager

The prereadout gain of a camera tube image intensifier modifies the SNR equation 4.5 as follows:

$$\text{SNR} = GS/\sigma, \tag{4.7}$$

where G = image intensifier gain factor

GS = amplified signal photoelectrons per pixel

σ_i = root mean square noise per pixel with intensifier, which is given by

$$\sigma_i = [G^2S + (\eta_q GS)^2 + G^2B + (\eta_q GB)^2 + N_{DC}t_{exp} + (N_{DC}t_{exp}\eta_{DC})^2 + (N_{sys})^2]^{1/2}, \tag{4.8}$$

where G^2S = amplified shot noise variance of signal photoelectrons per pixel

GB = amplified sky background photoelectrons per pixel

G^2B = amplified shot noise variance of sky background photoelectrons per pixel

η_q = imager quantum efficiency variation (percent root mean square)

N_{DC} = dark current (electrons per pixel per second)

η_{DC} = variation of dark current (percent root mean square)

N_{sys} = system readout noise (equivalent electrons per pixel root mean square)

The noise terms $G(S)^{1/2}$ and $G(B)^{1/2}$ in equation 4.8 amplify their respective photon-photoelectron conversion noises in the first photocathode by the intensifier gain G. The effect of intensifier gain, which can be a factor of up to 50,000 for an I-Ebsicon sensor in dark sky, is to make dark current and system noise much less significant compared to noise from B and S [15]. In camera operation, G is usually set to the highest value that avoids saturation in the prevailing sky brightness and that does not produce an excessive number of ionic scintillations in the image. In general, a value of G must be substantially greater than 1 for the image intensifier to form an adequate image.

The equation for the required number of photoelectrons S from a target image may be written from equations 4.7 and 4.8 as

$$S = (\text{SNR})[S + (\eta_q S)^2 + B + (\eta_q B)^2 + N_{\text{DC}} t_{\text{exp}}/G^2 + (N_{\text{DC}} t_{\text{exp}} \eta_{\text{DC}})^2/G^2 + (N_{\text{sys}})^2/G^2]^{1/2}. \tag{4.9}$$

The effect of increasing G is apparent in equation 4.9. As G is increased, the last three noise terms for dark current and system noise become smaller, and the required number of photoelectrons S becomes correspondingly smaller. When G is increased beyond where these noises become insignificant, the required S remains the same and sensitivity is not improved. An explicit solution for S from equation 4.9 may be obtained with the quadratic formula (see appendix A).

Equation 4.9 can be simplified for sensor evaluation when multiple measurements can be made on calibrated point sources, such as stars. The signal-to-noise ratio in this case may be defined as the average peak target signal divided by the root mean square of the background noise, which implies that repeated signal measurements of a calibrated star image can average out signal noise. The equation for the required number of target photoelectrons S in this case can be written directly from equation 4.9 as

$$S = (\text{SNR})[B + (\eta_q B)^2 + (N_{\text{DC}} t_{\text{exp}})/G^2 + (N_{\text{DC}} t_{\text{exp}} \eta_{\text{DC}})^2/G^2 + (N_{\text{sys}})^2/G^2]^{1/2}. \tag{4.10}$$

If the intensifier gain G can be set high enough to make dark current and readout noises insignificant, equation 4.10 may be written more simply as

$$S = (\text{SNR})[B + (\eta_q B)^2]^{1/2}. \tag{4.11}$$

Operating the sensor in dark sky with a short integration time, such as 33 milliseconds for the TV frame rate of 30 per second, is likely to allow a high enough gain setting for these conditions without camera saturation. To use equation 4.9, 4.10, or 4.11 to predict sensor sensitivity, it is necessary to calculate the required number of photoelectrons S for detection, which means that the other terms in the equations must be evaluated. For high intensifier gain such that dark current and readout noises are insignificant, as in equation 4.11, η_q is the only noise term that needs evaluation. If equation 4.10 does not apply, the noise terms in equations 4.9 and 4.10 may be evaluated by the procedures described in appendix B.

4.2.3.6 Signal-to-Noise Ratio Required for Detection

An adequate SNR must be determined that would result in a highly probable detection and a very low false-alarm rate. Analyses have shown that a signal-to-noise ratio equal to 6 results in an 85 percent probability of detection and a 10^{-6} false-alarm like-lihood, which are reasonable criteria both for a human operator looking at a TV screen and for an automatic image detector [18, 19].

4.2.3.7   Evaluation of Root Mean Square Quantum Efficiency Variation for High Intensifier Gain

The unknown term in equation 4.11 is the root mean square of the variation of the quantum efficiency, η_q. If $\eta_q = 0$, the sky background limited sensitivity condition exists, and the noise term in equation 4.11 becomes

$$\sigma = [B]^{1/2}. \tag{4.12}$$

In this case, video noise measured from the camera should increase as the square root of the increase in the background signal level, whether due to higher sky brightness or longer integration time. If the slope is steeper than equation 4.12 predicts, and intensifier gain is high enough for equation 4.11 to be valid, a value of η_q may be determined from the measurements. The procedure is to calculate background photoelectron numbers B from equation 4.3 or 4.4 for several sky background brightness intensities, and plot the noise component of equation 4.11, which is

$$\sigma = [B + (\eta_q B)^2]^{1/2}, \tag{4.13}$$

with a value of η_q chosen to make the noise plot have the same shape as the camera video noise measurement plot. The background brightness increment intervals should be the same for both the noise equation and the noise measurement plots.

4.2.3.8   Detection of a Moving Image

If the dwell time in a pixel of the image of a moving satellite, t_{dwell} (pixel width in arc seconds/satellite angular speed in arc seconds per second), is less than t_{exp}, then the required S in equations 4.10, 4.11, and 4.12 is increased by the factor

$$t_{exp}/t_{dwell}. \tag{4.14}$$

If t_{dwell} is greater than t_{exp}, no correction is needed.

4.2.4   Measured and Predicted Sensitivities of Experimental Test System I-Ebsicon Sensors

4.2.4.1   Experimental Test System 31-Inch f/5 Telescope and I-Ebsicon Camera

The sensor model was used to predict the sensitivity of the prime Experimental Test Site sensor, which was the 31-inch f/5 telescope with the I-Ebsicon camera. Figure 4.14 shows its predicted sensitivity operating at the standard TV rate of 30 frames per second as a function of night sky brightness under the conditions shown in a text box to the right of the plot.

Figure 4.14 also shows the atmospheric extinction-corrected m_v of solar-type stars which would produce a signal-to-noise ratio equal to 6 in the sensor as a function of sky brightness measured with a photoelectric photometer. Stars chosen for SNR measurement were of the solar spectral type and of brightness that produced a SNR of

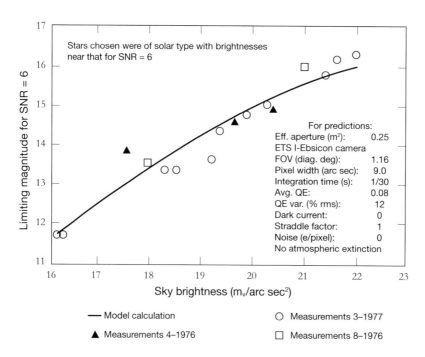

Figure 4.14   Plot of measured and predicted limiting magnitudes of the Experimental Test Site (ETS) 31-inch telescope and I-Ebsicon camera with a time exposure of 1/30 second (standard TV rate). e/pixel, electrons per pixel; FOV, field of view; QE, quantum efficiency; RMS, root mean square; SNR, signal-to-noise ratio.

about 6 [20, 21, 22]. The catalog magnitude of a star not producing a signal-to-noise ratio equal to 6 was adjusted for figure 4.14 to the value it would have to have to produce a SNR equal to 6.

Figure 4.14 indicates reasonable agreement of the predictions from equation 4.11 and measurements. Figure 4.15 shows a 0.18-degree portion of a calibrated star field in the galactic cluster NGC-188 [20] taken with the Experimental Test Site 31-inch sensor in the 0.5-degree zoomed mode in dark, clear sky (21 m_v per square arc second).

4.2.4.2   Limiting-Magnitude Summary for 31- and 14-Inch Experimental Test System Sensors

Table 4.2 summarizes the limiting apparent visual magnitude measurements of ETS sensors at the standard TV exposure of 1/30 second at the sky brightness indicated.

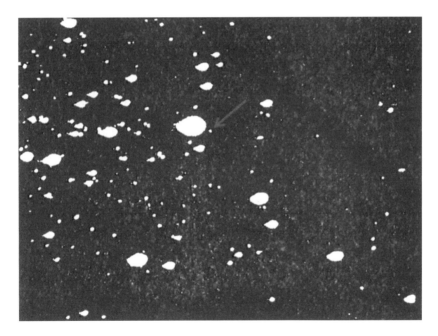

Figure 4.15   TV monitor display of a 0.18-degree (horizontal direction) portion of the galactic cluster NGC-188 taken with the Experimental Test Site 31-inch telescope and its I-Ebsicon camera in the 0.5-degree zoomed mode. The star designated by the red arrow is apparent visual magnitude 15.86 [20]. The bright star immediately to its left is magnitude 9.55 [20], or 334 times brighter, showing a larger diameter because of blooming in the camera. The integration time was 1/30 second for standard TV operation.

Table 4.2 Limiting apparent visual magnitude (m_v) measurements of Experimental Test Site sensors at time exposure of 1/30 second

Sensor with I-Ebsicon camera	31-inch f/5	31-inch f/5	14-inch f/1.7
Field of view (deg)	1.16	0.5	7.5
Limiting SNR = 6 m_v Dark sky (21.5 m_v/arcsec^2)	16.0 m_v	17.0 m_v	12.2 m_v
Limiting SNR = 6 m_v Bright sky (18.0 m_v/arcsec^2)	13.4 m_v	14.1 m_v	9.0 m_v

Calibrated solar-type stars were chosen for the measurements from those listed in Eggen and Sandage [20], Landolt [21] and Purgathofer [22].

4.2.4.3   Increased Sensitivity with Frame Integration

If greater sensitivity were required than what was achieved with a time exposure of 1/30 second, the readout beam in the cameras could be turned off to allow the integration of photoelectrons on the camera tube target. Sensitivity would increase by a factor roughly equal to the square root of the number of frames integrated. To achieve adequate integration times while avoiding saturation of the I-Ebsicon camera, it was necessary to accumulate frames on a commercial TV storage terminal/scan converter unit. The factor limiting the length of integration time was the buildup of fixed pattern noise from either quantum efficiency variation across the first photocathode or dark current variation across the target of the I-Ebsicon camera, depending on how high the gain could be set. The practical gain in sensitivity with integration was about 2.5 m_v (or a factor of 10) before the buildup of fixed pattern noise became very significant.

Typical signal-to-noise ratio values achieved with integration by the Experimental Test Site 31-inch telescope and I-Ebsicon camera in the zoomed 0.5-degree field of view are shown in table 4.3.

4.2.5   Sensor Implementation for Proof-of-Concept Demonstration

The Experimental Test Site telescopes, as supplied, could point at and track stars for camera testing, but considerable additional mount control hardware and software were required for deep-space surveillance. The deep-space surveillance mission was thought to comprise the three types of searches listed in table 4.4.

Table 4.3 Typical values of apparent visual magnitude (m_v) and signal-to-noise ratio (SNR) achieved with the Experimental Test Site 31-inch telescope in the 0.5-degree field of view with 7-second integration time

Sky brightness in m_v/arcsec^2	21.5 (dark)	18.3 (bright)
	m_v = 17.18, SNR = 24	m_v = 15.52, SNR = 9.3
	m_v = 17.41, SNR = 20	m_v = 15.60, SNR = 8.3
	m_v = 18.42, SNR = 6	

Table 4.4 Types of deep-space surveillance

Type	Description
1	A satellite was searched for using a known element set. If the satellite were found, its angular position in right ascension (RA) and declination (DEC) along with the time would be recorded, and the satellite would be reported to be where it was supposed to be. If the satellite were not found, a search around its predicted position would be conducted to try to find it. If the satellite were found then, its angular position and time would be recorded and reported. If not found, the satellite would be reported as lost. The term "maneuver query" could be applied for type 1 surveillance.
2	A specified region of space would be searched for particular classes of satellites normally associated with such regions (e.g., the equatorial geosynchronous belt and the Molniya belt)
3	The entire sky would be searched to find any satellites present without regard to satellite orbit type. As in all other searches, all satellites found would have their angular positions and times recorded and reported.

In addition, for all types of searches, any unknown satellite detected would be designated as an "uncorrelated target" and have its angular positions and corresponding times recorded. A decision would then be made whether or not to track the target for a time long enough to form a dead-reckoning vector or an approximate element set for acquisition at a later time. Sensor implementation for the satellite surveillance mission required a number of system functions, listed in table 4.5.

Figure 4.16 illustrates anticipated missions that the Experimental Test Site electro-optical sensors were designed to perform.

Table 4.5 System functions for the space surveillance mission

1	Driving the telescope axes at controlled rates and to positions corresponding to mission requirements (e.g., along the satellite trajectory or to specific areas in the sky for satellite searching). Such requirements included slewing to the starting point of a search and then executing step-and-settle motions in a specified field-of-regard to allow the field of view of the sensor to be searched for the presence of a satellite.
2	Performing an image-processing operation (i.e., a moving-target indication [MTI]) to indicate the presence of a point image which moves differently from stars. Initially, designation of a satellite image on a TV monitor was operator aided, but the goal was to develop an automated device (i.e., an AMTI) for the satellite designation..
3	Maintaining up-to-date catalogs of satellite element sets and of stars calibrated in position, brightness, and color.
4	Enabling a satellite's angular position to be recorded at specified times, and have its position corrected by comparison with calibrated star positions near the satellite at these times. Positional accuracy of 2 arc seconds was thought to be achievable.
5	Making a system to record the solar reflected light intensity and color from a satellite as a function of time, with time resolution as short as 1 millisecond.
6	Making a system that uses positional and brightness data taken on a satellite to confirm its identification or to designate it as an uncorrelated target.

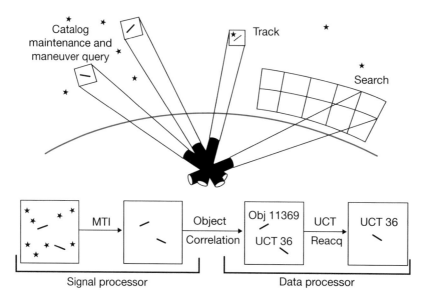

Figure 4.16 Diagram of anticipated missions and functions of an electro-optical sensor for satellite surveillance. MTI, moving-target indication; UCT, uncorrelated target.

4.2.5.1   Analog and Digital Moving-Target Indication Systems

Three electronic devices were developed at the laboratory to help an operator find point images that moved differently from stars in video images of the sky. The first device (employed in 1975) used a commercial broadcast analog TV storage tube and scan converter that could store and average a few successive video TV frames. After a time delay of a few seconds, the storage tube and converter would store a second average of the same number of frames and subtract the second frame average from the previously stored frame average. The result was a video frame of the time-delayed difference of the frames showing any image that had moved at least by one pixel during the time delay. The signal-to-noise ratio of such an image would be improved by averaging the stored frames. Star images, which were fixed by sidereal track, would be subtracted out and not be visible in the resultant frame. A moving-satellite image would thus become visible in a dark background. Figures 4.17 and 4.18 illustrate this

Figure 4.17   TV frame of nighttime sky in the Milky Way taken on day 95, 1978, with the Experimental Test Site 14-inch sensor in its 3-degree zoomed mode showing many stars and satellite 83553, a Kosmos 775 rocket body (arrow). A small video-generated rectangular box was manually positioned just below the satellite image.

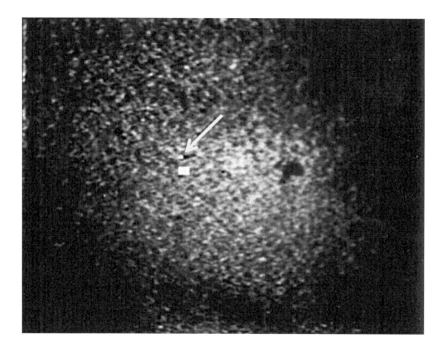

Figure 4.18   Video frame made from the difference of two stored frames containing the satellite image shown in figure 4.17, but taken a few seconds apart. The black-and-white image (indicated by arrow) is that of the satellite shown in figure 4.17 with an improved signal-to-noise ratio. A video-generated small rectangular box was manually positioned just below the satellite image.

operation. Figure 4.18 shows a subtracted image of two frames like that of figure 4.17, but taken a few seconds apart.

Although this technique helped operators find a satellite image, it did not automatically alert them of a possible satellite image being present. They had to carefully examine each frame to see if such an image were indeed present. The second device, built with a modified Westinghouse commercial 16-inch analog video disk recorder [23] and deployed in late 1976, produced an automated moving-target indication by detecting a moving image like that shown in figure 4.18 from its signal-to-noise ratio, drawing a box around it, and then sounding a detection alarm for operators. But they still had to manually examine the frame to verify that the image detected was that of a satellite. And even though they could see satellite images with signal-to-noise ratios considerably smaller than the moving-target indication SNR threshold of 6, they still had to know where to look in order to see them, which could be very difficult in a crowded star field. The early analog video systems were replaced by a third device, completed in 1980, that accomplished the same functions reliably with a digital signal-processing system.

4.2.5.2   Sensor Integration and Frame-Averaging Techniques to Detect Small Satellites

Figure 4.19 shows a faint satellite image detection with frame integration. In this case, the telescope was driven not at the sidereal rate but at an angular rate that approximated the topocentric motion of the satellite, based on an orbital element set. Improvement in detection sensitivity was about a factor of 10 with the increased effective integration time, which was limited by the buildup of video noise.

4.2.5.3   Taking Accurate Metric Data on Detected Satellites

The principal data taken on detected satellites were their topocentric angular positions of right ascension and declination in the sky as a function of time (i.e., metric data).

Figure 4.19   Detection of a faint satellite image by frame averaging. The satellite image is shown as a round point in the center, whereas surrounding star images are shown as elongated streaks.

Getting metric accuracy of 2 arc seconds within 1 standard deviation required measuring the positions of at least three position-calibrated stars as close to the satellite image as possible and then using triangulation to get the satellite's position relative to the stars. This local calibration procedure was partially automated by software algorithms, which included a graphical user interface.

4.2.5.4   Photoelectric Photometer System for Satellite Brightness and Color as a Function of Time

An important technique to help identify a detected satellite for resident space object identification is to observe and record the amplitude and color of its solar reflection as a function of time. Because a satellite can have many different surfaces made of different materials in different sizes and shapes, the satellite's solar reflection will change in both amplitude and time depending on the angle of the observer to the satellite, the angle of the Sun to the satellite, and the aspect angles of the satellite's body to the observer. It is possible that, under known observing conditions, a satellite could have repeatable or periodic light flashes in intensity and color. A record of these flashes would constitute a "signature" of the satellite, which could be used to identify the satellite on subsequent detections.

Another important function of a photometric photometer is the measurement of atmospheric extinction and sky brightness. Extinction is measured in units of m_v by recording signals from stars calibrated in brightness and spectral content from near zenith to the lowest elevation angle considered for use. An extinction coefficient in m_v per air mass could then be calculated (see subsection 4.2.1). Sky brightness is measured in m_v per square arc second from the signal of an area of sky in the angular aperture of the photometer that did not contain detectable stars.

Two methods of recording light from detected satellites and stars were implemented for Experimental Test Site sensors. The first (and simpler) method recorded the video signal amplitude of a detected image as a function of time using a commercial video signal integrator. This device would then provide the integrated video signal (of a designated box on the video screen), to a strip-chart recorder to record amplitude as a function of time. It had the advantage of being immediately usable without putting a satellite or star image out of view, but its limiting time resolution was 30 Hz (the TV frame rate), which was considered slow for resolving rapidly occurring solar reflections from satellites.

The second (and much more effective) method employed a photoelectric gallium-arsenide (GaAs) photometer, with color filter bands that had a time resolution of 1 millisecond to resolve rapidly occurring short light flashes. The photometer device was placed beside an I-Ebsicon TV camera on a telescope and was optically coupled to the telescope focal plane by a 40 percent reflecting and 40 percent transmitting pellicle that was swung in front of the I-Ebsicon camera focal plane during use. The optical system for the photometer was designed to spread the point image of a satellite or star uniformly across the photocathode of the gallium-arsenide photomultiplier tube so that any nonuniformity of response would not affect the output signal amplitude. An optical color filter, such as one used for the astronomer's UBVRI (ultraviolet-blue-visual-red-infrared) filter passbands [6], could be placed in the filter wheel to see whether the color of solar reflections from a satellite was close to one of these passbands. An automated system was implemented at the Experimental Test Site to calibrate the photometer and measure atmospheric extinction.

Figure 4.20 shows typical light-intensity plots as a function of time measured on two representative satellites. Although the plots seem different, even for the same satellite, they were taken under different geometrical lighting conditions, illustrating the challenge of identifying satellites by their temporal signatures.

How short in time could light flashes from satellites be? Some active satellites were spin stabilized, and rotating flat surfaces on them could make the observed duration of their flashes extremely short. For example, the twenty-six IDSP satellites (see figure 4.6a) have eight small flat solar panels and spin at about 150 revolutions per minute, making the duration of a flash due to specular solar reflection 0.55 millisecond as observed by the Experimental Test Site telescopes. The TV photometer technique, with its 30 Hz time limitation, would not show nearly as much information as a photometer with a 1 kHz rate. It did indeed seem that photometric data needed to be taken at least at the 1 kHz rate, but with the capability to average the data to longer times for slower and fainter flashes from other satellites.

Along with implementing the photometric systems, attempts were made to model the optical signatures of satellites from their materials and reflection geometry. But the number of variables made successful modeling possible only for simple shapes, such as specular and diffuse spheres, cylinders, and plates. The images of actual satellites were sometimes observed to get brighter as their range and solar phase angle increased. Consequently, the ETS operators recorded many satellite signatures under widely varying lighting conditions over several years in order to establish a satellite signature library that could be used to help identify a satellite. Examination of the Experimental Test Site optical signature database showed that the brightness of deep-space satellites varied from about 12 m_v to fainter than 18 m_v, with an average brightness of 14.5 m_v.

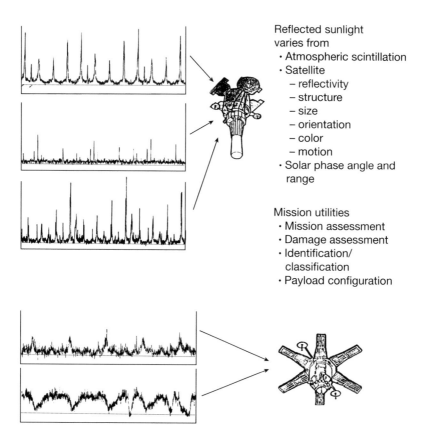

Reflected sunlight
varies from
  · Atmospheric scintillation
  · Satellite
    – reflectivity
    – structure
    – size
    – orientation
    – color
    – motion
  · Solar phase angle and
    range

Mission utilities
  · Mission assessment
  · Damage assessment
  · Identification/
    classification
  · Payload configuration

Figure 4.20   Time history plots of optical signatures for the Integrated Missile Early Warning Satellite (IMEWS; top) and for the second Russian communications satellite in the Molniya series (bottom).

4.2.5.5   Keeping Track of Newly Detected Satellites for Reacquisition

A principal part of the GEODSS proof-of concept demonstration, keeping track of newly detected satellites from angles-only metric data, required taking adequate and accurate metric data on a detected satellite to generate either a dead-reckoning vector or an initial orbital element set for near-term reacquisition and applying sophisticated algorithms to the data to then generate an accurate element set for inclusion in a satellite catalog (chapter 6 describes, in part, the algorithms developed by the laboratory to refine orbital element sets). Special techniques were developed for low-inclination satellites in geosynchronous orbit that did not move much during an observation period [24].

4.2.5.6   Rapid and Leak-Proof Search Techniques to Find Known and Unknown Satellites

Conducting such searches for deep-space satellites, while simultaneously keeping track of newly detected unknown satellites, required that two sensors be on at the same time, one to conduct a time-critical leak-proof search and another to track newly discovered uncorrelated targets, lost satellites, or maneuvered satellites in order to generate an orbital element set. In general, searches were conducted using step-and-stare search patterns around geostationary orbits for modeled satellites (i.e., points in the sky with fixed azimuth and elevation) spaced several degrees apart. The extent of the search pattern (field of regard) would cover a desired range of inclinations (usually ±5–10 degrees above and below the equatorial geosynchronous belt) to detect all satellites there. The common mode for the search was step and stare, with the telescope being driven at a sidereal rate during the stare time.

4.2.5.7   Scheduling Satellite Observations

The laboratory developed a sensor and an algorithm to efficiently schedule satellite observations, A panoramic sky-monitoring sensor (PANSKY) used a visible-band camera and zenith-pointing fish-eye lens to image the sky and a microprocessor to graphically designate, on a TV monitor, positions of calibrated stars that would be just visible in clear sky. And a dynamic scheduling algorithm designed and built for the Experimental Test Site had as input parameters satellite importance category, elevation angle, solar phase angle, and sky conditions. The heritages of this algorithm were of considerable use for GEODSS and, later on, for the laboratory-designed Space-Based Visible (SBV) sensor (a small visible-band surveillance sensor; see chapter 5).

4.2.5.8   Real-Time Detection of Satellite Orbital Maneuvers and Feature Deployment

Even though the electro-optical sensor developed for the Experimental Test Site could not resolve structural details of deep-space satellites, it could make real-time observations of events occurring on them, such as the deployment of solar panels (producing a brightness change), the ejection of hardware (e.g., lens covers), and the firing of a rocket engine. Camera operation at 30 frames per second could make such events easily observable to an operator. Figures 4.21 and 4.22 show examples of detectable

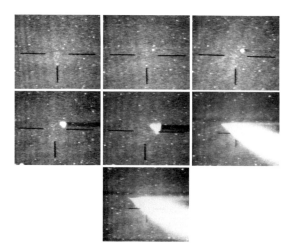

Figure 4.21   Sequence of seven video frames showing the approximately 20-second firing of a solid-fuel rocket on the transfer bus of the DSCS 2 and DSCS 3 satellites to put them into a geosynchronous circular orbit on day 303, 1982. Apparently, the aluminum-hydroxide exhaust plume is highly reflective of sunlight, and its angular extent is shown in the last two frames to be in excess of 1 degree, or 700 kilometers in distance at geosynchronous range. The satellite is seen drifting away from the rocket plume in the last frame.

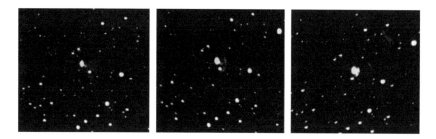

Figure 4.22   Sequence of three video frames showing the approximately 20-minute firing of a liquid fuel rocket to put SSC Object 15453 into a circular geosynchronous orbit on day 357, 1984. The first two frames just barely show the rocket exhaust upon firing, whereas the third frame shows the plume at full power.

rocket engine firings on deep-space satellites. Such firings are for maneuver and orbital transfer.

Whereas rocket engine firings and hardware ejections from satellites were relatively easy to observe, to draw correct conclusions about the observation of certain hardware deployments, such as the correct unfolding of solar panels, required detailed analysis of optical signatures and information from satellite owners/operators, to whom the utility of such observations was readily apparent, however. Indeed, providing near-real-time confirmation of deployments of hardware became a significant part of the GEODSS proof-of-concept demonstration at the Experimental Test Site. Based on the accomplishments described in this chapter, the laboratory helped the Air Force with specifying requirements for the multisite GEODSS system, with monitoring system procurement, and later with improving system techniques and technology.

4.2.6   Detection of Low-Altitude Satellites in Clear Daytime Sky in the Visible and Near-Infrared Band

Since low-altitude satellites were eclipsed by the Earth's shadow for most of the night, it was of interest to see whether they could be detected in clear daytime sky by a passive electro-optical sensor. Consequently, a project called "Daylight Satellite Acquisition and Tracking" (DAYSAT), was begun in fall 1980 [25, 26]. Since the daytime sky could be eight million times brighter than the nighttime sky, since solar phase angles in the daytime were large (larger than 90 degrees) and not as favorable as in nighttime, and since daytime brightness data on satellites were not available to derive the system parameters needed, developing DAYSAT technology proved to be quite a challenge.

Because it could be damaged by daytime light, the Experimental Test Site I-Ebsicon camera was rejected in favor of a silicon diode–array vidicon TV camera, which could operate in bright light with the desirable characteristics of low lag, low blooming, and high charge-storage capacity. In addition, the TV camera's spectral quantum efficiency was not peaked in the blue, and extended to the near-infrared spectrum [27]. Table 4.6 lists seven key technology issues that needed to be resolved for DAYSAT.

4.2.6.1   Demonstration of Daylight Satellite Acquisition and Tracking

A camera was developed at the laboratory, mounted on an Experimental Test Site 31-inch telescope in place of an I-Ebsicon camera and tested for limiting magnitude by star detections in the daytime. In figure 4.23, two data points are shown: one is for the

Table 4.6   Seven technology issues that needed resolution for DAYSAT

1	The largest silicon vidicon camera tube available had a 16-millimeter diagonal focal plane, compared to the GEODSS 80-millimeter focal plane. The small focal plane resulted in a small (0.25 degrees) field of view for an ETS 31-inch f/5 telescope.
2	A number of inexpensive well-engineered and reliable industrial surveillance cameras used silicon vidicons. However, these cameras were not designed for high signal levels, and their camera video amplifiers and beam current electronics needed to be modified to work with the high sky background signals [25, 26]
3	The high readout noise of the silicon vidicon camera tube meant that the camera had to operate near saturation to make sky background noise larger than readout noise. Variable optical attenuation was required to maintain this condition. The variable attenuation was done with long-pass Kodak Wratten Series 87c, 87, 88a, 89b, and 92 filters [28], which increased satellite image contrast relative to blue sky background .
4	Nonuniformity of response caused a steeply humped video pedestal with the high signal level, requiring high pass filter electronics to make a usable image [25, 26]
5	Fixed pattern noise from pixel-to-pixel quantum efficiency variation across the target was very high. The remedy for the noise from the quantum efficiency variation was to acquire an average of several TV frames of blank sky near (but outside of) the field of regard and to subtract it from each incoming TV data frame. It was necessary to do this if sky brightness changed by only a small amount; otherwise the subtracted frame was not effective. The average of several TV frames taken for the blank sky frame reduced temporal noise compared to that of a single TV frame.
6	Atmospheric seeing in the daytime was significantly worse (about 5 arc seconds) than at nighttime (about 2.5 arc seconds) because of atmospheric turbulence from solar heating.
7	Topocentric angular rates of low-altitude satellites varied from a few to several tenths of a degree per second and to nearly overhead rates of 1.5 degrees per second. The Experimental Test Site sensors were made primarily for deep-space satellites, which rarely went faster than 0.1 degree per second, hence some adjustment of the telescope mount control system was required.

Limiting magnitude of ETS DAYSAT sensor as a fuction of int. time

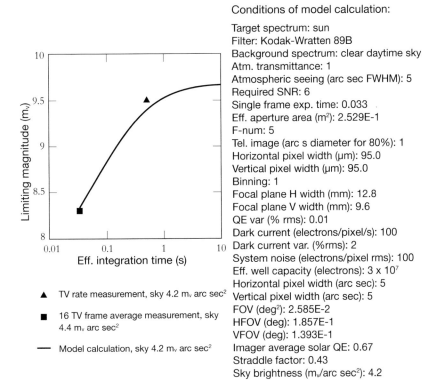

Conditions of model calculation:

Target spectrum: sun
Filter: Kodak-Wratten 89B
Background spectrum: clear daytime sky
Atm. transmittance: 1
Atmospheric seeing (arc sec FWHM): 5
Required SNR: 6
Single frame exp. time: 0.033
Eff. aperture area (m^2): 2.529E-1
F-num: 5
Tel. image (arc s diameter for 80%): 1
Horizontal pixel width (μm): 95.0
Vertical pixel width (μm): 95.0
Binning: 1
Focal plane H width (mm): 12.8
Focal plane V width (mm): 9.6
QE var (% rms): 0.01
Dark current (electrons/pixel/s): 100
Dark current var. (%rms): 2
System noise (electrons/pixel rms): 100
Eff. well capacity (electrons): 3 x 10^7
Horizontal pixel width (arc sec): 5
Vertical pixel width (arc sec): 5
FOV (deg^2): 2.585E-2
HFOV (deg): 1.857E-1
VFOV (deg): 1.393E-1
Imager average solar QE: 0.67
Straddle factor: 0.43
Sky brightness (m$_v$/arc sec^2): 4.2

▲ TV rate measurement, sky 4.2 m$_v$ arc sec^2

■ 16 TV frame average measurement, sky 4.4 m$_v$ arc sec^2

— Model calculation, sky 4.2 m$_v$ arc sec^2

Figure 4.23 Plot of measured and calculated detection sensitivity of a modified RCA Ultricon silicon vidicon camera and DAYSAT signal processor on an Experimental Test Site 31-inch f/5 telescope. FWHM, full width half maximum; HFOV, horizontal field of view; VFOV, vertical field of view.

TV rate integration time of 1/30 second in a clear daytime sky brightness of 4.2 m_v per square arc second and the other is for a 16-frame average (0.53-second integration time) in a sky brightness of 4.4 m_v per arc square second. Also shown in figure 4.23 is a sensor model plot for when the sky background brightness is very high, making readout noise and dark current insignificant, with detection sensitivity limited by the very small (0.01 percent root mean square) residual uncorrected quantum efficiency variation after background frame subtraction. Although this number seems small, it is significant when the effective photoelectron-well capacities of these daytime measurements are considered.

The ETS DAYSAT sensor was used to acquire and track low-altitude satellites (whose brightness ranged from 1 to 9.2 m_v) in the daytime. The brightness of most of the satellites tracked was 7 m_v or brighter. Some were tracked as close as 20 degrees from the Sun. Figure 4.24 shows the radio astronomy *Explorer* 38 spacecraft (SSC Object 3307) in track in the daytime with a brightness of 6 m_v. *Explorer* 38 was tracked out to a range of 9,000 kilometers with a brightness variation of 5.5–8.7 m_v.

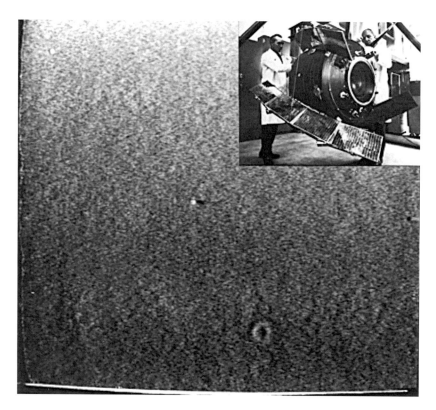

Figure 4.24   Radio astronomy *Explorer* 38 spacecraft (SDC Object 3307) in track from the Experimental Test Site in daytime on 12 February 1981, at a range of 7,100 kilometers. Its apparent visual magnitude then was 6 m_v. The inset photo is the satellite prior to its launch in 1968.

On 22 October 1981, a proof of concept was demonstrated for DAYSAT at the Experimental Test Site. Twenty-one satellites were acquired and tracked in the daytime, with accurate metric data on thirteen of them taken and sent to the North American Aerospace Defense Command (NORAD) [25, 26]. DAYSAT tests demonstrated that a sensor designed for deep-space satellite detection at night could, with a camera system change, be used to perform the GEODSS maneuver query mission on low-altitude satellites in the daytime even when their brightness was as faint as 9.5 m_v.

4.2.6.2   Atmospheric Clutter and Daylight Satellite Acquisition and Tracking

An unusual type of atmospheric clutter would occasionally be visible to the DAYSAT sensors. Sometimes, while searching for a low-altitude satellite with the camera system on an Experimental Test Site 31-inch telescope, a point image would drift in and move through the field just like a satellite image. Discrimination was made by observing whether the object remained in track using the orbital element set that applied to the satellite. A clutter object would not remain in track and would drift away. It was now important to identify the objects and find a way to quickly discriminate a satellite image from one or more clutter images. Typical ranges of clutter objects were used as the first discriminant. To estimate range, the researchers constructed a stereo parallax sensor system (figure 4.25), which consisted of two silicon vidicon sensors mounted on a steerable rigid bar and separated by a distance (limited by the width of the window) that could discriminate, by estimating the range by triangulation, between objects closer and farther away than 4 kilometers. The sensor system could measure the range of objects closer than 4 kilometers.

Figure 4.26 shows a particularly cluttered TV frame from the parallax sensor, indicating (since the clutter objects appear as two separated, side-by-side images) that the vast majority of clutter objects are closer than 4 kilometers.

It would, indeed, be hard to find a single point image in the frame that could be two superimposed images of an object farther away than 4 kilometers. The ability to discriminate a clutter object from a satellite was demonstrated in August 1981, when it was noticed that the Russian space station *Salyut* 6 occasionally passed within a few degrees of the star Polaris. Using Polaris as a sky reference point, the parallax sensor (with its 1.8-degree field of view) was aimed at an intersection point of the *Salyut* 6 trajectory. *Salyut* 6 was then detected three times as a single point image moving at the correct speed and direction in the presence of many closer point objects, which appeared as a pair of side-by-side images. The clutter objects were prolific all year around, but more so in the spring and fall. Consultation with Harvard

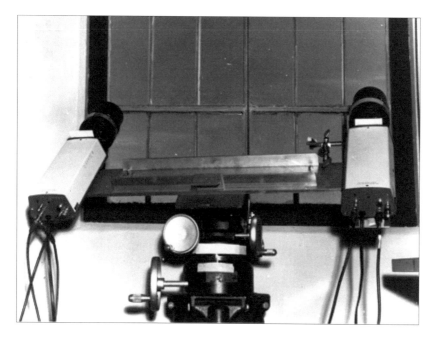

Figure 4.25   Parallax sensor aligned on the star Polaris through a window at Lincoln Laboratory, Lexington, Massachusetts.

University Biological Laboratory scientists confirmed that the atmosphere contains much biological material, such as plant seed transport vehicles and spider silks.[2] Spiders, which can survive freezing temperatures, can migrate over continents by spinning silks near the ground, which are then carried to high altitudes (a few kilometers) by convection. And, indeed, long, thin images were occasionally detected that most likely were spider silk.

The laboratory researchers concluded that, when not tracking an object with a known orbital element set, a passive electro-optical daytime sensor for satellite detection would need a discrimination method, such the use of two sensors in tandem to measure parallax, to verify that an image is far enough away to be that of a satellite and not of an atmospheric clutter object.

---

2. F.M. Carpenter and C.E. Wood, Biological Laboratories, Harvard University, Cambridge, Massachusetts, private communications, 1984.

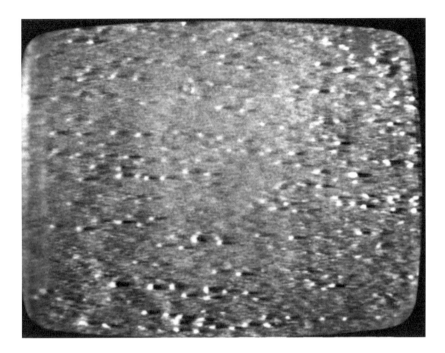

Figure 4.26 TV frame from the parallax sensor showing atmospheric clutter objects closer than 4 kilometers as two separate, side-by-side images.

4.2.7 Charge-Coupled Device Camera to Replace the GEODSS I-Ebsicon Camera

In the late 1970s, it was anticipated that solid-state charge-coupled-device (CCD) cameras would be developed that would eventually replace the I-Ebsicon vacuum-tube cameras. The Microelectronics Group at Lincoln Laboratory began a program to develop CCD imagers with properties suitable for space surveillance because commercially available focal planes were unsuitable.

The f-number of the GEODSS telescope was 2.15, with its 1-meter aperture covering a 2-degree-diameter field of view over its 80-millimeter-diameter flat focal plane. Table 4.7 shows five options to make a charge-coupled device camera for GEODSS.

Option 1 was not proposed because the additional observing time required would slow the scan coverage rate, and option 2, to make several four side abuttable charge-coupled devices and place them close enough together to eliminate the dead space

Table 4.7   Five options for making a charge-coupled device (CCD) camera for GEODSS

Option	Description
1	Use several CCD imagers spaced apart over the 80-millimeter GEODSS focal plane in a matrix pattern. Obtain multiple exposures and telescope movements to ensure that a satellite image falling in a dead space would be detected.
2	Closely abut several CCD imagers with negligible dead space between them to form a contiguous 80-millimeter-diameter focal surface.
3	Place a number of small CCDs side by side in rows, with spacing arranged so that continuous telescope motion perpendicular to the rows of images would have overlapping coverage for any satellite image that would appear on the focal plane. Adequate integration time would be achieved by clocking the CCD pixels with the same rate and direction of telescope motion. It would be essential to have smooth and steady telescope motion and low optical geometric distortion to keep both satellite and star images in the same column of pixels.
4	Couple CCD imagers to tapered fiber-optic bundles that would have their larger ends placed close enough to each other to form a contiguous 80-millimeter-diameter focal plane.
5	Develop and manufacture a single, contiguous CCD imager to cover the 80-millimeter-diameter focal plane.

between them to make a large contiguous focal plane, was unreasonably difficult. Option 3 was tested at the Experimental Test Site using a small 100×400–pixel CCD on the 31-inch telescope. The telescope was made to scan in declination at the same rate as the pixels were clocked along the 100-pixel direction. It was concluded, however, that this option would be difficult to use without resorting to exotic and expensive telescopes with appropriate drive capability and wide-field-of-view optics having negligible geometric distortion. Option 4, though it was expected to be difficult to accomplish, seemed possible to implement, and the only way to make a CCD imager for GEODSS until option 5 could be achieved.

4.2.7.1   Comparison of Charge-Coupled Device and Ebsicon Imagers

In every way, except for their small size, CCD imagers seemed very much better than Ebsicon imagers for space surveillance sensors. In listing relevant properties of Ebsicon and charge-coupled device imagers for comparison, figure 4.27 indicates three advantages for the CCD imager: (1) higher quantum efficiency, (2) more uniform response

80 mm Ebsicon tube

11 mm CCD

Technical Requirement	Tubes	CCDs	Potential CCD impact
Quantum efficiency	0.07	0.28	Improved sensitivity with bright background
Uniformity	Center much more sensitive than edges	5 percent edge to edge	Better search and UCT capability
Metric stability	Varies with Gain Power supply Orientation	Fixed by photolithography	Less frequent and simplified calibration
Ruggedness	Damaged by bright light	Damaged only by extremely bright light	Operational simplification improved lifetime
Photometric stability	Affected by previous exposures	Changes undetectable	Easier calibration, object discrimination
Focal plane size	80 mm	11 mm	Array of CCDs needed

Figure 4.27 Comparison of coupled-charge device and Ebsicon imagers (ca. 1980).

over a field of view, and (3) image positions fixed by photolithography. Figure 4.28 shows the spectral quantum efficiency advantage of an early (1980) front-illuminated CCD imager (photons must pass through the CCD electrode structure to get to the photoconductor), compared to the S-20 photoemitting photocathode of the Ebsicon imager.

The solar spectra that represent spectral shapes of white or gray satellites are closer in fit to the spectral quantum efficiency of the CCD than to that of the Ebsicon

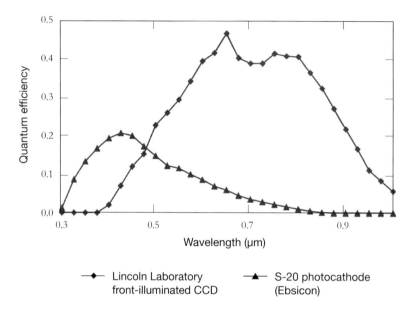

Figure 4.28   Plots of spectral quantum efficiencies of a front-illuminated charge-coupled device and S-20 (Ebsicon).

imager. On the other hand, the dark sky background spectra indicate that the CCD imager will see a brighter sky background than an Ebsicon sensor would see. These differing imager properties needed to be accounted for in predicting sensor performance for space surveillance.

4.2.7.2   Evaluation of Retrofit Camera Charge-Coupled Device Imagers

The camera to implement option 4 in table 4.7 was made with CCD imagers optically coupled to four fiber-optic tapered bundles (FOTBs) which were bonded together to form a contiguous 80-mm focal plane. However, one of the quadrant FOTBs contained four, two-side abutted CCDs to provide the narrower GEODSS zoom mode field of view with the same number of pixels as the full field of view. Consequently, a special camera was made by laboratory scientists to test the four, two-side abutted CCDs in an array component of the GEODSS Retrofit Camera, known as the "Quad Camera." Initially, images from the camera appeared quite noisy when compared to those from an I-Ebsicon sensor. Sources of noise (described in subsection 4.2.3) were subsequently evaluated as follows: the noise term σ for noise in background, in units of root mean square electrons per pixel (equation 4.6) with signal noise eliminated), is shown below to describe the possible noise sources.

$$\sigma = [B + (\eta_q B)^2 + N_{DC} t_{exp} + (N_{DC} t_{exp} \eta_{DC})^2 + (N_{sys})^2]^{1/2}, \qquad (4.15)$$

where B = shot-noise variance (or average number of) of sky background photoelectrons per pixel

η_q = imager quantum efficiency variation (percent root mean square)

N_{DC} = dark current (electrons per pixel per second)

η_{DC} = variation of dark current (percent root mean square)

N_{sys} = system noise (equivalent electrons per pixel root mean square)

The camera was then operated on a bench with no light input and with integration times varying from 0.5 to 10 seconds. Signal and noise were plotted as a function of integration time. The terms N_{sys}, N_{DC}, and η_{DC} were evaluated by fitting values for them in equation 4.15, with $B = 0$ for no light input. The charge-coupled device imager was then illuminated with a uniform light source, comparable to night sky brightness in intensity. The camera was operated again with the same integration times as before with new measurements being made of signal and noise. With the three terms N_{sys}, N_{DC}, and η_{DC} determined from the no-light input data, the noise term η_q was estimated from the new measurements. This procedure is illustrated in figure 4.29.

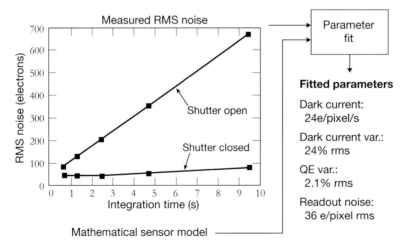

Figure 4.29 Plots of model parameters dark current, root mean square dark current and quantum efficiency variation, and root men square (RMS) readout noise as measured for the Quad Camera sensor.

It was apparent that the most significant noise was spatial, meaning that quantum efficiency or dark current variation noise or both were responsible. Members of the astronomy community indicated that they had encountered such sources of noise and had developed signal-processing techniques to reduce them [29, 30, 31, 32, 33]. These techniques were then adapted for use at the laboratory.

4.2.7.3   Large-Format Charge-Coupled Device Imager for Ground-Based Electro-Optical Deep-Space Surveillance

Late in 1992, Lincoln Laboratory scientists proposed to make a CCD imager with an imaging area of 61.4×47 millimeters with $2,560 \times 1,960$, 24-micron pixels as a suitable solid-state imager for GEODSS [34]. To fit in the 80-millimeter-diameter focal plane of GEODSS telescopes, the diagonal dimension of this CCD imager would be 78 millimeters, making a 2.05-degree diagonal field of view. The imager's pixel width on the telescope would be 2.27 arc seconds, or with 2×2–pixel binning, 4.53 arc seconds.

The proposed CCD imager would have frame-transfer readout, meaning that it would not require a frame shutter and would have no sensor dead time for operations. It would just fit on a 4-inch silicon wafer, which was the largest size available then for making silicon-based solid-state electronic devices. In addition, the large CCD would have back illumination so that light would not be attenuated by passing through the front-side metallic electrode structure to get to the photoconductive layer. The result of this design was greater effective quantum efficiency. Effective back illumination required developing a technique to make the silicon photoconductor layer thin enough that maximum absorption of backside-incident light would occur near the front-side electrode structure for efficient collection of photoelectrons.

The diagram in figure 4.30 shows the large, contiguous CCD imaging array with $1,960 \times 2,560$ 24-micron pixels and its eight separate frame-store arrays. Each of the four frame-store areas around the imaging array has two 980 (V) $\times$ 640 (H)–pixel frame-store arrays (where V = vertical; H = horizontal); the two serial-register readouts of each area are indicated by arrows. All frame-store arrays can read out simultaneously while the full imaging array is accumulating photoelectrons for the next frame. With frame-transfer readout and the highest data rate for low noise of 2 MHz, the maximum frame rate of the large imaging array is 3 frames per second with an integration time for the large imaging area of 0.33 second. Slower frame rates with longer integration times were possible.

A small 32×32–pixel frame-transfer charge-coupled device, used as a photoelectric photometer, was placed on each side of the imaging array, as also shown in figure 4.30.

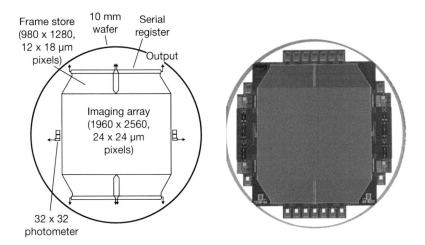

Figure 4.30 Diagram (left) and photograph (right) of Lincoln Laboratory's CCID-16 imager, developed to replace the GEODSS Ebsicon imager [34].

Each CCD has a single frame-store array with its readout register, indicated by an arrow in the figure, and operates at a 1 kHz frame rate, with a low-noise data rate of up to 2 MHz, to satisfy the GEODSS 1 kHz photometry requirement. To make a photometric measurement of a detected satellite image on the large array, the telescope would be moved to put the image on one of the 32×32–pixel CCDs.

Figure 4.31 shows the spectral quantum efficiency of the antireflection-coated (ARC), back-illuminated charge-coupled device along with the quantum efficiencies of a front-illuminated CCD and an Ebsicon (S-20 photocathode) for comparison. The tremendous quantum efficiency advantage of a back-illuminated CCD as compared to the other imagers is evident.

Lincoln Laboratory's Charge-Coupled Imaging Device 16 (CCID-16) was indeed a milestone in imaging science. It had the largest CCD focal plane ever made (until then), with a combination of high sensitivity, low noise, and a high enough frame rate for space surveillance applications. At the time, it had the world's largest integrated circuit. Since a camera to be used in the operational GEODSS system must eventually be a product that could be made by industry, it was decided to have the first camera for the new CCD imager made by industry. After competitive bidding, the Photometrics Company was selected to make the camera; with technology transfer from the laboratory, Photometrics made two cameras for the CCID-16 imager that fit into the GEODSS telescope.

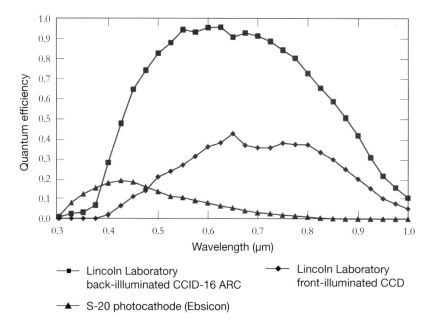

Figure 4.31   Plots of spectral quantum efficiencies of Lincoln Laboratory back- and front-illuminated charge-coupled devices, and an Ebsicon (S-20 photocathode).

Figure 4.32 shows a 2.05-degree diagonal dark sky image from a Photometrics camera with Lincoln Laboratory's CCID-16 imager on a GEODSS telescope. Stars are pin sharp, and image blooming of brighter stars is seen to be very low.

Artificial satellites, asteroids, and comets can be detected by using moving-target indication algorithms to compare successive frames to find any image that moved differently from star images between the frame times.

4.2.7.3   Noise Measurements for the Prototype GEODSS Charge-Coupled Device Camera

It was important to evaluate the sensor noise parameters described in subsection 4.2.3, and shown in equation 4.15, for the new cameras with the new imagers. As in the case of the earlier measurements on the Retrofit Camera CCDs, the CCID-16 camera was operated on an optical bench first with no light input, and then with a uniform floodlight on its imaging surface with intensity comparable to night sky brightness. The integration time was varied from 0 to slightly longer than 10 seconds. Signal and sources of noise were recorded and plotted, the signals and sources

Figure 4.32   Diagonal 2.05-degree dark sky image taken with the Photometrics camera with the Lincoln Laboratory's CCID-16 imager on a GEODSS telescope. Exposure time was 0.4 second.

of noise were fitted to equation 4.15, and the parameters in equation 4.15 were determined.

Figures 4.33 and 4.34 show plots of measured sources of noise for unbinned and 2×2–pixel binned cases and of equation 4.15 along with the models fitted to the data. The fitted parameters are listed in the text boxes. The plots are in units of electrons, but the camera output is in digital numbers, where camera gain is set to require a fixed number of electrons to be detected for each digital-number (DN) increase.

Figures 4.33 and 4.34 show great improvement in noise level compared to the earlier Quad Camera (see figure 4.29). In particular, the root mean square quantum efficiency variations of 0.33 and 0.35 percent for the unbinned and binned cases are much better than the 2.1 percent result for the quad camera. In spite of their very low percentages, as shown in figures 4.33 and 4.34, these variations in quantum efficiency have been found to be the most significant noise source limiting sensitivity.

A primary advantage of the large CCD imager is that it enables a GEODSS sensor to search the sky rapidly with a high sensitivity to detect small satellites in deep space.

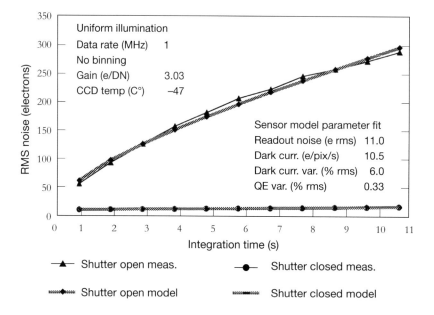

Figure 4.33 Plots of model parameters—dark current, root mean square dark current and quantum efficiency variation, and root mean square readout noise—for the case of no binning, as determined by the Photometrics camera and Lincoln Laboratory's CCID-16 imager. e/DN, electrons per digital-number increase.

Since limiting magnitude increases with integration time, there is an inverse relationship between sensitivity and search rate: a longer integration time requires a sensor to dwell in each telescope position for a longer time. Conducting a search requires more than one frame to be exposed at each telescope position because detection of a satellite image among stars depends on the relative motion of the satellite with respect to the stars.

Lincoln Laboratory subsequently transferred the CCID-16 technology to the Sarnoff Company and acted as consultants when Sarnoff built CCDs and camera electronics for the current GEODSS cameras [35].

Appendices

Appendix 4.A: Explicit Solution for Required Number of Photoelectrons (S)

An explicit solution for S from equation 4.9 may be written as follows:

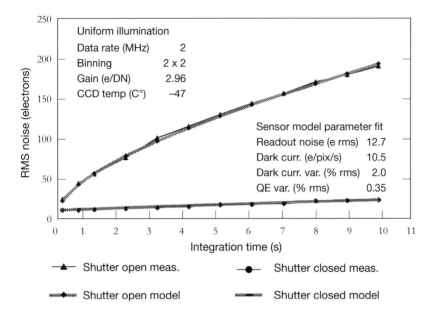

Figure 4.34   Plots of model parameters—dark current, root mean square dark current and quantum efficiency variation and root mean square readout noise—for the case of 2 × 2–pixel binning, as determined by the Photometrics camera and Lincoln Laboratory's CCID-16 imager.

$$S = (\text{SNR})[S + (\eta_q S)^2 + B + (\eta_q B)^2 + N_{\text{DC}} t_{\text{exp}}/G^2 + (N_{\text{DC}} t_{\text{exp}} \eta_{\text{DC}})^2/G^2 + (N_{\text{sys}})^2/G^2]^{1/2}$$
$$(4.9)$$

$$\text{Let } X_i = B + (\eta_q B)^2 + N_{\text{DC}} t_{\text{exp}}/G^2 + (N_{\text{DC}} t_{\text{exp}} \eta_{\text{DC}})^2/G^2 + (N_{\text{sys}})^2/G^2.$$

$$\text{Then } S = (\text{SNR})[S + (\eta_q S)^2 + X_i]^{1/2} \text{ and}$$

$$(1/(\text{SNR})^2 - (\eta_q)^2)S^2 - S - X_i = 0.$$

$$\text{Let } Y_i = (1/(\text{SNR})^2) - (\eta_q)^2.$$

Then, by the quadratic formula,

$$S = (0.5/Y_i)(1 + (1 + 4 Y_i X_i)^{1/2}).$$

Table 4.B.1.   Process for determining the first photocathode and video signal voltage relationship

Step	Procedure
1	Place the camera on a telescope or lens to look at a fixed sky brightness or artificial floodlight source calibrated in m_v/arcsec^2 units.
2	Fix the gain setting. Measure video voltage V above the video pedestal as a function of integration time t. Choose values of t that do not saturate the camera.
3	Calculate B (the number of photoelectrons from the first photocathode in the sensor) as a function of t, using equation 4.3 or 4.4.
4	Calculate the ratio B/V, which converts video signal voltage to photoelectrons for t and the gain setting used.

Appendix 4.B: Evaluation of Noise Terms in Equations 4.9 and 4.10

Sensor sensitivity can be estimated without knowing the numerical intensifier gain factor G for a fixed and repeatable, but unknown, gain value, but it is a complex procedure. Setting a fixed and repeatable gain value may mean setting a dial to a particular mark on it each time. Sensor sensitivity may also be estimated for the case of no image intensifier, or for $G = 1$.

Appendix 4.B.1: Relationship between Video Signal Voltage and Detected Photoelectrons

Before the noise terms in the equations 4.9 and 4.10 can be evaluated, the relationship between the number of detected photoelectrons and its corresponding video signal voltage for a fixed gain setting must be determined. In general, G is set high enough to make readout and dark current noises insignificant, but low enough not to cause camera saturation in the sky background or to cause a significant number of distracting ionic flashes to be visible in the display. A relationship between the first photocathode photoelectrons B or S and video signal voltage (V) is valid only for a particular gain setting. It can be determined by the sequence of steps shown in table 4.B.1.

Appendix 4.B.2: Required S for Fixed-Gain Setting or for Unity Gain

The equations for S, equations 4.9 and 4.10, are shown below.

$$S = (\text{SNR})[S + (\eta_q S)^2 + B + (\eta_q B)^2 + N_{\text{DC}}t_{\text{exp}}/G^2 + (N_{\text{DC}}t_{\text{exp}}\eta_{\text{DC}})^2/G^2 + (N_{\text{sys}})^2/G^2]^{1/2}.$$

$$(4.9)$$

$$S = (\text{SNR})[B + (\eta_q B)^2 + (N_{\text{DC}}t_{\text{exp}})/G^2 + (N_{\text{DC}}t_{\text{exp}}\eta_{\text{DC}})^2/G^2 + (N_{\text{sys}})^2/G^2]^{1/2}. \qquad (4.10)$$

The last three noise variance terms in the above equations will be more or less significant depending on the value of G. They can be most significant for a bare, unintensified imager with $G = 1$.

First, a conversion between video signal voltage and corresponding photoelectrons can be determined by the procedure of appendix B.1 for a fixed gain.

Then, numerical values for the noise terms $(N_{\text{DC}}t_{\text{exp}})^{1/2}/G$, $(N_{\text{DC}}t_{\text{exp}}\eta_{\text{DC}})/G$, and N_{sys}/G, can be evaluated as follows:

N_{sys}: The root mean square readout noise video signal voltage can be measured with the camera operating on a bench with no light input at the shortest integration times possible, and then extrapolating the noise back to what it would be with no integration time. The video noise in root mean square volts can be converted to first photocathode electrons per pixel root mean square (N_{sys}/G) with the voltage to photoelectron conversion ratio in appendix B.1.

N_{DC}: Dark current in a pixel can be measured by recording its video signal amplitude as a function of integration time t_{exp} with the camera operating on a bench with no light input. Electrons from dark current should increase linearly with integration time. Converting the video signal voltage to electrons per pixel for a t_{exp} value by the appendix B.1 procedure results in $(t_{\text{exp}}N_{\text{DC}})^{1/2}/G$ root mean square electrons per pixel, referred to the first photocathode, from dark current shot noise.

η_{DC}: The total noise recorded and converted to electrons per pixel as a function of integration times t_{exp} with no light input contains three root mean square noise sources: the previously determined temporal readout noise N_{sys}/G, the dark current temporal shot noise $(t_{\text{exp}}N_{\text{DC}})^{1/2}/G$, and the spatial dark current variation noise $(N_{\text{DC}}t_{\text{exp}}\eta_{\text{DC}})/G$. For a particular t_{exp}, the terms N_{sys} and $(t_{\text{exp}}N_{\text{DC}})^{1/2}$ can be root mean square subtracted from the total noise leaving the root mean square electron noise term $N_{\text{DC}}t_{\text{exp}}\eta_{\text{DC}}/G$ for equations 4.9 and 4.10.

η_q: With the readout noise and dark current noise terms determined above for the time t_{exp}, the spatial root mean square quantum efficiency variation term can be measured with the camera operating on a bench with a uniform flood light source input to the photocathode. As integration time is increased, the total noise increases, but with values of the dark current and readout noise terms known. They can be root mean square subtracted from the total noise to get a value for the noise term $\eta_q B$. Since B is known, a value for η_q can be calculated.

References

1. Grometstein, A. A., ed. 2011. *MIT Lincoln Laboratory: Technology in Support of National Security*. Lexington, MA: MIT Lincoln Laboratory.

2. Weber, R. 1978. The Ground-Based Electro-Optical Detection of Deep Space Satellites. *Proceedings of the Society for Photo-Instrumentation Engineers* 143:59–69.

3. Henize, K. G. (January 1957). The Baker-Nunn Satellite-Tracking Camera. *Sky and Telescope* 16:108.

4. Thekaekara, M. P. 1972. Evaluating the Light from the Sun. *Optical Spectra* 6 (3): 32.

5. Bair, M., et al. 1973. "Optical Properties of Satellite Materials," *Technical Report TR 194100-6-F*. ERIM.

6. "Satellite Systems," chapter 5, Los Angeles Air Force Base AFD-060912-025.

7. Bessell, M. S. 1990. UBVRI Passbands. *Proceedings of the Astronomical Society of the Pacific* 102:1181–1199.

8. W.E. Krag, "Visible Magnitude of Typical Satellites in Synchronous Orbit," MIT Lincoln Laboratory Technical Note 1974-23, 6 September 1974.

9. Allen, C. W. 1973. *Astrophysical Quantities*. 3rd ed. London: Athlone Press.

10. Broadfoot, A. L., and K. R. Kendall. 1968. The Airglow Spectrum, 3,100–10,000 Å. *Journal of Geophysical Research* 74:426.

11. Turnrose, B. E. 1974. Absolute Spectral Energy Distribution of the Night Sky at Palomar and Mount Wilson Observatories. *Proceedings of the Astronomical Society of the Pacific* 86:545.

12. Kneizys, F. X., et al. "Users Guide to LOWTRAN 7," Technical Report AFGL-TR-88–0177, Bedford, MA: Air Force Geophysics Laboratory, Hanscom Air Force Base, 16 August 1988.

13. Hardie, R. H. 1962. Photoelectric Reductions. In *Astronomical Techniques*, ed. W. A. Hiltner. Chicago: University of Chicago Press.

14. Konig, A. 1962. Astrometry with Astrographs. In *Astronomical Techniques*, ed. W. A. Hiltner. Chicago: University of Chicago Press.

15. Weber, R. (April 1976). Predicted and Measured Capabilities of the Lincoln ETS, Photon-Noise-Limited Electro-Optical Systems. *MIT Lincoln Laboratory Technical Note* 1976-9:26.

16. *Photomultiplier Tubes*. 1971. Harrison, NJ: RCA Electronic Components Division.

17. Baum, W. A. 1962. The Detection and Measurement of Faint Astronomical Sources. In *Astronomical Techniques*, ed. W. A. Hiltner. Chicago: University of Chicago Press.

18. H.L. Van Trees. 1968. *Detection, Estimation, and Modulation Theory, part 1*. New York: John Wiley & Sons.

19. *RCA Electro-Optics Handbook*. 1968. Harrison, NJ: RCA Corporation.

20. Eggen, O. J., and A. Sandage. 1969. New Photometric Data for the Old Galactic Cluster NGC-188. *Astrophysical Journal* 158:669.

21. Landolt, A. U. 1973. Photoelectric Sequences in the Celestial Equatorial Selected Areas 92–115. *Astronomical Journal* 78:959.

22. Purgathofer, A. Th. 1969. UBV Sequences in Selected Star Fields. *Lowell Observatory Bulletin* 7:98.

23. G.T. Flynn, "An Electro-Optical MTI System for the Detection of Artificial Satellites (the Nith Finder)," MIT Lincoln Laboratory Technical Note 1977-32, 27 June 1977.

24. Taff, L. G., and J. M. Sorvari. 1982. Differential Correction for Near-Stationary Satellites. *Celestial Mechanics* 26:423.

25. Rork, E. W., S. S. Lin, and A. J. Yakutis. "Ground-Based Electro-Optical Detection of Artificial Satellites in Daylight from Reflected Sunlight," MIT Lincoln Laboratory Project Report ETS-63, 25 May 1982.

26. Rork, E. W., R. J. Bergemann, S. S. Lin, J. M. Sorvari, and A. J. Yakutis. "Ground-Based Electro-Optical Surveillance of Satellites in Daylight by Detection of Reflected Sunlight," in *IEEE Eascon-83 Conference Proceedings*, p. 103, 19 September 1983.

27. "RCA Electro-Optics and Devices Camera Tube 4532/U Series Specifications," Lancaster, PA: RCA Corporation, June 1980.

28. Kodak Wratten Filters for Scientific and Technical Use. In *Publication No. B-3, 3*. Rochester, NY: Eastman Kodak Company, 1966.

29. Tyson, J. A., and P. Seitzer. 1988. A Deep Sky Survey of 12 High Latitude Fields. *Astrophysical Journal* 335:552.

30. Lauer, T. R. 1989. The Reduction of Wide Field/Planetary Camera Images. *Proceedings of the Astronomical Society of the Pacific* 101:445–469.

31. Tyson, J. A. 1986. Low-Light-Level Charge Coupled Device Imaging in Astronomy. *Journal of the Optical Society of America. A, Optics and Image Science* 3:2131.

32. Walker, G. A. H., R. Johnson, C. A. Christian, P. Waddell, and J. Kormendy. 1984. The CHFT CCD Detector. *Proceedings of the Society for Photo-Instrumentation Engineers* 501:353–358.

33. Baum, W. A., B. Thomsen, and T. J. Kreidl. 1981. Subtleties in Flat Fielding of Charge Coupled Device (CCD) Images. *Proceedings of the Society for Photo-Instrumentation Engineers* 290:24–27.

34. Burke, B. E., J. A. Gregory, R. W. Mountain, B. B. Kosicki, E. D. Savoye, P. J. Daniels, et al. "Large-Area Back-Illuminated CCD Imager Development," in *Optical Detectors for Astronomy: Proceedings of an ESO CCD Workshop Held in Garching, Germany, October 8–10, 1996*, J.W. Beletic and P. Amico, Eds., pp. 19–28, Dordrecht: Kluwer Academic, 1998.

35. Tower, J. R., P. K. Swain, F. L. Hsueh, R. M. Dawson, P. A. Levine, G. M. Meray, et al. 2003. Large Format Backside Illuminated CCD Imager for Space Surveillance. *IEEE Transactions on Electron Devices* 50:218–224.

5

Technology Development for Space-Based Electro-Optical Deep-Space Surveillance

Jayant Sharma

5.1 Introduction

Optical space surveillance from the ground started with the detection of Sputnik 1 in 1957. MIT Lincoln Laboratory has been involved in the field of space surveillance from the very beginning, with radar tracking from its Millstone Hill radar in Westford, Massachusetts. Since then, both optical and radar tracking techniques have developed and matured. Capabilities exist today to measure positions of low Earth orbit (LEO) resident space objects (RSOs) as small as 10 centimeters in size and geosynchronous orbit (GEO) RSOs as small as 1 meter [1]. The U.S. Air Force has invested considerable resources in ground-based electro-optical systems. In the late 1980s, Lincoln Laboratory proposed to build an electro-optical system for space surveillance (primarily deep-space surveillance) that would operate in a low Earth orbit. There are several advantages in basing a sensor in space rather than on the ground:

1. A space-based sensor would be above the atmosphere and thus immune to the effects of atmospheric weather, unlike a ground-based sensor, whose performance can be adversely affected by nighttime clouds as much as 40–50 percent of the time. This immunity can ensure access to high-interest objects orbiting the Earth, such as new launches and maneuvering satellites in the geosynchronous belt.

2. Due to the sky brightness from the scattering of sunlight by the atmosphere, ground-based sensors are adversely affected by the background noise in the focal plane, thus substantially limiting their data collection on resident space objects in the daytime. A space-based sensor, which is constrained only by how close to the Sun the sensor can point without focal plane damage, can reduce the duration of that background noise effect by several hours per day,

depending on the design of its telescope. and because the sky is always dark for a space-based sensor, the background noise on the focal plane is low when it is pointing away from the Sun or the Moon, which enhances the sensor's detection sensitivity

3. At least three ground-based sensors, geographically distributed around the Earth, are required to detect and track all the resident space objects in the geosynchronous belt, whereas a single space-based sensor orbiting in low Earth orbit would be able to view the entire geosynchronous belt of satellites (figure 5.1) over one orbit, limited only by its telescope design.

On the other hand, basing a sensor in space rather than on the ground does have certain disadvantages:

1. Because building a sensor system that can survive a launch and operate semiautonomously in space for several years requires a large amount of development and testing, a space-based sensor of modest aperture will cost significantly more than a larger ground-based one.

2. Even though basing a sensor in space removes the effect of sky brightness, the sensor is still limited by how close it can be pointed to the Sun to avoid excessive illumination of, and consequent damage to, the focal plane. Moreover, other practical considerations dictate where a space-based sensor can point. It usually points away from the Earth and at targets that are under favorable illumination conditions. A more favorable condition is defined by a smaller phase angle, as shown in figure 5.1. As the phase angle becomes smaller, the target gets brighter in the focal plane of the sensor as more of the satellite illuminated by the Sun can be observed. As the phase angle becomes larger, however, the illuminated parts of the satellite start facing away from the sensor and the satellite appears dimmer to the sensor.

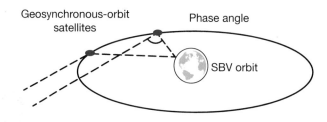

Figure 5.1 Definition of phase angle.

3. Unlike a ground-based sensor, a space-based sensor must be operated remotely, and its hardware cannot generally be modified once the sensor is launched.

4. Finally, the data acquired by a space-based sensor have to be brought down to the ground for full processing. Typically, the data are stored on board the sensor's satellite and are downloaded as the satellite passes over a ground station. Commands for the satellite are also uploaded during these ground-station contacts. Unless the space-based sensor is designed from the outset to make use of relay satellite communication channels, its tactical, quick-response operation is therefore quite limited, as compared to that of a ground-based sensor.

5.2 The Electro-Optical Space-Based Visible Sensor

The first steps in demonstrating the potential of space-based space surveillance began with the Midcourse Space Experiment (MSX) program in 1987. Funded and managed by the Department of Defense's Ballistic Missile Defense Organization (renamed the "Missile Defense Agency" in 2002), the MSX satellite was a long-duration, "observatory"-style measurement platform that was designed to collect high-quality data on terrestrial, Earth-limb, and celestial backgrounds; ballistic missile–like targets; and resident space objects (RSOs). To collect phenomenology data in support of ballistic missile defense objectives, the MSX satellite gathered optical data from the far ultraviolet (110 nanometers) to the very-long-wave infrared (28 microns) with fully characterized and calibrated sensors [2]. Its Space-Based Visible (SBV) sensor, a small visible-band surveillance sensor built by Lincoln Laboratory, was designed to perform above-the-horizon surveillance experiments and acquire broadband optical data (450–950 nanometers) on targets and backgrounds [3, 4].

Measuring 17 feet in length and weighing 6,000 pounds on the ground, the MSX satellite is shown in figure 5.2 during its final integration and test at Vandenberg Air Force Base in California, from which it was launched. The core of the satellite contained a long-wave infrared (LWIR) sensor, called the "SPIRIT III," with a focal plane cooled to less than 10° K by solid hydrogen cryogen stored in the dewar visible at the center of the spacecraft. The sunshade for the SPIRIT III sensor and apertures of all of the MSX sensors were located at the "top" of the satellite (figure 5.2). As also shown in the figure, the SBV sensor was composed of a 73-pound telescope, which was boresighted with all of the other MSX sensors at the top, and a 100-pound electronics assembly toward the bottom of the satellite.

The SBV sensor, as the first proof-of-concept, space-based electro-optical sensor dedicated to space surveillance, demonstrated three key technologies:

Figure 5.2 MSX satellite showing Space-Based Visible (SBV) sensor components prior to the satellite being mounted on its booster.

1. High off-axis stray-light-rejection optics to allow detection of faint targets near the sunlit Earth limb;
2. Advanced staring charge-coupled device (CCD) focal plane arrays to allow for sensitive searches of large areas of the sky; and
3. Onboard signal-processing capability to reduce the large volume of focal plane data to a manageable set (on stars and resident space objects) that could be downloaded easily with the available communication channel

The sensor's high-stray-light-rejection design allowed the detection of faint targets in high-background-light environments near the sunlit Earth limb. This characteristic was essential for the collection of phenomenology data on missile targets and low-altitude resident space objects. The SBV sensor employed an off-axis optical design to accomplish this objective, hence its "boxy" structure, as shown in figure 5.3 (left). Also, to minimize scattered light caused by reflections from internal contamination, the telescope and the mirrors were kept extremely clean during integration, launch, and operations, and they continued to remain so until the system's end of life about ten years after launch. Stray-light rejection can be quantified by the minimum detectable object that can be seen in the presence of stressing backgrounds. At the SBV Critical Design Review, the goal was set to establish the capability of detecting a 68-centimeter-diameter specular sphere with a reflectivity of 0.8 at a range of 3,000 kilometers at a tangent height of 100 kilometers above the sunlit Earth. The detection-sensitivity results shown in figure 5.4 illustrate that the threshold capability of this goal was substantially exceeded; minimum detection capability was actually equivalent to a 22-centimeter-diameter sphere under the conditions described above. Figure 5.4 also shows that the performance of the sensor did not degrade during its construction, launch, and operation of the telescope.

Figure 5.3 Major technology demonstrations of SBV sensor include high stray-light rejection optics (left), the focal-plane array (center), and the onboard signal processor (right).

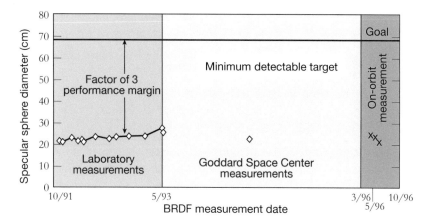

Figure 5.4   Graph of SBV sensor's detection sensitivity for space-based targets as function of specular sphere diameter and bidirectional reflectance distribution function (BRDF) measurement data.

The second key technology incorporated into the design of the SBV sensor was that of its low-noise CCD focal plane arrays (figure 5.3; center). These four abutting 420×422–pixel arrays, each with a frame-store region for rapid readout, were designed and fabricated by the Semiconductor Division of Lincoln Laboratory in the late 1980s. CCD focal planes can be characterized by their dark current, nonuniformity, readout noise, charge-transfer efficiency, well depth, and percentage of damaged pixels, all of which are affected by on-orbit radiation. The SBV sensor's focal plane exceeded performance expectations with respect to each of these measures. Focal-plane dark current and nonuniformity appeared to be increasing slowly because of radiation damage after almost three years in orbit; focal-plane readout noise was also affected slightly on orbit but was not significant even at the end of the sensor's life. There was also no detectable change in sensor's charge-transfer efficiency. Thus the SBV sensor performed for eight years after the Critical Design Review well above the detection thresholds set at that review.

The third key technology demonstrated by the SBV sensor was the onboard signal processor (figure 5.3; right). During routine space surveillance operations, the SBV gathered sensor data by staring at a chosen location in the sky and collecting the image data over a sequence of frames, referred to as a "frame set." A typical frame set included as many as 16 frames, resulting in almost three million pixels of information. The signal processor analyzed these three million pixels of information per field and retained only the information most vital for space surveillance, thus reducing the data volume by as much as a factor of 1,000. The retained information consisted of selected stars needed to determine the pointing of the telescope boresite and any streak signatures left by resident space objects moving through the sensor's field of view. Figure 5.5 illustrates the SBV's signal processing by showing the superposition of 16 raw frames (left) and the signal-processed image of these 16 frames (right). In these images, the stationary point sources are star detections, and the streaks are satellite detections.

Apart from these key technologies, the SBV sensor also demonstrated the utility of derivative technologies and techniques. Its off-axis design and chosen focal length–to–diameter (f/d) ratio resulted in a large field of view, which proved highly effective in demonstrating search techniques for efficient data collection on resident space objects. The sensor's orbital location enabled it also to demonstrate the capability of a single sensor to survey the entire geosynchronous belt—a very important region for both commercial and military satellites. Ground-based optical space surveillance sensors must be deployed in groups of at least three or even four in order to provide comparable coverage.

5.2.1   Experimental Phase of the Space-Based Visible Sensor: Space Surveillance Demonstrations

The metric and photometric accuracy of the SBV sensor and its capabilities to perform wide-area searches and geosynchronous-belt surveillance were the key technological

 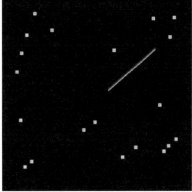

Figure 5.5   SBV sensor raw full-frame (left) and signal-processor (right) images.

developments that it demonstrated during its life as a contributing operational plat-form for space-based space surveillance for the Air Force Space Command [5]. The experience gained by the Air Force Space Command and Lincoln Laboratory contrib-uted quite significantly to the operations of the follow-on Space-Based Space Surveil-lance Program (SBSS). Some of the demonstrations during the operational test phase of the SBV sensor are discussed in the following section.

5.2.1.1   Space-Based Visible Metric and Photometric Processing

To assess the ability of the SBV sensor to conduct space-based space surveillance, it was first necessary to determine the metric and photometric characteristics of the sen-sor on orbit. The metric positioning of targets in the SBV sensor's field of view re-quired knowledge both of the precise pointing of the sensor's boresight and of the Earth-centered position of the MSX/SBV satellite in orbit at the time the data were gathered. By using the pointing information, the start points and endpoints of the streaks detected on the focal plane (figure 5.5) could be transformed into two angular measurements on the sky, such as right ascension and declination. Each endpoint and the precise Earth-centered position of the SBV sensor constituted a metric observa-tion of the target. This observation was then merged with other ground-based optical and radar observations on the same target in order to establish the target's trajectory. In some cases, SBV sensor data were used exclusively for determining the target's trajectory.

The process of producing a metric observation involved several steps. With the SBV sensor operating in a staring mode, the raw sensor data were gathered on board and sent to the signal processor, which extracted the pixel values and intensities associated with a preselected number of star detections (note that in the normal or sidereal mode of operation, the star positions on the focal plane were invariant over a frame set). In addition, a moving-target indication algorithm within the signal processor identified any objects moving relative to the stationary background. Pixel intensities and focal-plane coordinates for both the selected stars and the moving targets were downloaded in a signal processor report. Software at the ground-based SBV Processing and Opera-tions Control Center (SPOCC) further processed each signal processor report. The star detections were centroided, and the pattern and exact positions of detected stars were matched to a catalog of known stars. This process allowed for a highly accurate determination of the SBV sensor's pointing, to within a few tenths of an arc second, without the use of instruments such as onboard gyroscopes. Once pointing was deter-mined, the information was used to map the endpoints of the streaks on the focal plane to absolute angular positions on the sky, thus producing a metric observation.

The Earth-centered position, or ephemeris, of the MSX satellite at the time the data were gathered was needed for further processing. As part of an independent processing pipeline, Lincoln Laboratory maintained the MSX ephemeris to an accuracy of 6 meters (limited by the quality and quantity of tracking data available). The determination of the MSX ephemeris was accomplished by processing ranging data from the Space Ground Link System, a network of ground telemetry sensors used by the Air Force to track its space assets and the tracking data from the Millstone Hill and other radars in the Space Surveillance Network. A diagram illustrating the components of the metric-data-generation process is shown in figure 5.6.

To assess the metric performance of the SBV sensor, routine on-orbit metric calibration was conducted by observing satellites for which the positions were well established and by comparing these satellites' known positions with their positions as observed by the sensor. During the design phase of the SBV program, the goal was set for the sensor to produce metric observations of resident space objects with an accuracy of 4 arc seconds (1 σ); the goal was chosen to be approximately one-third of the field of view of 12 arc seconds of each pixel in the focal plane. Confirmation of achieving the 4-arc-second metric accuracy is shown in figure 5.6. The inner circle of this diagram shows 4 arc seconds; for comparison, the outer circle is 10 arc seconds, which was the accuracy requirement for the GEODSS (Ground-Based Electro-Optical Deep-Space Surveillance) system (see chapter 4) at the time of these experiments.[1] As a consequence, the metric accuracy of the SBV sensor exceeded that of the GEODSS system by a factor of 2.5. Further validation of the metric performance of the SBV sensor is provided in figure 5.7, which shows a large number of residuals of SBV sensor observations on satellites in the Global Navigation Satellite System (GLONASS), the Russian equivalent of the U.S. Global Positioning System. The sensor determined the positions of these satellites to a high accuracy (within a few meters). These results demonstrate that the SBV sensor collected metric data at the accuracy level of 4 arc seconds.

The SBV sensor acquired metric and photometric data on resident space objects simultaneously; intensity and pixel coordinates for streaks detected on RSOs were downloaded into the SPOCC as part of the signal processor report. The intensity

---

1. The Ground-Based Electro-Optical Space Surveillance (GEODSS) system consists of three 1-meter-class telescopes located on the White Sands Missile Base in New Mexico, the island of Maui in Hawaii, and the island of Diego Garcia in the Indian Ocean. The GEODSS system grew out of Lincoln Laboratory technology first developed at the Experimental Test Site in Socorro, New Mexico. In September 1998, the Space Surveillance Network was augmented with the Transportable Optical System (TOS), another Lincoln Laboratory system, located in southern Spain. All these systems contribute to deep-space surveillance.

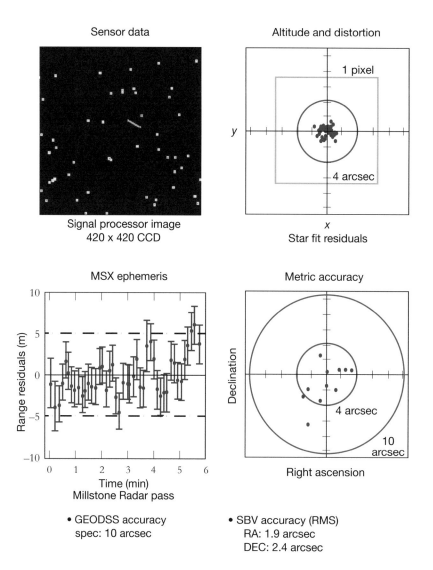

Figure 5.6 SBV sensor metric data processing. DEC, declination; RA, right ascension; RMS, root mean square.

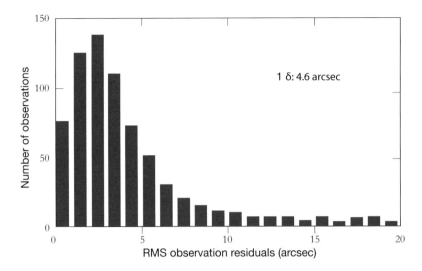

Figure 5.7 Histogram of SBV sensor Global Navigation Satellite System (GLONASS) observation residuals.

information was used to determine an average brightness for the target during the data collection. The brightness of an object was quantified as "SBV magnitude" on a logarithmic scale, with larger values representing dimmer objects.[2] Figure 5.8 shows an example of a detected satellite signature and a histogram displaying photometric data collection as a function of brightness. It can be seen from this diagram that the SBV sensor's overall sensitivity was at SBV magnitude 15, with saturation on the focal plane occurring around magnitude 7. An SBV sensor database of substantially more than 100,000 observations on 2,300 known objects had been established in the first three years of operation, with metric and photometric information available on active and inactive payloads, rocket bodies, upper stages, and space debris.

5.2.1.2 Space-Based Visible Wide-Area-Search Capability

The field of view of the SBV sensor was 1.4×1.4 degrees for each of its four charge-coupled devices. Because of the significant distortion inherent in the design of off-axis optical systems, the total field of regard for all four CCDs was 6.6×1.4

---

2. SBV magnitude differs slightly from the conventional astronomical magnitude because of the spectral response of the SBV sensor's CCDs, enhanced in the red as compared to the astronomical V filter.

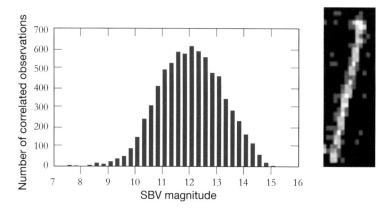

Figure 5.8 Histogram of SBV sensor's detection sensitivity (left) and color intensity map of resident space object (RSO) streak (right). The colors indicate intensity, ranging from blue (dim) to green to yellow to red (bright).

degrees (as compared to 2 degrees diagonal for the GEODSS telescopes' field of view). However, given that the onboard signal processor could only process the data from one CCD at a time, the instantaneous field of view was limited to that of one focal-plane array. The capability of the SBV sensor's wide field of view when applied to search applications is shown in figure 5.9, with data gathered during a single data acquisition, or "look," of the SBV sensor on one of the four focal planes. The signatures of five correlated satellites are evident in the signal processor image in the left panel of the figure. The right panel of the figure illustrates the same detected signature data but presented along with the predicted positions of those same targets, as based on the Air Force Space Command resident space object catalog (the detected stars have been omitted for clarity). Although these five satellites are seen in the same focal plane of the SBV sensor, they represent members of each of three orbital regimes: a geosynchronous orbit, a geosynchronous transfer trajectory, and a low-altitude orbit.

The wide-field-of-view search capability of the SBV sensor was extremely useful in addressing the problem of "lost objects." Resident space objects can become lost from the RSO catalog if their predicted positions differ significantly from their actual positions because of the paucity of tracking data or maneuvers—a concept illustrated in figure 5.10. For an RSO whose element set has been updated recently, such as is the case of the Intelsat VII, the predicted position and the position detected by the SBV sensor were quite close. But, as the element sets grew older, as for the NATO's III-D

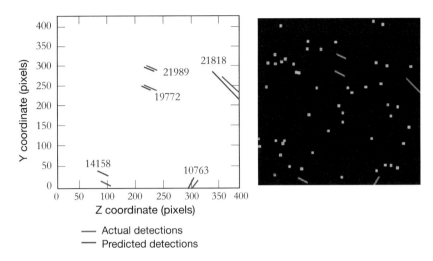

— Actual detections
— Predicted detections

RSO	SBV-RSO range (km)	Description	Orbit
10763	2794	NOAA 5 debris	LEO
14158	41247	Inactive galaxy I	GEO
19772	41317	Active IntelSat V-F15	GEO
21818	11259	Ariane Spelda debris	GEO transfer
21989	41334	Active IntelSat K	GEO

Figure 5.9 Application of wide field of view of SBV sensor to RSO detections.

satellite, the deviations began to grow. In the case of the Meteosat rocket body, which had not been tracked in more than one month, the predicted position and the actual position differed significantly. Nevertheless, the SBV sensor was able to detect and correlate Meteosat successfully. This capability of the SBV sensor was exercised extensively; the sensor found nearly 400 lost RSOs during its operational lifetime.

5.2.1.3 Space-Based Visible Surveillance of the Geosynchronous Belt

The SBV sensor had two unique characteristics that made it well suited for surveillance of the geosynchronous belt. The first was its wide field of view, which allowed for

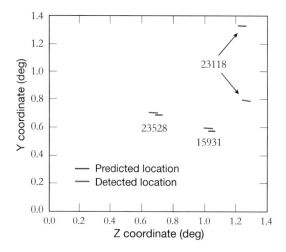

RSO	Description	Observed SBV magnitude	Age of element set
23118	Meteosat rocket body	14.5	38 days
15391	NATO IIID	13.3	7 days
23528	IntelSat VII	11.7	< 1 day

Figure 5.10   SBV sensor's correlation of lost RSOs.

multiple targets to be seen in one look. The other was its orbital location, which gave it the capability to survey the entire geosynchronous belt, as illustrated in figure 5.11. Evidence of this capability is shown in the lower portion of the figure, with the positions of all the geostationary satellites observed by the SBV sensor during the technology-demonstration phase.

Since geostationary satellites remain fixed over a given location on the equator, they are of considerable value to both military and commercial enterprises. One of the applications of the large field of view of the SBV sensor, in combination with its capability to survey the entire geosynchronous belt, is illustrated in figure 5.12. The left panel of the figure shows a signal-processed frame set from the SBV sensor with detections on five objects. It is frequently the case in the geosynchronous belt that satellites are maintained in clusters. The detected cluster shown in figure 5.12 consisted of five direct-broadcast and mobile telecommunications satellites located at 259 degrees east

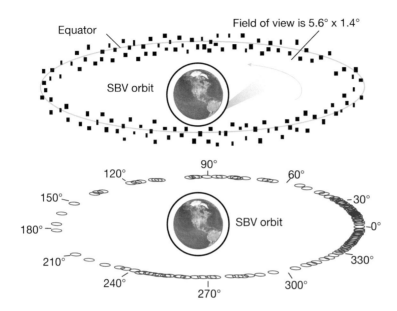

Figure 5.11 Coverage of geosynchronous belt from SBV orbit (top). Locations of all station-kept geosynchronous satellites observed by SBV sensor (bottom) as of November 1997.

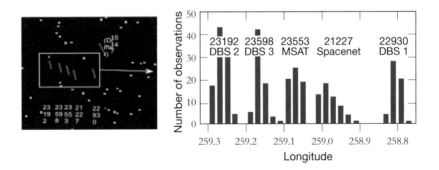

Figure 5.12 SBV sensor's observation of geosynchronous cluster. DBS, Direct Broadcast Satellite; MSAT, Mobile Satellite.

longitude, as well as one "serendipitous" object, a Russian payload passing through the field of view. A histogram showing the total number of sensor detections as a function of longitude is displayed in the right panel of the figure and clearly shows that these satellites were located quite close to one another. Because the SBV sensor, with its large field of view, could acquire data on an entire cluster simultaneously, the problem of properly correlating each member of the cluster to the RSO catalog was greatly eased. Images such as those in figure 5.12 show that a wide-field-of-view, high-accuracy, space-based electro-optical sensor can both support and enhance geosynchronous belt surveillance.

5.2.2   Operational Phase of the Space-Based Visible Sensor: Enhancement of Sensor Productivity

Space surveillance is separated into two classes: low-altitude and deep-space surveillance. The problems of acquiring and tracking objects in low altitude are addressed quite adequately using ground-based phased-array radars, such as the FPS-85 Spacetrack radar or the future Space Fence radar. Although these radars serve low-altitude surveillance well, their lack of detection sensitivity makes them unable to address the problems of deep-space surveillance, unlike the SBV sensor, whose coverage of the entire geosynchronous belt made it well suited to do so. Following its experimental demonstration phase, the SBV sensor was operated as a contributing component of the Space Surveillance Network (SSN), with a focus on developing operational strategies for efficient observations of geosynchronous satellites. Three techniques were developed to enhance the sensor's data collection in this capacity.

First, during geosynchronous search, data were collected and processed in sequence from all four of the sensor's charge-coupled devices (CCDs) before the satellite was reoriented. This technique was an extension of the wide-area-search capability (described earlier) to multiple focal planes in sequence.

Second, and most significant, an important piece of software for choosing some optimal scheduling for observing resident space objects was redesigned with the strengths of the SBV sensor in mind. The conjunction-optimized look-ahead (COLA) scheduler took the tasking list of objects to observe on any given day and sought out regions of space in which RSOs were in apparent conjunction with any two of the four CCDs. With this design, the SBV sensor typically saw at least two RSOs per CCD.

Third, the SBV sensor demonstrated "pinch-point" operations, wherein the sensor monitored specific points or areas—"pinch points"—in the geosynchronous belt that were rich with crossing satellites.

5.2.2.1   Conjunction-Optimized Look-Ahead Scheduler

As described earlier, the SBV sensor was an instrument with as inherently wide field of view, each of its four CCDs covering a field of regard of approximately 1.4×1.4 degrees and all four together covering a total of 6.6×1.4 degrees. It was evident that the SBV sensor could use this strength to increase its productivity when tracking the previously mentioned geosynchronous clusters of resident space objects. Further, a study [6] indicated that such clustering occurred in the field of regard for transient, nongeosynchronous deep-space RSOs as well. An example of simultaneously observing multiple satellites is shown in figure 5.9, in which the SBV sensor detected one satellite in a low Earth orbit (LEO), another in a highly elliptical orbit (HEO), and three more in geosynchronous orbits (GEO) within the same field of view—all in apparent angular conjunction even though they were significantly distant from one another in range. The COLA scheduler was specifically written to take advantage of such conjunctions in scheduling observations in order to maximize the sensor's productivity and enhance its operational utility. The time series shown in figure 5.13 plots the history of the mission-planning productivity for the SBV sensor during these experiments. The first noticeable improvement in the productivity occurred

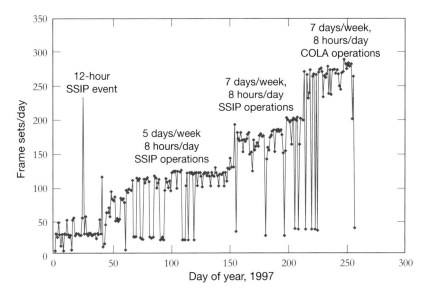

Figure 5.13   Effects of conjunction-optimized look-ahead (COLA) scheduler on productivity enhancement. SSIP, Space Surveillance Interface Processor.

when the sensor began operating eight hours a day, five days a week. A second significant increase in productivity occurred when the sensor started operating seven days a week. The Space Surveillance Interface Processor (SSIP) was the standard scheduler used in the early days of operation, with no optimization for conjunctions. The distinct improvement in SBV sensor productivity with the COLA scheduler is evident in figure 5.13 around day 220.

Clearly, the use of temporal, spatially apparent conjunctions of satellites in a large field of view is an effective technique for enhancing the productivity of the SBV sensor and, by inference, any wide-field-of-view sensor.

5.2.2.2   Pinch-Point Operations

An interesting example of the use of a space-based sensor arose from pinch-point operations, which focused sensor searches on densely populated regions of the geosynchronous belt. Satellites launched into geosynchronous orbit conform to well-established guidelines, driven by orbital evolution and the necessity for conserving station-keeping fuel, that cause the satellites to be clustered in the right ascension of the ascending node (figure 5.14). The important feature to note is the structure of the geosynchronous population. During the active life of these satellites, their inclinations are maintained near zero degrees and thus their ascending nodes between –90 and +90 degrees. When a satellite's fuel is depleted or when it can no longer be maneuvered, its ascending node will rapidly evolve to 90 degrees clockwise, and the satellite's trajectory will slowly evolve along the curve shown in figure 5.14, completing the cycle over approximately fifty-three years. Because geosynchronous satellites have only been launched into orbit around the Earth for about thirty years, the satellites have evolved, at most, a little more than halfway through the cycle. Satellites that have inclinations of 15 degrees and ascending nodes of near-zero degrees, as seen in the rightmost data in figure 5.14, are some of the first geosynchronous satellites ever launched.

The results shown in figure 5.14 illustrate that, though an active geostationary satellite is maneuvered so that it remains in its position on the equator, once it becomes inactive, its inclination and ascending node evolve in predictable ways. Figure 5.15 shows the distribution of ascending nodes for this set of satellites in figure 5.14; the satellites with inclinations less than 0.5 degree have once again been removed from the plot.

Thus a productive search for geosynchronous satellites can be conducted by examining the region around the appropriate right ascension, where the satellites cross the equator. The effectiveness of pinch-point operations is illustrated in figure 5.16, which

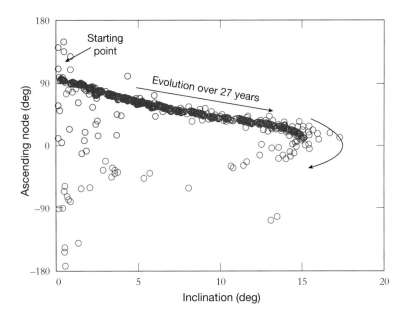

Figure 5.14 Graph of data showing the correlation between the inclination and right ascension of an ascending node for a large sample of geosynchronous satellites (with inclination greater than 0.5 degree) during the orbital evolution due to geopotential and lunisolar forces. Note the clustering at any given time in the evolution.

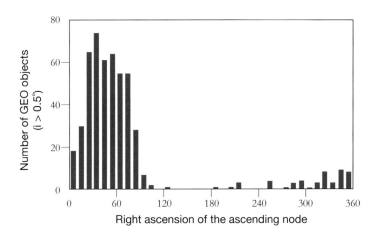

Figure 5.15 Histogram of right ascension of ascending node for geosynchronous satellites. The clustering of ascending nodes between 0 and 90 degrees is the pattern referred to as the "geosynchronous pinch points" (satellites with inclinations less than 0.5 degree are excluded in this figure).

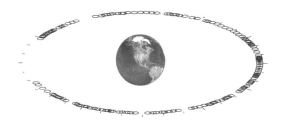

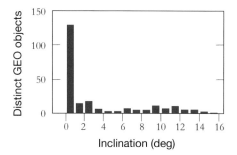

Figure 5.16   Example of data collected over 24 hours in a pinch-point operation, demonstrating success in detecting geosynchronous satellites in high-inclination orbits. (Top) Longitude distribution of detected satellites. (Bottom) Histogram of inclinations of detected satellites.

shows the distribution in both longitude and inclination of all geosynchronous satellites detected during a single day. Data from only search operations are shown in these results. The top panel in figure 5.16 shows the global coverage, with a gap over the Pacific Ocean that is due to a reduced number of satellites and limited coverage; the bottom panel shows geosynchronous objects, with both low and high inclinations, that were detected simultaneously at the pinch points.

5.2.2.3   Space-Based Visible Photometry

The SBV sensor produced information on the brightness of a resident space object at the same time that the sensor established the object's metric position. Figure 5.8 shows that the brightness is determined on the basis of the pixel intensities detected by the sensor and quantified as SBV magnitude. Although the SBV sensor was originally designed to achieve photometric sensitivity of SBV magnitude 14.5, the sensor's actual demonstrated sensitivity on orbit was shown to exceed magnitude 15. Although one of the principal contributions of any space surveillance sensor is metric information, a great deal of information can be gleaned from the photometric signature of an RSO as well. Indeed, not only is the photometric signature instrumental to the ground processing of the SBV sensor data in order to achieve high metric accuracy, but it can also be used to discriminate multiple RSOs on the basis of their size, configuration, reflectivity characteristics, and status, provided that measurements are made over a large range of phase angle (a common measure describing the relative geometry between a target, the observing sensor, and the Sun).

A diagram depicting this phase angle geometry is shown in figure 5.17. If a sensor repeatedly observes a given satellite over a range of phase angles, a characteristic "phase curve" can be determined. The phase curve of satellites will display distinctive features that can be used to classify them photometrically or to discriminate them within classes. For example, the stability of a three-axis stabilized satellite can be determined with this technique, as shown in figure 5.18, in which photometric data on both a stable and an unstable Hughes 601–class satellite are given as a function of phase angle. Active, three-axis stabilized satellites typically have large solar panels that track the Sun. When observations are made at small phase angles, the solar panels of stable satellites provide a bright reflection, as can be seen in their decreasing magnitude (brighter) signatures with decreasing phase angles. Since the solar panels of unstable satellites no longer track the Sun, however, they show no such trend. It would thus be possible to draw conclusions regarding the status of the vehicle from this example.

As the results shown in figure 5.18 indicate, data from the SBV sensor may be used to obtain information on the operational status of satellites and can aid in their identification. Both of these capabilities are useful to the Air Force Space Command in conducting space surveillance operations.

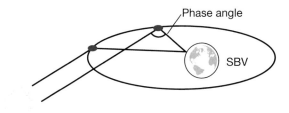

Figure 5.17   Geometry of space-based photometric observations.

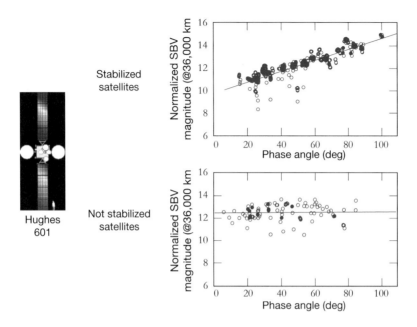

Stabilized satellites

Hughes 601

Not stabilized satellites

Figure 5.18   Photometric phase functions of stabilized (top) and unstabilized (bottom) Hughes 601 satellites, showing detectable differences.

5.3   Summary

Because the SBV sensor was this country's first space-based electro-optical space surveillance sensor, both its technology and the techniques needed to operate it had to be invented. And because the SBV sensor was riding on the MSX multisensor, multiuse satellite, the constraints of the other MSX sensors had to be taken into account, at least until they finished their experiments and were shut down. Among the techniques for space-based space surveillance developed by Lincoln Laboratory researchers were calibration of sensors for metric and photometric accuracy, dynamic scheduling of observations with conjunction optimization for enhanced throughput, and creation of photometric databases for observed satellites.

The experience in operating the SBV sensor from 1996 to 2008 was leveraged to support the Space-Based Space Surveillance System (SBSS), which also benefited from lessons learned in developing the sensor's hardware and software. Indeed, the ground-based processing software developed by Lincoln Laboratory for the SBSS was built on the experience of the SBV sensor's processing software. In addition, through the SBV/

MSX program, the Air Force Space Command gained significant experience in tasking and utilizing space surveillance data.

References

1. Pensa, A., and R. Sridharan. "Monitoring Objects in Space with the U.S. Space Surveillance Network," in Mission Design and Implementation of Satellite Constellations: Proceedings of International Workshop Held in Toulouse, France, November 1997, J.C. van der Ha, Ed., pp. 305–315, Dordrecht: Kluwer Academic, 1998.
2. Mill, J. D., and B. D. Guilmain. 1996. The MSX Mission Objectives. *Johns Hopkins APL Technical Digest* 17 (1): 4–10.
3. Harrison, D. C., and J. C. Chow. 1996. The Space-Based Visible Sensor. *Johns Hopkins APL Technical Digest* 17 (2): 226–236.
4. Stokes, G. H., C. von Braun, R. Sridharan, D. Harrison, and J. Sharma. 1998. The Space-Based Visible Program. *Lincoln Laboratory Journal* 11 (2): 205–238.
5. Sharma, J., G. H. Stokes, C. von Braun, G. Zollinger, and A. J. Wiseman. 2002. Toward Operational Space-Based Space Surveillance. *Lincoln Laboratory Journal* 13 (2): 309–334.
6. Sridharan, R., B. Burnham, and A. Wiseman. "Performance Improvements of the SBV," in Proceedings of 1998 Space Control Conference, pp. 71–80, Lexington, MA, 14–16 April, MIT Lincoln Laboratory Project Report STK-253, 1998.

6

Technology Developments in Catalog Discovery

Ramaswamy Sridharan and George Zollinger

Fundamental to all activities in space surveillance has been the ability to create and maintain an accurate, orderly digital catalog of all resident space objects (RSOs) orbiting Earth. And essential to the establishment of such a catalog are the following:

1. Detection and tracking of new RSOs and their addition to a catalog;
2. Discrimination of the new RSOs from one another, using various metrics to identify them uniquely in the catalog.

New RSOs appear in space through two types of events:

1. New launches, which typically deliver at least two and sometimes as many as thirty new objects into space;
2. RSO fragmentations, which add from one to several thousand new objects, depending on the mechanism of fragmentation.

Whenever a new resident space object with an orbital element set is added to the catalog, it is important to characterize the RSO, that is, to identify whether it is a payload, a rocket body, or a piece of inert debris and to provide information as to its size, function, and provenance. Such information, along with the element set, enables the RSO to be unequivocally identified whenever it is tracked by a sensor.

Identification of an RSO as a payload is necessary for the monitoring of its operational status. Identification of rocket bodies is necessary since they may carry residual fuel, which has often been observed to explode and cause the bodies to fragment. Debris pieces, which are more numerous than payloads, may collide with them, causing loss of control or fragmentation; or with rocket bodies, causing them to explode and fragment; or, more likely, with other debris pieces, causing further fragmentation. Collisions are orbit dependent; for example, there is a finite probability of collisions between payloads in geostationary orbit [1]. And, of course, collisions between payloads and rocket bodies can also occur, but are much less likely than collisions with the much more numerous debris pieces.

This chapter will first describe special techniques, including search strategies, developed over the years to ensure the discovery of new RSOs both in low Earth orbits (LEO) and in deep-space orbits. It will then give several examples of the application of the search strategies and techniques of discrimination to facilitate unique identification of RSOs and will cover "catalog discovery," as defined below, to include the cataloging of debris pieces from fragmentation events among RSOs.

A United Nations agreement [2, 3] mandates that all nations announce their satellite launches and provide some information on the initial satellite orbits. Because this information is generally incomplete and applies only to the immediate aftermath of the launches, however, considerations of national security have made it incumbent on the U.S. Department of Defense to deploy a system to

- discover newly launched RSOs;
- discriminate and identify them;
- track and determine their orbits and any changes thereof throughout their lifetimes;
- analyze the status of the important functioning satellites;
- assess the potential for collision in space between functioning satellites and other objects; and
- predict the reentries of the RSOs resulting from atmospheric drag or intentional maneuvers.

These activities constitute catalog discovery and orderly catalog maintenance and are a significant part of space situational awareness, which is a major mission for the U.S. Air Force Space Command.

6.1   Launch Activity and Catalog Discovery

Figure 6.1, which presents the world's satellite launch activity by year and country from 1957 through 2010, illustrates the wide diversity of satellite owners at present, although the Soviet Union/Russia and the United States have clearly dominated launch activity since the beginning of the Space Age in 1957.

The Orbital Debris Program Office of NASA's Johnson Space Center has been generating estimates of the number of RSOs in low Earth orbit [4]. Figure 6.2 plots the RSO population from data in a NASA report [5]. Although it reflects all Earth-orbiting objects that are larger than 10 centimeters in characteristic size at

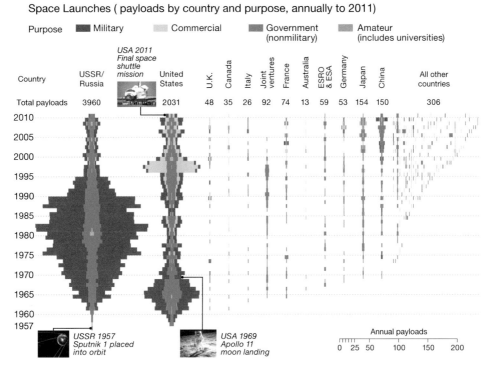

Figure 6.1   Launch activity in the Space Age (redrawn with permission from the website Universe Today). ESRO, European Space Research Organisation; ESA, European Space Agency

altitudes below 2,000 kilometers and larger than about 50 centimeters at higher altitudes, a great number of space-debris objects smaller than 10 centimeters are also in orbit. Estimates by NASA's Johnson Space Center indicate that the detectable RSO population might soar to as many as 100,000 objects if the threshold size of detection were reduced to 1–2 centimeters. All such RSOs constitute a collision hazard to active satellites and other large objects in orbit.

6.1.1   Detection of Newly Launched Resident Space Objects

Techniques for detecting RSOs newly launched into orbit differ somewhat depending on whether adequate prelaunch information is supplied by the launch agency (cooperative launches) or not (noncooperative launches). The time of launch from a launch

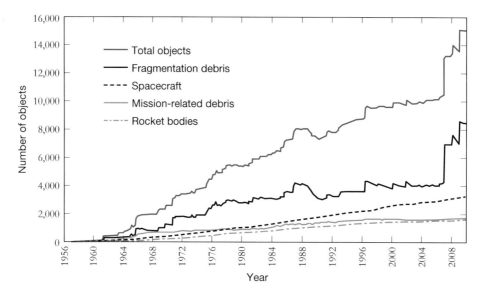

Figure 6.2   Cataloged resident space object population by object type and numbers per year (from NASA's Johnson Space Center).

site is dependent on what type of orbit is desired and whether the satellite is joining an existing constellation of satellites or not. A common type of orbit is Sun-synchronous at low altitude and high inclination, such that the daily precession of the orbital plane around the Earth as driven by the dominant Earth perturbation is equal to the average daily angular motion of the Earth around the Sun. This type of orbit ensures that the orbital plane faces the same sunlight geometry from orbit period to orbit period [6]. Other common types of orbit for satellites launched into existing constellations are low Earth orbit for constellations such as Earth resources or meteorological satellites and medium Earth orbit for constellations such as the U.S. Global Positioning System (GPS) or the Russian Global Navigation Satellite System (GLONASS). For these orbits, the time of launch is determined by the desired right ascension of the orbital plane, and typically the launch takes place a few minutes prior to when the launch site passes under the orbital plane. The time of a launch into a very high eccentricity orbit is chosen such that the perigee altitude increases initially to avoid high atmospheric drag and consequent demise of the satellite [7, 8]. The time of a launch into geosynchronous orbit is often chosen such that the control of the satellite's orbital inclination during its lifetime requires a minimum of fuel—every pound of fuel on board the satellite is one less pound of useful payload. Essentially, knowing the type of orbit desired allows the launch agency to reliably predict time of launch within any given day.

All search strategies in this chapter depend on the availability of a consistent and complete catalog of existing resident space objects so that a new RSO will not be mistaken for an existing one. Once a new RSO is detected, it is important to track it to establish its orbit and characterize it using the various metrics enumerated earlier (location, orbit, size, type, communications characteristics, etc.). An orderly RSO catalog should contain all this information.

6.1.2   Search Strategies for Resident Space Objects Newly Launched into Low Earth Orbit: Radar Systems

A search strategy for newly launched RSOs is based on whether the object was launched into a low-altitude or a deep-space orbit. Historically, most satellites in low Earth orbit are confined to altitudes below 3,000 kilometers, except for two laser geodetic satellites launched into 6,000-kilometer-altitude orbits in 1976 and 1992. A very effective search strategy for detecting and cataloging RSOs in low Earth orbit is to use a fence-type radar system, which searches a defined volume of space in its field of regard (generally in a leakproof manner) and detects the new RSOs as they transit this volume. An example is the oldest radar in the Space Surveillance Network, called the Air Force Space Surveillance System (AFSSS), known as the "Naval Space Surveillance (NAVSPASUR) System" until it was handed over to the Air Force in October 2004 [9]. The NAVSPASUR radar consisted of three transmitters and six receivers. The transmitters created an east–west "fence" of continuous-wave radar energy along a great circle of the Earth at about 30 degrees north latitude (figure 6.3) extending across the continental United States.[1] All RSOs newly launched into low Earth orbit above 28 degrees but below 152 degrees inclination (as limited by the location of the NAVSPASUR radar and the orientation of the radar fence)[2] and above a certain size (approximately 1 sq. meter, limited by the sensitivity of the radar) were detected and their orbits established within, at most, 24 hours of launch.

A second strategy of assured detection of RSOs launched into low Earth orbit is from the phased-array radars that deploy leakproof fences. Examples are the PAVE

---

1.   The Air Force Space Surveillance System (AFSSS) was shut down in 2013. It is slated to be replaced after 2017 by two geographically separate fence-type radars with detection and track capability, to be called the "Space Fence Radar System" [10].

2.   The new Space Fence is located on the Kwajalein Atoll at about 8 degrees north latitude. Hence the new lower and upper limits in inclination are 8 and 72 degrees.

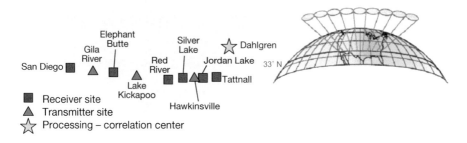

Multistatic radar interferometer zenith fence at 33° N latitude
Detection range (1.0 m² target): 18,500 km
Detects LEO objects > 33° inclination
200 m metric accuracy

No tracking data: Detection fence only

Figure 6.3   Navy Space Surveillance (NAVSPASUR) fence-type radar of the present-day Air Force Space Surveillance System (AFSSS).

PAWS (PAVE Phased Array Warning System) missile defense radars, which erect a leakproof fence at low elevation; the Cobra Dane radar, with a similar leakproof fence; and the FPS-85 (figure 6.4) and Cavalier radars, which tend to have higher-elevation fences that are not leakproof [11,12]. Because of their locations and sensitivities, these radars are more limited than the new Space Fence in their detection sensitivity and inclination coverage, but they will be extremely useful to supplement the data from the Space Fence to improve the orbital element sets of the new resident space objects.

A third search strategy is to use narrow-beam radar in a fence-search mode to detect new RSOs launched into low Earth orbit. An excellent example of this is the ALTAIR (Advanced Research Projects Agency [ARPA] Long-Range Tracking and Instrumentation Radar; see chapter 3), which simulates a low-elevation fence whose azimuth is about 75 degrees wide by continuously moving its antenna. The radar exploits its wide beamwidth at VHF (about 3 degrees) to ensure a leakproof search over the wide-azimuth fence. The center point of this fence is defined by the time when and the azimuth where the newly launched RSO is expected to break the horizon [13].

A fourth strategy, which works for cooperatively launched RSOs, is to use a narrow-beam radar in an along-orbit search mode to detect the new RSOs. This

Location: Eglin AFB, Florida
Frequency: UHF 437–447 MHz
Peak power: >30 MW

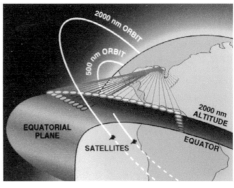

Detection range (1.0 m² target): 2500 nmi (single pulse)
Detects LEO ROSs > 15° inclination
20 m metric accuracy

Figure 6.4   Example of phased-array radar—the FPS-85.

approach is demonstrated in an example on detection and tracking of picosats in subsection 6.1.4.

Thus detection is guaranteed for all new RSOs in low Earth orbit by using a fence-type radar, subject only to its sensitivity limits.

6.1.3   Search Strategies for Resident Space Objects Newly Launched into Low Earth Orbit: Optical Systems

Ground-based optical sensors have been used to detect resident space objects in low Earth orbit (e.g., the Baker-Nunn system deployed by the Smithsonian Astrophysical Observatory [14]), but they can reliably operate only in cloudless nighttime conditions and only when the RSOs are sunlit. Hence, by the very nature of the low-altitude orbit and the geometry needed, the time available for search by such sensors is limited. And because they are unreliable for the task of timely detection of new RSOs in low Earth orbit, ground-based optical sensors have been replaced by more expensive, but more reliable radars.

In principle, space-based optical sensors (see chapter 5) are not restricted to nighttime operation, but they suffer from a few limitations compared to ground-based fence-type radars:

1. Creating a large optical fence to be searched in a leakproof manner is difficult because of the limited field of view of most space-based telescopes.

2. Optical detection is dependent on the resident space objects being sunlit, but RSOs in low Earth orbit may spend a significant part of each orbit in Earth's shadow.

3. Space-based optical systems require near-real-time communications in order to be tactical and responsive to any changes in launch times; such communications capability is an expensive feature not typically included in these systems.

6.1.4   Radar Detection, Tracking, and Discrimination of Picosats Newly Launched into Low Earth Orbit

The following example illustrates the use of a narrow-beam radar (the fourth strategy for radars described in subsection 6.1.2) to detect and characterize multiple new satellites cooperatively launched into low Earth orbit [15].

Miniaturization of electronic components has led to a proliferation of small satellites capable of significant missions. For example, a team at the California Polytechnic Institute developed a design for a CubeSat, a 10-centimeter cube satellite that can be built for less than $50,000 [16]. Variously called "minisats," "microsats," or "picosats," small satellites pose a significant cataloging problem for the U.S. Space Surveillance Network because of both their small size and their number (anywhere from a few to many tens of picosats can be injected into orbit from a single launch). If many are deployed from the same booster, it can be difficult to discriminate between them, particularly because smaller satellites tend not to have unique scatterers. A challenging example was the launch of eleven small satellites, including several picosats, from the first test of a Minotaur launch vehicle [17, 18]. The Minotaur placed five minisatellites into a Sun-synchronous orbit at an altitude of about 700 kilometers. Later in the mission, one of these, the OPAL (Orbiting Picosat Automated Launcher), deployed six picosats. The Millstone Hill radar was instrumental in early detection, tracking, and, most important, discrimination of these satellites.

The following is a brief physical description of the satellites deployed during this mission and launched from Vandenberg Air Force Base on 26 January 2000, listed in the order of their separation from the fourth stage of the rocket:

1. ASUSat, 14-sided cylindrical can, measuring 24.5 × 32 centimeters in diameter;

2. OPAL, 6-sided, carrying six smaller satellites;

3. Orbital Calibration Sphere, 3.5-meter-diameter inflatable sphere;

4. FalconSat, 17- × 18-inch (about 43- × 45.5-centimeter) box;

5. JAWSAT, 35- × 35- × 42-inch (about 89- × 89- × 106.5-centimeter) box; and

6. Pegasus (the cylindrical fourth stage of the Minotaur launch vehicle), about 1 meter in diameter.

OPAL then deployed six small satellites: the JAK satellite (measuring $2.5 \times 7 \times 10$ inches, or about $6.5 \times 18 \times 25.5$ centimeters), Stensat (about 12 cubic inches, or about 196.5 cubic centimeters), Thelma and Louise (each measuring $2.5 \times 7.5 \times 20$ inches, or about $6.5 \times 19 \times 51$ centimeters), and a pair of picosats (each measuring $3 \times 4 \times 1$ inch, or about $7.5 \times 10 \times 2.5$ centimeters, connected by a 100-foot or about 30.5-meter tether).

The process of discriminating among the RSOs involved the following steps:

1. A laboratory-developed radar cross-section software suite, RCS_TOOLBOX, was used to predict the expected RCS of the satellites and their expected radar signature, which enabled the radar analysts to identify them.

2. Laboratory-developed software was used to predict the detailed launch and satellite separation sequence.

RCS_TOOLBOX, which estimated the average radar cross section of a target, could also predict the radar signature if given physical details of the target. In the case of the Minotaur mission, the rocket body was large and rose above the radar's horizon about 1 minute earlier than did the cluster of its payloads, making it easily discernible. JAWSAT, although not as large as the Orbital Calibration Sphere, was much larger than any of the other targets, so it, too, was easy to identify. ASUSat was the smallest satellite, and FalconSat was a cube with a repeating lobe structure in its RCS signature.

In conjunction with RCS_TOOLBOX, the time sequence of events, as provided by the launch agency, was used to discriminate the various satellites, and, in particular, to identify the OPAL, which subsequently deployed the picosats. The key to the success of this process was the Orbital Calibration Sphere's inflatable payload. The radar signature of a sphere (in principal and orthogonal polarizations) is unique, making it easy to identify. Once the sphere was identified, the Millstone Hill radar used an along-orbit search strategy to detect all the other satellites. The RCS predictions and the deployment information were used to discriminate them.

Because of the picosats' small size, detecting them was the most significant challenge of the project. Radar cross-section prediction for a picosat was −22 dBsm average (actual tracking showed the RCS to vary between −14 and −30 dBsm). At a range of about 2,000 kilometers from the radar, the estimated signal-to-noise ratio was about 15 dB per pulse for the picosat at its average RCS. Coherent integration of a sequence of pulses was used to enhance the SNR both for detection and, more important, for tracking in order to collect metric data with as high an accuracy as possible. All objectives—detection, discrimination, and metric data collection for cataloging—were achieved by the Millstone Hill radar in this example project.

6.1.5   Deep-Space Search Strategies

A consistent theme in this book is that deep-space surveillance is different from and much harder than low Earth orbit surveillance. The challenges in the acquisition and track of the resident space objects launched into a deep-space orbit, as compared to a low Earth orbit, are driven by the widely varying orbits of the RSOs in various phases of the launch and by the long time sequence (on the order of several hours to several days) from launch to the final orbit of the payload. A deep-space launch consists of

1. A low-altitude portion, in which the second stage of the rocket may be left in orbit for a short period of time, typically hours to days, before it reenters the atmosphere and, presumably, disintegrates.
2. A high-eccentricity orbit, into which the third stage of the launching rocket and the attached payload are injected. The inclination of the orbit may be different from that of the low-altitude stage.
3. A second deep-space orbit (with or without a fourth stage of the rocket), which may be a high-eccentricity orbit or a near-circular orbit into which the payload gets injected. The near-circular orbit typically is in medium Earth orbit at an altitude of approximately 16,000 kilometers or in a near-geosynchronous orbit at an altitude of approximately 36,000 kilometers. The inclination of the final orbit may be different from that of the initial high-eccentricity orbit and the initial low Earth orbit.
4. Possibly, a small set of intermediate orbits transitioning the payload from the second deep-space orbit in item 3 to the final orbit desired.

The time span for the complete launch and injection sequence is typically 6–9 hours for Hohmann transfers [19] and can be a day to several days when intermediate orbits are used.

Detection and tracking of satellites launched into a deep-space orbit are difficult because

1. The most obvious strategy for assured detection and tracking of a deep-space launch is, of course, close coordination with the satellite owner or launch agency and the availability of all the staging information. Unfortunately, given

the number of countries or organizations launching satellites, such cooperation is often difficult to elicit.

2. Most of the deep-space orbits of satellites are at long ranges, necessitating integration of a large number of pulses from any radar to achieve detection, best achieved by a narrow-beam, high-sensitivity radar system. Because a fence-type radar is designed to detect with a single pulse in order to facilitate the widest leakproof fence possible, such a radar is not normally useful for deep-space surveillance, although it can, of course, readily detect the low Earth orbit portion of the launch sequence.

Search and acquisition by a ground-based optical sensor requires that satellites in deep-space orbit be sunlit and that the sensor itself be in Earth's shadow with clear skies. Such conditions cannot be guaranteed even with geographically distributed sensors like those of the current GEODSS (Ground-Based Electro-Optical Deep Space Surveillance) system. Search and acquisition by space-based optical sensors also pose geometry problems. Typically, deep-space satellites are launched such that they are sunlit through most of their initial deep-space orbits to enable solar panels to produce power. If the phase angle (i.e., the angle between the Sun, center of the Earth, and satellite) is low, the space-based sensor may be unable to point at the satellite for fear of damage to its focal plane from direct solar illumination, a pointing problem that can be solved in the future with a network of space-based sensors and near-real-time tasking of the same. Over the course of days after a launch, space-based sensors are invaluable in discovering and tracking deep-space satellites.

There are three possible strategies for detection and tracking satellites newly launched into deep-space orbits [20]. Whereas two of these strategies are for timely acquisition, the third is not but is used when the first two fail. An essential difference from a launch into low Earth orbit is that, for any of these deep-space search strategies to be successful, there must be a predicted set of orbital elements for all phases of the launch, based either on prior history, an expectation of the staging, or, when possible, on actual detection and tracking of the staging.

6.1.5.1   Deep-Space Search Strategy 1: Tracking through Maneuvers

This strategy is the best but also the most difficult to execute; it requires tracking the deep-space launch during all the critical phases when there are likely to be significant changes in orbit resulting from rocket thrusts or satellite separation. Simple astrodynamics shows that certain parts of the orbit are suitable for efficient maneuvers of a newly launched satellite. For example, an apogee maneuver is ideal to change the satellite's inclination or orbital plane, whereas a perigee maneuver is more efficient to alter its eccentricity or orbital period. A typical launch sequence into a deep-space orbit entails the following:

1. A generally vertical launch to escape the dense atmosphere and consequent high drag as rapidly as possible, followed by an injection into a low-altitude orbit.

2. An insertion burn into a high-eccentricity orbit at the appropriate location. For example, a Molniya type of satellite is inserted into a high-eccentricity orbit at a Southern Hemisphere location corresponding to the argument of perigee required in the final orbit. A satellite destined for geosynchronous orbit, on the other hand, is injected into a high-eccentricity orbit near one of the equator crossings of the low-altitude orbit.

3. Subsequent maneuvers to adjust the satellite's orbit to the final configuration. Such maneuvers occur at the perigee of the high-eccentricity orbit for adjustment of orbital period or at the apogee for changes in the orbital plane.

4. Separation of the satellite and deployment of appendages. Separation of a satellite from the stage of the rocket it is attached to occurs prior to or immediately after the maneuvers in items 2 and 3. Deployment of appendages (solar panels, antennas, etc.) occurs at various times.

Hence, once a launch destined for deep space has been detected and an initial orbit for the low-altitude portion of the launch sequence established, predictions can be made for the subsequent maneuvers with high confidence. Surveillance sensors can be tasked for search and acquisition at the appropriate locations and times. Discontinuous tracking is adequate as long as the locations of maneuvers can be predicted and the satellite is tracked through the maneuvers so that a postmaneuver orbit can be established. The most efficient location for a maneuver to raise apogee is at the orbit's perigee; the most efficient location for a maneuver to change orbital inclination or to raise the perigee is at the orbit's apogee. A combination of ground-based radars and space-based optical sensors will be the desired mix in the future for implementation of this strategy—the radars will provide tactical (day and night) availability, and the optical systems will cover regions and maneuvers that are either out of view of the radars or are difficult to detect because of the detection sensitivity of the radars.

A successful example of deep-space search strategy 1—tracking through maneuvers—is the tracking of a Defense Support Program (DSP) satellite launched into geosynchronous orbit, one of the types of launches routinely supported by the

Lincoln Laboratory radars on the Kwajalein Atoll and in Westford, Massachusetts.. Similar search strategies have been successfully used in a wide variety of both cooperative and noncooperative launches.

The Defense Support Program satellite system was the principal source for U.S. detection of missile and satellite launches worldwide. Like every domestic launch, the launch of a DSP satellite was supported by radar and, when possible, optical tracking. If the launch was routine, the radars and optical systems provided confirmatory data. If, on the other hand, something failed in the launch sequence, the data from the radars became invaluable in diagnosing the causes of failure.

The DSP 20 satellite was launched on 9 May 2000 from Cape Canaveral, Florida, and was successfully inserted into geosynchronous orbit. Figure 6.5 shows the launch sequence.

The satellite was launched atop the interim upper stage of a Titan/IUS launch vehicle, which put the IUS/DSP into low-latitude orbit; the IUS then fired several times to put itself and the satellite it was carrying into transfer orbit and finally into a geosynchronous orbit, at which point the interim upper stage separated from the satellite and maneuvered away. DSP 20 went through a series of maneuvers to achieve operational configuration. The unfurling of the solar panels and the ejection of the lens cover of the infrared telescope were important elements of this series.

Figure 6.5 of the launch sequence and the coverage shows that ALTAIR on Kwajalein had good visibility of the DSP 20 complex through the first thrust of solid rocket motor 1 (SRM 1 ignition), which served to put the interim upper stage/DSP satellite (IUS/DSP) into a Hohmann transfer orbit. The most important products of ALTAIR's coverage of the launch were

1. Confirmation of the separation of the Titan upper stage and the IUS/DSP in the low-altitude orbit;

2. Confirmation of the solid rocket motor 1's thrust of the IUS/DSP into transfer orbit; and

3. An element set of the transfer orbit based on the tracking of the IUS/DSP post-thrust; and handoff of the element set to sensors with subsequent coverage.

The radars in Westford, Massachusetts, called the "Lincoln Space Surveillance Complex" (LSSC), consisted of the Millstone Hill narrowband tracking radar and the Haystack and the Haystack auxiliary wideband imaging radars. The Millstone Hill radar used the element set forwarded by ALTAIR to acquire the interim upper stage/ DSP 20 complex and tracked the complex through to insertion into geosynchronous

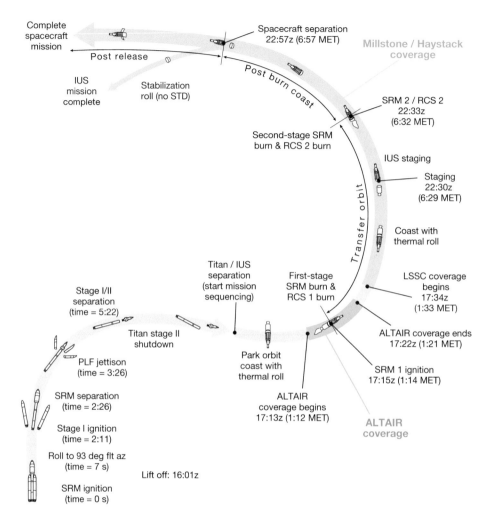

Figure 6.5 Launch sequence and times of support from the Lincoln Laboratory radars (note that the time sequence is expanded or contracted as necessary). flt az, flight azimuth; IUS, inertial upper stage; LSSC, Lincoln Space Surveillance Complex; MET, Mission Elapsed Time; PLF, payload fairing; RCS, reaction control system; SRM, solid rocket motor; z, Zulu Time (Greenwich Mean Time).

orbit, separation of DSP 20 from the interim upper stage, and the satellite's acquisition of Earth and Sun.

Figure 6.6 shows the Millstone Hill radar display of the tracking of the interim upper stage and DSP 20 after separation of the two in geosynchronous orbit. The radar coherently integrated a large number of pulses (at approximately 30 pulses per second) to create a tracking display of the magnitude of signal (in arbitrary units) along the *y*-axis against the relative Doppler along the *x*-axis. Normally, the target being tracked would be centered at zero Doppler. A separate return at about −1 Hz Doppler (about 0.12 meter per second difference in slant range speed) indicated that the IUS and DSP 20 had successfully separated in orbit, thus demonstrating the ability of radar tracking to follow the injection into geosynchronous orbit of the complex and the separation of its two component objects. The radar also demonstrated the ability to detect closely spaced objects in orbit. The detections were achieved within a few coherent integration cycles of the radar—a few minutes at most.

The next significant event in the launch sequence was the unfurling of the solar panels, shown in the Doppler-time-intensity (DTI) display of the tracking at the Millstone Hill radar (figure 6.7; see the DTI discussion in chapter 3). About a quarter of the time into the display, the intensity of the target (shown as color in the display), and the Doppler width of the target increased significantly. These increases indicate an event that was inferred to be consistent with the unfurling of the solar panels of DSP

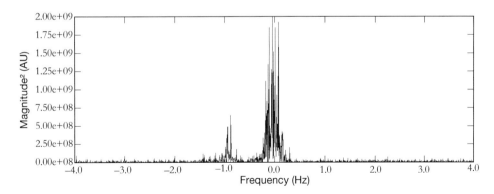

Figure 6.6 Spectral representation of the tracking gate showing the Defense Support Program (DSP) satellite at about 0 Hz relative frequency and the interim upper stage at about −1 Hz relative frequency (about 0.1 meter per second relative range rate along the radar line of sight); *x*-axis: relative Doppler; *y*-axis: magnitude of signal in arbitrary units ($\times\ 10^8$ AU). Magnitude^2, magnitude squared.

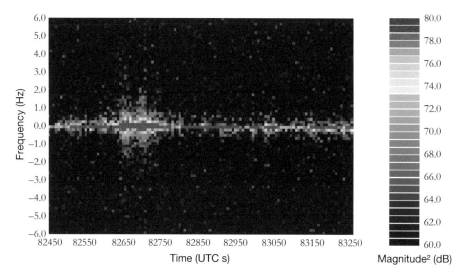

Figure 6.7   Doppler-time-intensity (DTI) display at the Millstone Hill radar showing the Doppler (*y*-axis) broadening of the radar return of the Defense Support Program (DSP) satellite at the time when its solar panels were deployed. UTC, Coordinated Universal Time (Greenwich Mean Time).

20, as confirmed by the launch deployment sequence. This event was also detected within a few integration cycles of the radar—a few minutes at most.

During the launch sequence, the final event that was susceptible to easy confirmation at the Millstone Hill radar was the ejection of the lens cover of the infrared telescope. The lens cover is a relatively large (about 1-meter-diameter) circular plate (for protection of the aperture of the infrared telescope), which was ejected by springs at a velocity of approximately 1 meter per second relative to the parent satellite in a direction toward the center of the Earth and, as a result, almost along the radar's line of sight. The consequence was that, in the Doppler-time-intensity display at the radar, a second target appeared almost instantaneously, offset about 10 Hz from the parent target, as shown in figure 6.8. Note the second horizontal line at a relative Doppler of about 10 Hz beginning halfway through the time interval on the *x*-axis and persisting for the rest of the tracking interval. The measured intensity of the infrared telescope cover is significantly smaller than the intensity of the parent satellite. The occurrence and confirmation of this event were vital to the functioning of DSP 20.

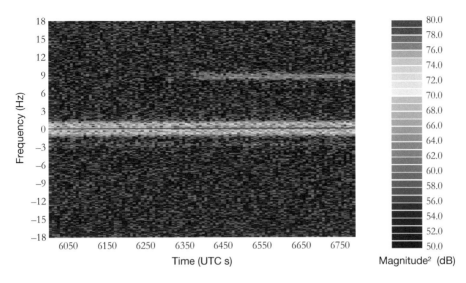

Figure 6.8   Doppler (y-axis)-time (x-axis)-intensity (color) display at the Millstone Hill radar when the infrared telescope cover was ejected.

6.1.5.2   Deep-Space Search Strategy 2: Forecast Postmaneuver Orbit Search

If sensor nonavailability precludes the use of search strategy 1, an alternative strategy can be used, wherein the expected orbits postmaneuver are predicted and radars, optical systems, or both are requested to search these predicted orbits. Generally, these predictions are based on a history of such launches (satellite owners being conservative people and satellites often being deployed into constellations for particular purposes). Also, given some tracking of the initial low Earth orbit and a prediction of the objective of the mission, it is possible to generate a set of expected orbits postmaneuver that can then be provided to the sensors as search orbits. Success of this strategy has been repeatedly demonstrated, particularly at the Millstone Hill radar on noncooperative launches from the former Soviet Union as well as other countries.

As an example, consider the standard Molniya launch from the Plesetsk launch site in Russia [20]. As the ground traces in figure 6.9 show, the payload gets injected into a high-eccentricity orbit in the Southern Hemisphere. Lacking coverage of this injection, a radar must initiate a search based on the expected orbit with the following possible assumptions:

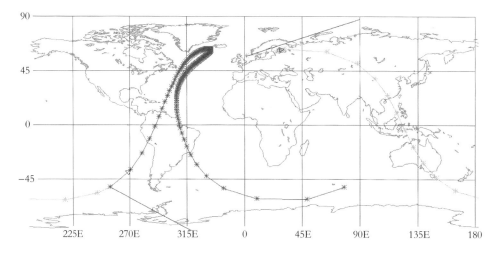

Figure 6.9   Ground trace of a Molniya launch. E, east longitude.

1. An argument of perigee of the high-eccentricity orbit of either 280 or 288 degrees, based on historical data;

2. A time of injection into the high-eccentricity orbit that differs by approximately 2 minutes, consistent with the argument of perigee variation;

3. An initial orbital period of the high-eccentricity orbit of either 710 or 730 minutes, again consistent with historical data (note that the orbital period is adjusted to 718 minutes, i.e., a semisynchronous period, typically after 4 days);

4. A semimajor axis consistent with the estimated orbital period; and

5. The inclination and right ascension of the ascending node assumed to be the same as the values in the low-altitude portion of the orbit.

Using these orbital guesses, a deep-space radar like the Millstone Hill Radar can conduct a series of searches—essentially one search for each orbital guess, plus an along-orbit time search to account for uncertainties. This method has been developed and applied with great success in the real-world operations of the laboratory's deep-space radars.

Another example of a forecast postmaneuver orbit search is the search for the Russian Gorizont 45 satellite on its launch into a geosynchronous orbit (figure 6.10).

As figure 6.10 illustrates, the following sequence happened on this launch:

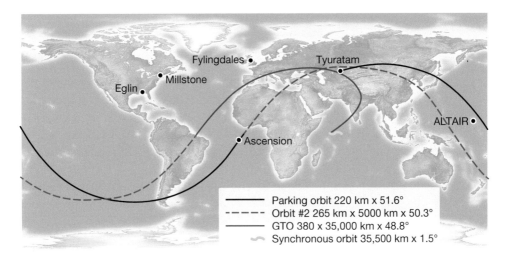

Figure 6.10   Ground trace of the launch of Gorizont 45 into geosynchronous orbit.
GTO, geostationary transfer orbit.

1. Launch took place from the Tyuratam launch complex in Kazakhstan into the parking orbit shown in black in the figure. ALTAIR tracked the Gorizont launch vehicle from the launch complex but did not detect any separation and maneuver into a high-eccentricity transfer orbit.

2. The rocket stage carrying the Gorizont satellite maneuvered near Ascension Island, close to its equator crossing, into an intermediate eccentric orbit represented by the blue line in the figure. The phased-array radar at Fylingdales station in Great Britain detected and tracked the rocket stage in this orbit.

3. The high-eccentricity transfer orbit to geosynchronous altitude was initiated at the south-to-north equatorial crossing on the northeast coast of South America. It is represented by the green line in the figure.

4. The satellite was injected into a near-geosynchronous orbit off the east coast of Africa where the green line terminates. The geosynchronous orbits of the rocket body and the payload are represented by the pink dots.

The only method to track this launch was to use the parking orbit as tracked by ALTAIR and fit it to the intermediate transfer orbit as detected by Fylingdales radar. It was then necessary to predict, first, the possible injection into a geosynchronous transfer orbit and, second, the geosynchronous orbit. Finally, the predictions were used to search with available sensors. The strategy was quite successful, though it took a few days to execute.

6.1.5.3   Deep-Space Search Strategy 3: Predicted Location or Orbit Search

Occasionally, search strategies 1 and 2 both fail for lack of adequate information or good predictions. A fallback strategy, which is neither timely nor assured of success, can be useful in many such cases. This strategy consists of a search near a location or along an orbit predicted on the basis of at least some information, however incomplete.

Common examples of a predicted location search are searches for satellites in geosynchronous (especially geostationary) orbits. A launch into geosynchronous orbit involves several major maneuvers, and sensors often fail to track the launch vehicle and payload through these orbit changes. In such a case, a predicted injection point into geosynchronous orbit can be calculated from the transfer orbit (or perhaps from some external information) and a localized search of the geosynchronous region can be conducted. This procedure has been used successfully many times by the Millstone Hill radar and other deep-space sensors of the Space Surveillance Network. The procedure is also applicable to the case where a geostationary satellite maneuvers in orbit and moves to a different location. The search for Gorizont 45, discussed earlier, constituted, at least to some degree, an example of a predicted location search.

A good example of a predicted orbit search is the search for the Soviet Prognoz 5 satellite by the Millstone Hill radar using "serendipitous" data. Prognoz 5 was one of a series of scientific research satellites launched by the former Soviet Union. A major objective of the Prognoz series was to examine the interaction of the Earth's ionosphere with the solar wind. These satellites had unusual orbital characteristics—a very high eccentricity, on the order of 0.9–0.95, an orbital period of about 4 days, an initial perigee height of about 500 kilometers, an initial apogee height of about 200,000 kilometers, and an initial orbital inclination of about 63 degrees (note that all these parameters vary from orbit to orbit because of the significant effects of lunisolar perturbations). The launch time and the consequent right ascension of the ascending node of the satellites were severely constrained by the requirement that the lunisolar perturbations must raise the initial perigee height in order to prevent the satellite from reentering the atmosphere within a few orbits. The orbital evolution of such (high-eccentricity) orbit satellites has been extensively studied [7, 8].

The Prognoz's launch sequence is similar to the Molniya's except that the period of the Prognoz's predicted orbit is about 4 days instead of 710 to 730 minutes. Without prior information on the argument of perigee and the orbital period of the satellite, only the orbital plane is known, based on detection data from the Defense Support Program satellite and Space Surveillance Network on the low Earth orbit portion of the launch. Search with a narrow-beam radar (namely, the Millstone Hill radar) has a low probability of success because the range to the satellite increases rapidly, and small uncertainties in the search element sets result in large search volumes. Optical sensors are often not available in time to support such a launch. However, in the case of Prognoz 5, a simple calculation based on the predicted orbital elements showed that the satellite (or the rocket body, which also ends up in a similar orbit) was likely to pass through the fence-type radar coverage of the Navy Space Surveillance (NAVSPASUR) System (figure 6.3) [9] within the radar's detection range in the satellite's fourth and fifth orbits. Keen-eyed NAVSPSUR analysts picked up detection data on an uncorrelated target that seemed to have the appropriate orbital characteristics. Since the NAVSPASUR System is a static-zenith radar fence, knowing the position of the satellite and the time at detection enabled the Millstone Hill radar analysts to recompute an orbit for the satellite, conduct a search based on the new information, and detect and track the satellite, thus adding it the satellite catalog.

6.2   Catalog Discovery of Resident Space Objects Generated by Fragmentation Events

According to the Orbital Debris Program Office of NASA's Johnson Space Center, fragmentation events are the leading cause of space debris. One of the best ways to detect the fragmentation of resident space objects in low Earth orbit is to regularly process the detection data for uncorrelated targets (UCTs) from phased-array radars, such as the FPS-85, and missile warning radars. In the future, detection data for UCTs from the Space Fence radar system will also need to be processed. These various systems will provide a complete picture of fragmentation events in low Earth orbit in a near-tactical time frame (within 24 hours of the event). Depending on the altitude and location of perigee of orbiting fragmentation debris, the systems may also detect fragmentation events in highly elliptical orbits but cannot be relied on to be complete. In medium Earth and geosynchronous orbits, fragmentation events can be detected only by using optical surveillance instruments to repeatedly search those orbits or orbit regions. GEODSS systems do conduct regular searches of the geosynchronous belt and will detect uncorrelated targets, some of which can then be recognized as arising from a fragmentation event and perhaps even associated with a parent satellite. An excellent example of such a search and detection is the discovery of high-area-to-mass-ratio geosynchronous resident space objects by

Thomas Schildknecht, Reto Musci, and Tim Flohrer [21]. Space-based optical systems like the Space-Based Space Surveillance (SBSS) satellite, and ground-based systems, like the Space Surveillance Telescope (SST) and GEODSS system, can also conduct searches of the geosynchronous orbit region and discover fragmentation events.

The highly elliptical and medium Earth orbit regions of space can be searched by ground-based optical sensors such as those of the GEODSS system for new fragmentation events, and by radars for fragments where a possible fragmentation event has been identified. The Millstone Hill radar has frequently been used to search and detect debris from the Russian Kosmos series of satellites in Molniya-like, high-eccentricity, semisynchronous orbits. The first few such satellites suffered catastrophic fragmentation of unknown cause.

The Fragmentation Event Detection System (FEDS) was designed and implemented as part of the Space Surveillance Software System at Lincoln Laboratory to detect such events by using uncorrelated target data from the FPS-85 radar [22]. Located on Eglin Air Force Base in Florida, as part of the Space Surveillance Network, the FPS-85 operates a surveillance fence continuously and is thus able to detect new resident space objects in low-altitude orbits fairly rapidly. The radar system automatically tries to associate each detection with a cataloged resident space object. If unassociated, the detection is termed an "uncorrelated target" (UCT). Observational data and element sets on UCTs are collected and automatically calculated by the radar system, then automatically sent to the Millstone Hill radar once per day. The Fragmentation Event Detection System uses these uncorrelated-target data from the FPS-85 to automate the detection of breakups of resident space objects in orbit.

6.2.1   Analytical Description of the Fragmentation Event Detection System

Using historical data on uncorrelated targets from the FPS-85 radar, the FEDS generates a model of the statistical likelihood of a fragmentation event. Based on this model, data are evaluated daily in a three-stage process. Element sets meeting the criteria of this process are considered possible evidence of a fragmentation event. When such an event is actually detected, three types of reports are generated:

1. A list of the element sets involved in the event;
2. Information about the location of the event; and
3. VRML (Virtual Reality Modeling Language) code for the three-dimensional graphic display of the event.

The Fragmentation Event Detection System has been successful at detecting fragmentation events ranging in size from small events resulting in a few debris pieces, such as the breakup of NASA's COBE (Cosmic Background Explorer) satellite, to large events resulting in several hundred debris pieces, such as the breakup of a Pegasus satellite. These and other detected fragmentation events are summarized in table 6.1.

The FPS-85 phased-array radar detects fragmentation events in both low-altitude and high-eccentricity orbits as uncorrelated targets. The FEDS algorithm assumes that when a fragmentation event occurs, there will be a significant increase in the number of objects detected for a given day in a small region of orbit space. Dividing orbit space into forty rectangular "microbins," defined by inclination and right ascension, the FEDS computes the historical mean and standard deviation for the number of objects detected per day in each of these microbins. If the number of objects detected for a given day exceeds the mean detection rate for that bin by three standard deviations, then the bin may contain a candidate fragmentation event, and this information is passed on to the next stage. The first step in this process is to consider the set of all uncorrelated targets and, from this information, to form the subset of all UCTs excluding the last 30 days. Mathematically,

$$I_T = \{\text{all UCTs}\},$$

where I_t is the set of all UCTs;

Table 6.1 Table of fragmentation events

Event	First detection	Number of fragments	Days detected
GLONASS	Day 340, 1996	4–9	7
Pegasus	Day 156, 1996	16–151	40
Ariane/Cerise	Day 206, 1996	1	0
Ariane	Day 129, 1996	4–9	3
SL-12	Day 82, 1996	19–154	2
Delta II rocket body	Day 6, 1996	6–8	3
GLONASS debris	Day 349, 1995	8–22	3
GLONASS debris	Day 256, 1995	18–28	3
SL-19 rocket body	Day 194, 1995	4–14	3
GLONASS debris	Day 132, 1995	3–8	3
COBE	Day 63, 1995	6–21	10

$$I = \{x \mid x \in I_T \wedge x.\text{epoch} < (\text{Current date} - 30 \text{ days})\} \, ,$$

where x is a subset of UCTs seen in the last 30 days.

The fragmentation model must exclude the most recent data from the FPS-85 radar system because these have been compared against the statistical model. The FEDS then divides this subset into ten nearly equal population "inclination bands," defined by

$$I_J = \{x \mid x \in I \wedge [\theta_J] < x.\text{inclination} \leq [\theta_{J+1}]\} \, ,$$

where $[\theta_J]$ and $[\theta_{J+1}]$ are angles chosen for all $J = 0, 1, 2, \ldots 9$. These two angles are generated by using the rule that the population of I_J is roughly one-tenth of the overall population of UCTs distributed by inclination.

Each of the inclination bands is then divided into four right-ascension microbins by a similar process. The right-ascension bins are defined such that

$$r_{JK} = \{x \mid x \in I_J \wedge [\psi_K] < x.\text{right ascension} \leq [\psi_{K+1}]\} \, ,$$

where $[\psi_K]$ and $[\psi_{K+1}]$ are equally distributed UCTs grouped by right ascension chosen for all $K = 1, 2, 3, 4$. These two angles are generated by using a rule similar to the one above, except for the following differences: (1) the angle used is right ascension, (2) each bin is constructed solely from an inclination band and not from the overall population; and (3) the target percentage is 25 rather than 10.

There are thus ten inclination bands, each with four right-ascension microbins, or a total of forty microbins. The FEDS calculates the mean $M_{r_{JK}}$ and standard deviation $\sigma_{r_{JK}}$ for the expected number of objects detected daily in each of these bins on basis of the training data.

New uncorrelated targets from a day that is not part of the training set are then sorted into the forty microbins defined earlier. For each microbin, the FEDS performs the following check:

$$|D_{JK}| \geq M_{r_{JK}} + 3 \times \sigma_{r_{JK}} \, ,$$

where D_{JK} is the population of UCTs in the JKth bin.

If the number of uncorrelated targets detected in a microbin exceeds the mean by three standard deviations, then this microbin progresses to the second stage. Most bins have means and standard deviations slightly below one UCT per day, so this test typically isolates bins with four or more UCTs.

The second stage attempts to normalize the data and to overcome system irregularities in the performance of the radar. Although the FPS-85 radar system attempts local catalog maintenance of uncorrelated targets, it typically loses track of them after a few days. Because these then reappear as "new" UCTs, the UCT detection rate of each bin will be periodic. Another long-term aspect of the data is that uncorrelated targets will move across bins because of naturally evolving right ascension. Bin movement also occurs across inclination bands, but typically at a low rate of change relative to that for right ascension. Still another factor is that the hours of operation of the surveillance fence vary. The final factor affecting FEDS results is the occasional operation of a second debris fence. All these factors affect the detection rates for all bins. Since the cumulative effect should be close to uniform, each bin should have the same ratio of UCTs detected in the bin to total UCTs (D) detected that day. Because there are forty microbins, each microbin represents 2.5 percent of the overall historical data.

The FEDS then performs the following check:

$$\frac{|D_{JK}|}{|D|} \geq \frac{M_{r_{JK}}}{M} + 3 \times \frac{\sigma_{r_{JK}}}{M} .$$

This equation means that if the percentage of uncorrelated targets detected in a microbin exceeds the expected percentage of detections by three standard deviations, then this microbin progresses to the third stage.

At this stage, the FEDS has determined that the abnormal number of UCTs detected in one or more bins is not symptomatic across all bins. But the fact that a set of UCTs has similar inclination and right ascension is not sufficient to establish a fragmentation event. For example, the objects detected may not be near one another: a group of UCTs may contain both deep-space and near-Earth objects with similar inclination and right ascension. Hence the following check is made:

1. Sort D_{JK} by semimajor axis; and
2. Check if three or more UCTs have semimajor axes within 600 kilometers of one another.

If three or more uncorrelated targets meet the criterion in item 2, then the FEDS reports this cluster as a possible fragmentation event. The microbins used in this process are, on average, 18 degrees inclination by 90 degrees right ascension, a measure that encompasses most fragmentation events. But if an event occurs close to or on a bin boundary, then the resulting pieces may be scattered over several bins and, as a result, may not be detected as a fragmentation event. This problem is addressed by establishing "macrobins," each of which combines four adjacent microbins. Macrobins R $_{JK}$ are defined by the following relation:

$$R_{JK} = r_{JK} \cup r_{J+1K+1} \cup r_{J+1K} \cup r_{JK+1}.$$

Macrobins do wrap around. Hence

if $K\,4$, then set $K + 1 = 0$, and if $J = 9$, then set $J + 1 = 0$.

This substitution is necessary to make the macrobins wrap around in inclination and right ascension. A sample macrobin is depicted in figure 6.11. The circle represents a fragmentation event that has occurred on the intersection between the inclination and right-ascension boundaries. If the event is small, spreading out across the adjacent four microbins will make it undetectable on the microbin scale, but because it is encompassed by the macrobin, it will be detectable on the macrobin scale. And because every intersection of two boundary lines defines a macrobin, the FEDS contains a total of forty macrobins. On average, a macrobin will cover 36 degrees inclination by 180 degrees right ascension. Macrobins, like microbins, have a mean and standard deviation associated with each bin. Once the macrobins are created, the FEDS evaluates each macrobin according to the same three-stage process applied to microbins.

Large breakups (such as that of a Pegasus) frequently give rise to so many detections that the fragmentation event is detected both in its microbin and in the four macrobins that share that microbin.

Detection by the FEDS only signifies that in a ring-shaped region of space defined by inclination, right ascension, and semimajor axis, there has been a significant increase in the number of uncorrelated targets detected. The next step is to determine how the UCTs are distributed in the cluster ring. At this point, an analysis must be

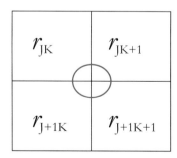

Figure 6.11   Sample macrobin.

performed on each cluster to determine whether a fragmentation event has occurred. If the UCTs are distributed evenly throughout the ring, then those detections are unlikely to be the result of a recent fragmentation event. If a new fragmentation event has occurred, then the UCTs should be in a dense region in that ring.

The Fragmentation Event Detection System provides several tools to assist in the analysis of a cluster. The first is a list of element sets for each uncorrelated target in a cluster. The system also keeps information about all previously detected clusters. This information can be useful in determining whether a cluster is a result of an older breakup. Operational experience has shown that in nearly one out of four cases, the increase in the number of UCTs is not the result of a recent fragmentation event but is, instead, a false alarm. The system also maintains a corresponding Virtual Reality Modeling Language (VRML) file for every detected event. By viewing these VRML files, analysts can determine the distribution of uncorrelated targets through a given cluster ring. Because each UCT tends to have an epoch close to its time of observation, the UCTs in a given cluster will have different epochs. Hence, to accurately determine the distribution of uncorrelated targets, the FEDS propagates all UCTs to a common epoch. The system then determines the geocentric position of each UCT and generates the VRML file. Presented in figure 6.12 is a sample VRML file showing the initial results of a Pegasus breakup. In this particular example, the forty-three

Figure 6.12   Virtual Reality Modeling Language (VRML) file of a Pegasus breakup.

distinct uncorrelated targets contained in small section of a cluster ring are clearly the result of a fragmentation event.

6.2.2   Performance Evaluation

The performance of the Fragmentation Event Detection System was analyzed over a test period, with the following results:

- Eleven out of twelve fragmentation events in the data set were detected.
- Event sizes ranging from 4 to 154 objects were detected.
- All events detected were visible over several different days.
- A false alarm rate of 22.6 percent (early in its operation—before adequate training data) was recorded.

Table 6.1 shows twelve fragmentation events that occurred in the data set. The columns represent the object(s) associated with the event, the date of first detection by the FEDS, the number of uncorrelated targets detected for a given event, and, finally, the number of different days on which the system recognized the fragmentation event. The one fragmentation event that went undetected resulted from the collision between the French military satellite Cerise and a debris piece from an Ariane rocket stage. This collision was notable because it was the first proven collision between a satellite and space debris. In this case, only one detectable UCT was created (there were probably more that were too small to be detected), unusual for an orbital impact at about 10 kilometers per second. The FEDS only detected the one UCT because the two parent objects were correlated targets, and one UCT is insufficient to pass the first stage.

Analysis of clusters detected by the FEDS showed that roughly a quarter of the fragmentation events reported by the system were, in fact, false alarms, most of which were quickly identified by examining either the VRML files or the element sets. And most of the false alarms occurred at the beginning of the historical data set. As the FEDS acquired more data, the probability of false alarms decreased. Figure 6.13 shows a plot of valid and invalid events reported by the FEDS. The number of bins represents the total number of micro- and macrobins that registered an event. For example, in 1996 on day 156, there were two microbin and seven macrobin detections of a fragmentation event. The event in question was a Pegasus breakup.

Finally, the response times of the FEDS to actual fragmentation events are shown in table 6.2.

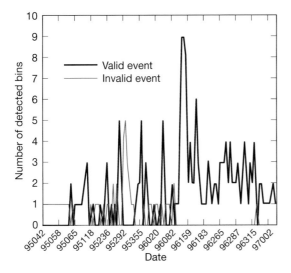

Figure 6.13   Plot of valid and invalid event detections (dates in abbreviated year–day number format).

Table 6.2   Response time chart

Fragmentation event	Space Control Center number	Breakup date	FEDS detection date
GLONASS debris	18347	Day 336, 1996	Day 340, 1996
Pegasus	23106	Day 155, 1996	Day 156, 1996
Ariane	21057	Day 122,* 1996	Day 129, 1996
SL-12	23797	Day 50, 1996	Day 52, 1996
Delta II rocket body	22232	Day 5,* 1996	Day 6, 1996
GLONASS debris	20957	Day 348, 1995	Day 349, 1995
GLONASS debris	23174	Day 255,* 1995	Day 256, 1995
SL-19 rocket body	23440	Day, 193,* 1995	Day, 194, 1995
GLONASS debris	23338	Day 131, 1995	Day 132, 1995
COBE	20332	Day 62,* 1995	Day 63, 1995

* Estimated breakup date

It is important to note that, although the FEDS analysis was done with FPS-85 data because they were easily available, the system can be used to analyze uncorrelated-target data from any fence-type radar engaged in surveillance of space.

6.3   Orderly Catalog Maintenance

The next major task for space surveillance was to maintain and publish an orderly catalog of all detectable resident space objects orbiting Earth and, occasionally, of RSOs in Earth-escape orbits. Because objects in such orbits do not constitute any kind of threat to the country or to other RSOs and are of interest only to the scientists and the particular launch agency involved, by and large, data on their orbits are maintained by the individual agency (most often, NASA, the Jet Propulsion Laboratory, Johns Hopkins University Applied Physics Laboratory, National Association of State Departments of Agriculture, or European Space Agency). The task of RSO catalog maintenance is handled by Air Force Space Command with supporting data and analyses supplied by the sensors and analysts of the Space Surveillance Network. Some interesting aspects of this task are discussed below.

The catalog database must contain the following information:

1. An identification of each resident space object—a unique sequence number, information on the date of launch, the launch agency, the type of object, and, where relevant, the date of the object's demise (usually caused by loss of critical system such as power, reentry into the atmosphere or other causes).

2. A regularly updated set of orbital elements, state vectors, or both for each RSO at a recent time, along with information on the orbital model that was used to generate the elements or the state vectors. The orbital model information is needed by users to compute compatible predictions from the elements or the state vectors.

3. Information on photometric or radiometric measurements of the RSO.

4. Any other available ancillary identifying information about the RSO—images, communication frequencies and protocols, and so on.

The earlier discussion of catalog discovery has outlined the necessity for discovering and tracking all resident space objects newly launched or resulting from fragmentation events. This information, combined with the descriptions furnished by the launch agencies, forms the basis of the identification of RSOs. Such information is often not available on a newly detected resident space object. The following forensic techniques can then be applied to ascertain the provenance of such an RSO, generally labeled "UCT analysis" [23].

An "uncorrelated target" (UCT) is defined as a resident space object that has been tracked, but whose element set does not correspond to any of the known RSOs in the catalog. In such a case, the following three possibilities exist:

1. The uncorrelated target is a resident space object that has maneuvered in or out of its orbit;
2. The UCT signals a breakup event of a known RSO; or
3. The UCT is a newly detected RSO with no obvious provenance.

The first step to determine which of these possibilities is actually true is to request additional tracking in order to refine the orbital element set of the uncorrelated target. The next step is to run an orbital correlation against the catalog based on the fact that, once in orbit, the unidentified RSO's orbital plane or orbital energy is not changed much (no more than a few degrees or a small fraction) by a maneuver or a fragmentation. Once a prior RSO with a similar orbit is identified, the following questions can be answered:

1. Is the uncorrelated target a result of a maneuver by a prior resident space object in a similar orbit? If a conjunction point for the two RSOs can be determined and a plausible maneuver can be deduced, and if the prior RSO is no longer found in its original orbit, then the maneuver hypothesis is most likely true.
2. Is the uncorrelated target a result of a fragmentation event involving a prior resident space object? If a conjunction point for the two RSOs can be determined but the prior RSO is confirmed to be still in its (perhaps slightly altered) orbit, then the fragmentation hypothesis is most likely true. In such a case, there are likely to be additional uncorrelated targets as a result of the fragmentation event. The examples in the FEDS analysis earlier illustrate such events.

If there is no obvious provenance in the existing catalog, then the following questions need to be addressed:

1. Is the uncorrelated target a result of a recent launch? Given both the identification of launches by agency announcements and the ubiquitous coverage by non–space surveillance launch detection systems, it is easy to prove or disprove this possibility.
2. Is the uncorrelated target an orphan? If there is no provenance and no recent launch with which to identify it, the UCT is a previously untracked resident space object and must be added to a provisional catalog without full identifying information. In such a case, radar or optical signature analysis, or both, should

be used to fully characterize the RSO. The full set of characteristics can then be compared with known RSOs from launches that have never been tracked or prior RSOs that have been "lost" to the catalog because of the lack of tracking data. A knowledgeable analyst is vital to this process.

Ultimately, the catalog database must contain all resident space objects, whether identified completely or not, to ensure that newly detected RSOs are determined to be not in the catalog before adding them to it.

6.4   Summary

As stated at the outset of this chapter, the creation and maintenance of an orderly catalog with unique identifying information on all man-made objects in space is fundamental to all activities in space surveillance. Although the problem of cataloging satellites in low Earth orbit in the early days of space surveillance was solved by the use of fence-type radars, the problem of catalog discovery in deep space, like that of deep-space surveillance itself, has been historically difficult and has necessitated significant technological developments.

Section 6.1 described the techniques and technologies developed at Lincoln Laboratory to address this problem; these have been successfully applied at all deep-space radars and optical systems of the Space Surveillance Network. Noncooperative foreign launches were the major challenge encountered and met. From the early days on, Lincoln Laboratory analysts working on U.S. government and commercial launches have developed, and demonstrated the success of, technologies for deep-space surveillance. As a result, a robust set of strategies is now routinely applied at all deep-space sensors. The FEDS algorithm described in section 6.2 was developed and demonstrated by the laboratory to identify fragmentation events, trace them back to the parent objects, and develop catalog entries for all of the debris objects created.

References

1. Leclair, R., and R. Sridharan. "Probability of Collision in the Geostationary Orbit," presented at the ESA Space Debris Conference, Darmstadt, Germany, March 2001.
2. United Nations. Office for Outer Space Affairs, "Convention on Registration of Objects Launched into Outer Space," http://www.unoosa.org/oosa/en/ourwork/spacelaw/treaties/introregistration-convention.html, 1974.

3. Tranchetti, F. 2013. *Fundamentals of Space Law and Policy*, 11–13. New York: Springer.

4. Orbital Debris Program Office at NASA's Johnson Space Center. http://www.orbitaldebris.jsc.nasa.gov, last updated April 2016.

5. *Orbital Debris Program Quarterly*, vol. 14, no. 1, January 2010.

6. *Wikipedia*, "Sun-Synchronous Orbit," last modified 11 May 2016.

7. Lidov, M. L. 1962. The Evolution of Orbits of Artificial Satellites of Planets under the Action of Gravitational Perturbations of External Bodies. *Planetary and Space Science* 9:719–759.

8. Sridharan, R., and M. L. Renard. (March 1975). Non-Numeric Computation for High Eccentricity Orbits. *Celestial Mechanics* 11 (2): 179–194.

9. "Naval Space Surveillance System (NSSS): Baseline Systems Description," San Diego, CA: SPAWAR System Center, Intelligence, Surveillance and Reconnaissance Department, 6 April 2000.

10. Lockheed Martin. "Space Fence: How to Keep Space Safe," http://www.lockheedmartin.com/us/products/space-fence.html, 2016; *Wikipedia*, "Air Force Space Surveillance System," http://en.wikipedia.org/wiki/Air_Force_Space_Surveillance_System, last modified 11 April 2016.

11. Camp, W. W., J. T. Mayhan, and R. M. O'Donnell. 2000. Wideband Radar for Ballistic Missile Defense and Range-Doppler Imaging of Satellites. *Lincoln Laboratory Journal* 12 (2): 267–280.

12. J. Emory Reed. (March 1969). The AN/FPS-85 Radar System. *Proceedings of the IEEE* 57 (3): 324–335.

13. Hall, T. D., G. F. Duff, and L. J. Maciel. 2012. The Space Mission at Kwajalein. *Lincoln Laboratory Journal* 10 (2): 48–63.

14. Henize, K. G. (January 1957). The Baker-Nunn Satellite-Tracking Camera. *Sky and Telescope* 16:108.

15. Bougas, W., T. A. Cott, S. E. Andrews, and R. Sridharan. "Detection and Discrimination of PICOSATS by Radars: An Example," presented at IEEE Core Technologies for Space Systems Conference, Colorado Springs, CO, November 2001.

16. Woellert, K., P. Ehrenfreund, A. J. Ricco, and H. Hertaveld. 2011. CubeSats: Cost-Effective Science and Technology Platforms for Emerging and Developing Nations. *Advances in Space Research* 47 (4): 663–684.

17. *Wikipedia,* "Minotaur I," https://en.wikipedia.org/wiki/Minotaur_I, last modified 31 January 2016.

18. Orbital Sciences Corporation. "Minotaur," https://www.orbitalatk.com/flight -systems/space-launch-vehicles/minotaur/, 2016.

19. Curtis, H. *Orbital Mechanics for Engineering Students*, p. 301, Elsevier Aerospace Engineering Series, Oxford: Elsevier, 2014.

20. R. Sridharan and A. Pensa, "Search Strategies for Deep Space Launches," MIT Lincoln Laboratory Technical Note 1977-21, 15 June 1977.

21. Schildknecht, T., R. Musci, and T. Flohrer. 2008. Properties of the High Area-to-Mass Ratio Space Debris Population at High Altitudes. *Advances in Space Research* 41 (7): 1039–1045.

22. Zollinger, G. R., and R. Sridharan. 2000. A Fragmentation Event Detection System. *Acta Astronautica* 46 (7): 477–489.

23. R. Sridharan, W. P. Seniw, and A. Freed, "Correlation Techniques for Deep Space Uncorrelated Targets," MIT Lincoln Laboratory Technical Note 1979-24, 13 March 1979.

7

Characterization of Resident Space Objects

Ramaswamy Sridharan and Richard Lambour

The need to characterize resident space objects (RSOs) goes back to the earliest days of space surveillance, to 1957 with the launch of Sputnik 1. Characterization at that time implied the discrimination of an active payload from the rocket body and other objects associated with a new launch; this was essential for early tracking and cataloging of a payload so that intelligence assets could be devoted to monitoring it. Over the years, RSO characterization requirements have expanded significantly to include assessments of orbital location, mission, functional status, provenance, size, shape, surface properties, attitude stability, spin period, axis orientation, and any short- and long-term behavior of any of the monitored RSO attributes. Although the primary consumers of such information are intelligence agencies, the information is also extremely useful to the operators of the Space Surveillance Network (SSN), who must select and schedule surveillance sensors (as discussed in section 2.1 of chapter 2).

Generally, information on an RSO mission is available for cooperative launches, but even in the case of noncooperative launches, an RSO mission can often be identified by heritage, that is, by similarity in orbit and structure to a previous RSO whose mission is known, primarily because satellite builders historically have been conservative.[1] And the status of an RSO mission can often be monitored by identifying whether the RSO is on location and in an attitude that permits the performance of its mission.

Although analysis of data collected by wideband radar imaging sensors is, without a doubt, the best method for characterizing resident space objects (as discussed in

---

1. Such conservatism does not, however, apply to the space surveillance experimental community—DARPA (Defense Advanced Research Projects Agency), AFRL (Air Force Research Laboratory), or the CubeSat experimenters.

chapter 3), analysis of metric and radiometric data from narrowband radar sensors and low-resolution optical sensors more commonly available in the Space Surveillance Network is used extensively for limited, but tactically useful RSO characterization. Section 7.1 illustrates the variety of characterization information that can be derived by using narrowband radars. Section 7.2 briefly discusses the use of broadband optical sensors for RSO characterization. Section 7.3 describes an interesting experiment in data fusion from radar and optical sensors for detailed characterization of a class of space-debris objects in low Earth orbit (LEO).

7.1   Narrowband Radar Characterization of Deep-Space Resident Space Objects

Rather than develop a theoretical construct (as in a Bayesian approach [1, 2]), this section will describe real-world examples that demonstrate the utility of several techniques for deep-space RSO characterization.

7.1.1   Example 1: Estimation of Size, Shape, and Attitude Variation of a Deep-Space Resident Space Object

Figures 7.1 and 7.2 illustrate sample displays at the Millstone Hill Radar for coherent and noncoherent tracking. The top panels of the figures show the radar cross-section (RCS) signature (temporal variation) in both principal and orthogonal polarizations. Note that the signal-to-noise ratio (SNR) is shown in dB and the RCS in dBsm along the y-axis. The x-axis shows time in seconds past a fiducial mark.

The measured radar cross section of a resident space object has an approximate relationship to its size [3, 4], but it is also dependent on the transmitter frequency of the radar; the measured polarization ratio has an approximate relationship to the RSO's shape, but with significant caveats. The repetitive variation of the RSO's radar cross section over an extended period of time implies an attitude motion or spin about an axis whose period can be determined by autocorrelation and used in further analysis.

Figures 7.1 and 7.2 illustrate the real-time assessment of these metrics, though they obviously need to be confirmed by extended offline analyses of the data. The tracking of the resident space object in figure 7.1 is coherent; the object's large radar cross section (lower panels of figure) implies that it is sizable, which is typical of communications satellites having large antennas, and attitude stable, whereas the apparently periodic variation in its RCS (top panel of figure) implies that there is probably a slow variation in the RSO's attitude, a precession, which is quite normal

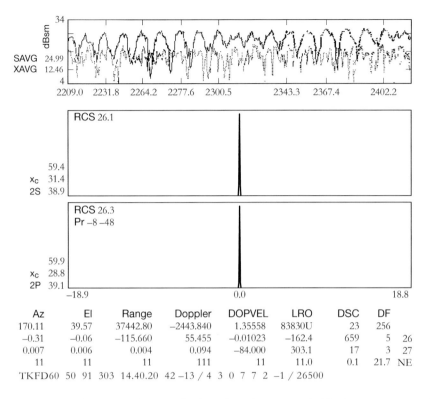

Az	El	Range	Doppler	DOPVEL	LRO	DSC	DF	
170.11	39.57	37442.80	−2443.840	1.35558	83830U	23	256	
−0.31	−0.06	−115.660	55.455	−0.01023	−162.4	659	5	26
0.007	0.006	0.004	0.094	−84.000	303.1	17	3	27
11	11	11	111	11	11.0	0.1	21.7	NE

TKFD60 50 91 303 14.40.20 42 −13 / 4 3 0 7 7 2 −1 / 26500

Figure 7.1 Millstone Hill radar display of a coherent track. The bottom two graphs (polarimetric and sum channels—see chapter 2) illustrate the integrated signal (in dB along y-axis) versus frequency offset from estimated Doppler along the x-axis. Coherence of the target can be inferred from the signal energy being concentrated in one or a very few spectral bins. The top graph presents the temporal variation of radar cross section in the principal (solid line) and orthogonal (dotted) polarization channels, with the y-axis in dBsm and the x-axis in time in seconds relative to an epoch. Even though the RCS has a periodic variation, the target remains coherent, as seen in the bottom graphs, because the period is at least an order of magnitude larger than the integration interval. Az, azimuth; EL, elevation; RANGE, DOPPLER, range to target and Doppler of target; Pr, Polarization ratio; RCS, measured radar cross-section of target; all other acronyms on the figure are internal variables used in processing and display and not relevant to the text

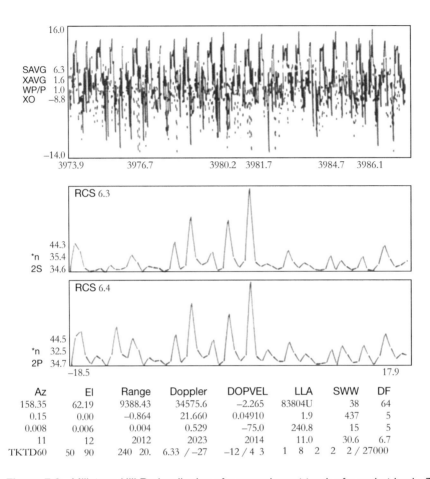

Az	El	Range	Doppler	DOPVEL	LLA	SWW	DF
158.35	62.19	9388.43	34575.6	−2.265	83804U	38	64
0.15	0.00	−0.864	21.660	0.04910	1.9	437	5
0.008	0.006	0.004	0.529	−75.0	240.8	15	5
11	12	2012	2023	2014	11.0	30.6	6.7
TKTD60	50 90	240 20.	6.33 / −27	−12 / 4 3	1 8 2	2 2 / 27000	

Figure 7.2 Millstone Hill Radar display of a noncoherent track of a rocket body. The top panel represents the temporal variation in the RCS in dBsm in the principal (*solid line*) and orthogonal (*dotted line*) polarizations along the *y*-axis and the time of track along the *x*-axis; the bottom two graphs represent the spread spectrum of the received energy with the *y*-axis being the signal in dB and the *x*-axis being the frequency offset from expected Doppler.

in all stabilized objects in space (e.g., the Earth). Such behavior is common with spin-stabilized cylindrical satellites and, occasionally, with three-axis-stabilized satellites having extended solar panels. The tracking of the RSO in figure 7.2 is noncoherent; the object is suspected to be a rocket body spinning about its axis of maximum inertia, which is common after a rocket body's end of thrust and separation from its satellite. A size estimate can be inferred from the radar cross section, and a spin period can be inferred from the periodic specular peaks in the RCS, although, of course, a detailed analysis with recorded data is necessary to confirm these. Clearly, considerable information on an RSO's size, attitude, and attitude variation, and some information on its shape, can be derived from the RSO's measured RCS signature, as this example illustrates.

7.1.2   Example 2: Status of a Satellite Derived from Its Radar Cross-Section Signature

Telstar 401 is a classic example of an active satellite in geostationary orbit that failed unexpectedly. Although more will be said in chapter 8 on the conjunction and collision avoidance activity that was spawned by this failure, here we note one of many methods to recognize a satellite's failure using radar data.

Figure 7.3 captures Telstar 401's radar cross-section signature as measured by the Millstone Hill radar before and after failure of the satellite. Telstar 401 was a communications satellite with large solar arrays pointed at the Sun and an antenna pointed at

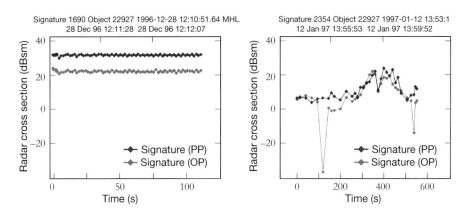

Figure 7.3   Millstone Hill radar cross-section signature of Telstar 401 (SCN 22927) before (20 December 1996) and after (12 January 1997) failure of the satellite in orbit. MHL, Marshall Islands Local Time; OP, orthogonal polarization; PP, principal polarization.

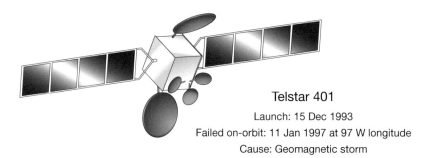

Telstar 401

Launch: 15 Dec 1993
Failed on-orbit: 11 Jan 1997 at 97 W longitude
Cause: Geomagnetic storm

Figure 7.4   Artist's sketch of Telstar 401 in its deployed configuration in orbit.

the Earth (see figure 7.4). When the satellite was active and controlled in attitude (prior to failure), its radar cross section was nearly invariant over short periods of time (minutes), as seen in the left panel of figure 7.3. When the satellite failed, however, it lost attitude control and started tumbling [5, 6]. This led to a temporally varying RCS (on short timescales), as seen in the right panel of figure 7.3. Thus monitoring the RCS signature on satellites has been an effective technique for assessing their status and detecting major changes. This capability is useful particularly in the case of satellites for which owner information is lacking, and it has been used extensively at both the Millstone Hill radar and ALTAIR for this purpose.

A different method of monitoring the status of Telstar 401, using the satellite's longitudinal location in the geostationary belt over several years, is shown in figure 7.5. Active, functioning geostationary satellites tend to be held in a narrow-longitude box assigned by the International Telecommunications Union (ITU), as Telstar 401 was in the period between January 1996 and January 1997, after which the satellite started an uncontrolled oscillation, as also shown in figure 7.5, over a wide-longitude region—an unlikely behavior if the satellite were functioning normally.

7.1.3   Example 3: Status of a Satellite Derived from Solar Panel Speculars in Its Radar Cross-Section Signature

Serendipity occasionally played a part in the development of techniques during the early days of deep-space radar surveillance. The detection of "speculars," that is, large spikes in a satellite's radar cross section over a short period of time due to changes in the orientation of its solar panels with respect to the radar-to-satellite line of sight, is one such serendipitous example; detecting these speculars led to a technique for assessing the operational status of satellites in geostationary orbit [7].

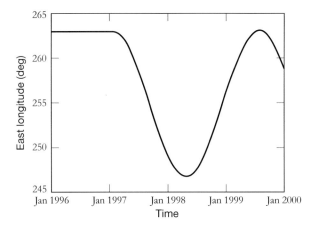

Figure 7.5 Variation in the east longitude location of Telstar 401. Note the abrupt change in behavior in January 1997.

Figure 7.6 illustrates the typical orbital deployment of a geostationary satellite. It is common for a three-axis-stabilized satellite to orient its solar panels in a plane perpendicular to the plane of its orbit (in this case, the Earth's equatorial plane). The panels rotate around their long axis (parallel to the Earth's polar axis) to follow the Sun in order to maximize the solar illumination received, and the electrical power generated from it.

Figure 7.7 illustrates the "anomalous" behavior of the radar cross section signature of the RCA-A communications satellite, as measured by the Millstone Hill radar. This behavior was deemed to be anomalous because the measured RCS of communications satellites is normally invariant with time over short time periods (on the order of minutes). The RCA-A's enhanced peak radar cross section of 25 dBsm, as compared to a normal 12 dBsm, and its well-behaved lobe seemed to indicate that a flat panel was being rotated around the radar-to-satellite line of sight.

The repetition of this anomalous RCS signature on successive days and the fact that the radar-satellite line of sight was oriented approximately toward the Sun during those days conclusively proved that the satellite's radar cross section was being substantially enhanced by the flat back of its solar panel (the front was pointing toward the Sun). This effect was verified by examining the RCS signatures of a number of geostationary satellites. Note that, on more recently deployed geostationary satellites, because the Earth-pointing parabolic antenna may dominate the measured RCS, the enhancement due to the solar panels may be barely noticeable.

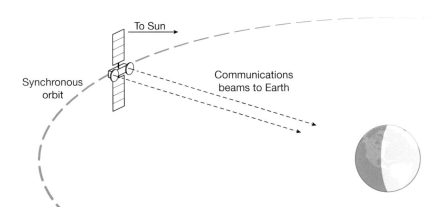

Figure 7.6   Typical configuration of three-axis-stabilized satellite in geosynchronous orbit.

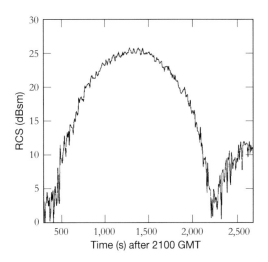

Figure 7.7   "Anomalous" behavior of measured radar cross section of RCA-A satellite on day 250, 1980.

Three significant observations emerged from this investigation:

1. The lobe-like behavior of the measured radar cross section for three-axis-stabilized geostationary satellites is repetitive from day to day, and the times of its occurrence can be predicted.
2. The existence of this behavior distinguishes such satellites from the class of geostationary cylindrical spin-stabilized satellites, which do not have a flat-panel structure.
3. The presence of this behavior, if it can be measured at predicted times, is an indicator that the satellite is operationally active, and because the proper illumination of the solar panels is essential to generate the power needed to operate the satellite, its absence indicates a possible failure of the satellite.

Occasionally, two lobes have been seen in the RCS signature of a communications satellite, which may indicate that the north and south solar panels are offset with respect to each other. An example is shown in figure 7.8.

The solar panel "specular" lobe effect was predicted and demonstrated at radars operating at about 420 MHz (UHF) frequency (ALTAIR, Pirinclik), as well as at about 1,300 MHz (L-band) frequency (Millstone Hill radar). Analysts at ALTAIR and the Millstone Hill radar have used this effect for discrimination and status assessment

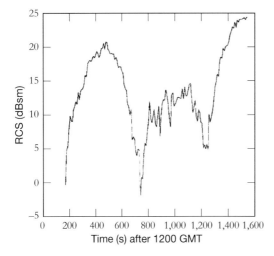

Figure 7.8 Radar cross-section "specular" behavior of a geosynchronous satellite showing two lobes.

very effectively over the years [8]. The effect has also been used to discriminate between a satellite and its rocket body soon after their injection into geosynchronous
orbit—the rocket body lacks solar panels and does not exhibit the RCS behavior
described here. The same "specular" lobing effect can be seen on satellites in other
orbits—Global Positioning System (GPS) satellites in medium Earth orbits and
satellites in low Earth orbits—when the necessary radar-satellite-Sun geometry is
visible to a given ground sensor.

GPS satellites are in circular orbits of 55 degrees inclination with a period of about
12 hours. They form a constellation such that, for any point on the Earth, at least
four active satellites are visible for the purpose of navigation. Figure 7.9 shows the approximate orbital configuration of a GPS satellite with the solar panels pointed toward
the Sun and the antennas pointed toward the Earth. Given the satellite's orbit, it is
feasible to calculate the time of occurrence of near alignment of the radar line of sight
with the satellite-Sun vector. Data were collected on the GPS-6 satellite on three successive days over the same time interval in 1989; the specular features in RCS are
shown in figure 7.10 [8].

The Russian Molniya communications satellites have highly eccentric orbits of 63
degrees inclination with a period of about 12 hours, perigee altitudes of about 600
kilometers, and apogee altitudes of about 36,000 kilometers. Since these satellites also
have solar panels, it was feasible to calculate the time of occurrence of their solar-
panel speculars and to measure them at ALTAIR [8]. Figure 7.11 displays the results
on three days in 1989. The orbital configuration of a Molniya satellite, with multiple

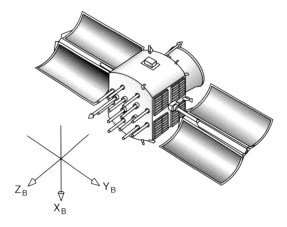

Figure 7.9   Drawing of Block 1 GPS satellite.

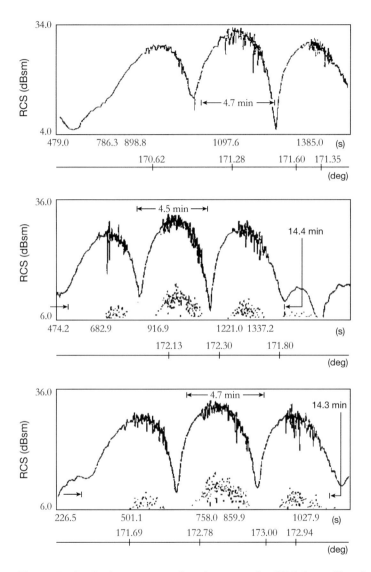

Figure 7.10   Radar cross-section signatures for GPS-6 satellite observed on three successive days in 1989.

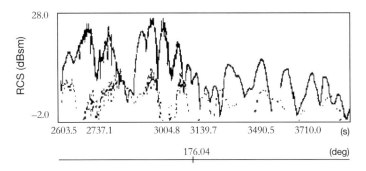

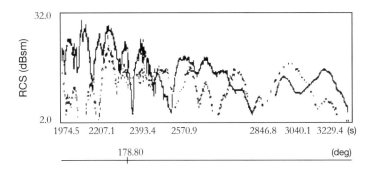

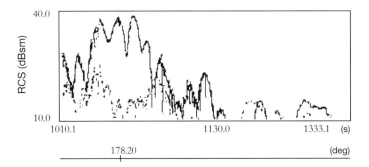

Figure 7.11 Radar cross-section signatures of the Molniya 1-70 satellite observed on 28 February, 3 March, and 7 March 1989.

solar panels, is more complex than that of a standard geostationary satellite, as shown in figure 7.12. It is suspected, though not verified, that the radar cross-section signature of a Molniya satellite is a result of sequential occurrence of lobes on several solar panels.

Finally, figure 7.13 displays the RCS peaks due to lobing on another geostationary satellite—Raduga 18—as measured at ALTAIR [8]. The occurrence of the peaks 12 hours apart demonstrates the lobing phenomenon.

Thus the active status of satellites in deep-space orbits can be ascertained by monitoring the proper orientation of their solar panels, that is, whether they are pointed toward the Sun. The absence of the speculars would indicate one of the following two possibilities:

1. The solar panels are not generating power because they are not pointed at the Sun, hence the satellite is inactive
2. The satellite does not have extended solar panels, hence cannot be monitored in this manner.

7.1.4 Example 4: Status of a Satellite Derived from Attitude Control in Its Radar Cross-Section Signature

There is a class of satellites whose attitude (hence functioning) is apparently actively controlled from orbit to orbit—a feature that has been inferred from radar signature

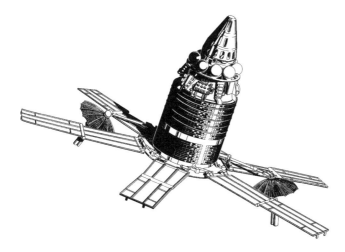

Figure 7.12 Illustration of a Molniya 1 satellite (reprinted by permission of Aerospace Corporation).

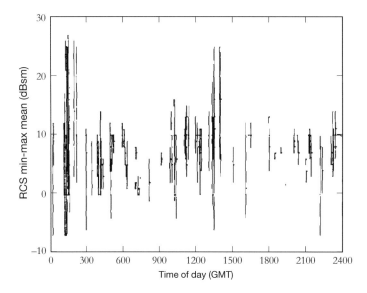

Figure 7.13   Radar cross-section behavior versus time of day over several days for Raduga 18 satellite. Note specular signatures, 12 hours apart, from front and back of solar panels.

data. Assessing the status of this class thus requires predicting the times of the changes in attitude and monitoring the satellite's radar cross section at those times. An example of such monitoring using the Millstone Hill radar is given in figure 7.14. The satellite is attitude stable early in the track but appears attitude unstable to the narrowband radar past about 1755 GMT. Unlike the case of a satellite's loss of attitude control, as demonstrated in an earlier example, this pattern is repetitive from orbit to orbit, implying that the satellite is under active attitude control to perform its mission.

7.1.5   Example 5: Characterization of a Satellite Malfunction with Narrowband Radar

A particularly interesting example of the application of several techniques of narrowband radar characterization in support of the analysis and recovery of a malfunctioning satellite occurred in 1981. The summary of the effort below is drawn from Seniw and Spence [9].

FLTSATCOM V was a communications satellite launched on 6 August 1981 from the Kennedy Space Center in Florida. The fully deployed configuration in its intended

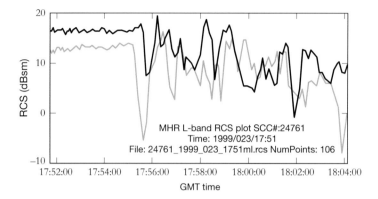

Figure 7.14   Track of a deep-space satellite at the Millstone Hill radar (MHR) showing the transition in the attitude of the satellite as inferred from the change in its radar signature. ml.rcs, maximum likelihood radar cross section.

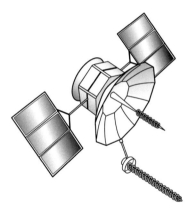

Figure 7.15   Drawing of planned operational configuration of FLTSATCOM 5 satellite in geostationary orbit.

geostationary orbit is shown in figure 7.15, with the solar panels and the UHF transmit and receive antennas deployed and the antennas' axis (z-axis) pointing toward the center of the Earth to provide Earth coverage. In the high-eccentricity transfer orbit and in the early portion of the geosynchronous orbit, the solar panels were folded around the satellite's hexagonal body, the UHF parabolic transmit antenna was folded inward to form an inverted cone around the z-axis, and the UHF receive antenna was folded over the cone in sections.

The launch took place at 0816 GMT on 6 August 1981; and by that afternoon, Millstone Hill radar was informed of a satellite anomaly and was requested by the Air Force to support the satellite owner with tracking and analysis. The radar tracked the satellite through the rocket motor thrust at apogee from its transfer orbit into a super-geosynchronous orbit, where the satellite drifted westward in sublongitude by 30 degrees per day. Monitoring the satellite for any indication of anomalous behavior and rapidly providing orbital updates to the satellite's owners, Millstone continued to observe it during its many maneuvers until it was stopped at 93 degrees west sublongitude in a near-geostationary orbit.

The Millstone Hill radar used many techniques, along with metric support, to assess the anomalies in the configuration of the satellite over the next 40 days. The first question was whether the solar panels were intact and stowed around the hexagonal sides. The peak radar cross section of the hexagonal sides was calculated to be 640 square meters; that of each of the six solar panel sections, to be 3,070 square meters. These calculations were based on the radar line of sight being perpendicular to each side and each section and assuming that the side of the satellite and the solar panel section reflected the radar energy like metallic flat plates. Although the RCS difference was large enough for discrimination, it depended on the line of sight being near perpendicular to the sides of the satellite.

The satellite was spinning on its z-axis at about 60 revolutions per minute; its spin-axis alignment in inertial space was obtained from its owners. Thus it was possible to calculate a favorable geometry for the "glints" off the sides of the satellite when a near-peak radar cross section was detectable. Figure 7.16 shows the spin direction and the nutation (or wobble) of the satellite, as determined by the owners, and the consequent scattering geometry.

The Millstone Hill radar operated with a 1-millisecond pulsewidth at approximately 30 pulses per second. The satellite was spinning at about 60 revolutions per minute with a nutation angle (θ) of about 10 degrees. Complex geometrical calculations were thus involved in determining possible times for glints to occur. Further consideration of the geometry led to the expectation that, in the near-perpendicular position, there should be one glint every spin period or two glints one-half or one-sixth of a spin period apart. In addition, the fact that glints were observed at all the predicted times confirmed the direction of the spin axis, its orientation, and θ as the nutation angle.

Figure 7.17a illustrates the case in which the glints were observed one-half of a spin period apart, and figure 7.17b shows the case one hour later, when the glints were one-sixth of a spin period apart. In both cases, the glints were detected by collecting RCS

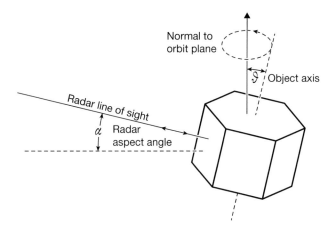

Figure 7.16   Scattering geometry of the sides of the satellite.

data pulse by pulse (with the pulse repetition frequency being an integer multiple of the satellite's spin frequency) and noncoherently averaging the data over 100 spin periods.

Note that, in figure 7.17, the measured radar cross section is shown on the y-axes in square meters rather than in the conventional decibels per square meter (dBsm). The radar cross sections shown on the figures are undoubtedly lower than the actual radar cross sections because the averaging of the 100 periods of radar returns could not be precisely aligned with the satellite's dynamics, to say nothing of what the actual radar and satellite geometry, hence radar cross section, were at the time of the observation. As such, the measurements could be consistent with scattering from either the sides of the satellite or the solar panels stowed against its sides. Given the observation of the expected solar panel signature discussed next, however, and the fact that the glint signatures were as predicted and that they evolved smoothly without additional components, it is likely that the scattering was from the stowed panels.

The solar panels were subsequently deployed into an operational configuration. Again the radar cross section of the satellite was observed at a geometry where the radar, the satellite, and the Sun were nearly collinear (see subsection 7.1.3), as shown in figure 7.18 (note that the RCS on the figure's y-axis is in dBsm: 10 dBsm =10 square meters, 20 dBsm = 100 square meters, 30 dBsm = 1,000 square meters). The figure shows that the motion of the solar panels traced out the scattering pattern expected for a cut along a flat plate, with the RCS being consistent with that of the solar panels. This pattern is additional evidence of the solar panels having been intact on the sides of the satellite and subsequently deployed.

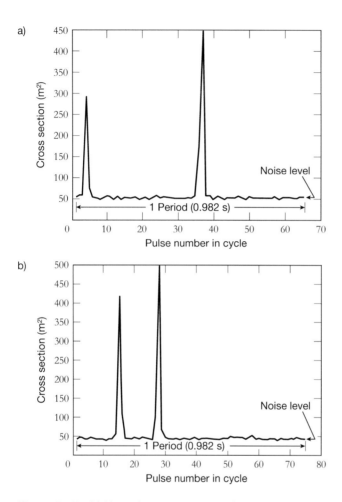

Figure 7.17   (a) Noncoherent averages of 100 periods of radar returns, 9 August 1981. (b) Noncoherent averages of 100 periods of radar returns, 9 August 1981, one hour after those shown in figure 7.17a.

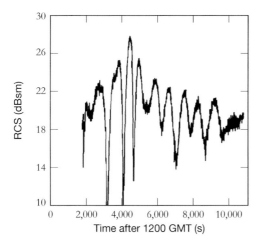

Figure 7.18   FLTSATCOM 5 solar panel radar cross section in near-geostationary orbit.

On 11 September 1981, the owners of FLTSATCOM V attempted to deploy the sections of the parabolic UHF antenna. The Millstone Hill radar observed this maneuver but did not detect a significant increase in the satellite's radar cross section, which would have occurred if the antenna had unfurled and pointed at the Earth. The RCS measurements are shown in figure 7.19.

The variations in its radar cross section are normal for a satellite in a 6-degree-inclination orbit with some attitude motion. In particular, there is no significant change in RCS at or after the time the unfurl command was issued. Thus the analysts at Millstone concluded that the unfurl command failed.

A small piece of space debris was detected by the FPS-85 radar and tracked by the Millstone Hill radar in the transfer orbit of FLTSATCOM V. Metric assessment indicated that this piece was probably related to the satellite in its initial transfer orbit. Assessment of the radar cross-section signature using orthogonal polarization data indicated a significant linear aspect to this piece, which was small, with a small RCS, as shown in figure 7.20. The provenance of the piece, which apparently broke off from the satellite during its first orbit, was determined by the satellite's manufacturers using Millstone Hill radar analysis, imaging data from the Haystack wideband radar, and satellite manufacturing data.

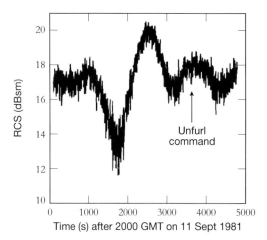

Figure 7.19 Radar measurements of FLTSATCOM 5 at the time of the attempted unfurling of the parabolic UHF antenna, 11 September 1981.

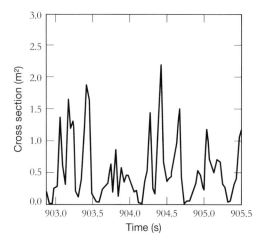

Figure 7.20 Two cycles of radar cross-section measurement of the space-debris piece in transfer orbit. Note that RCS on y-axis is in square meters.

7.1.6   Example 6: Detection and Discrimination of Closely Spaced Satellites in Geostationary Orbit

Because they are very popular for a variety of applications, geostationary orbits are, by international agreement, controlled by the International Telecommunications Union (ITU), which allocates specific orbital "slots' or locations to satellites intended to assume a geostationary orbit. There is substantial crowding of desirable slots—locations that provide good connectivity to countries and continents, and locations that extend the lifetimes of satellites by minimizing the number of maneuvers they need to make to stay within their assigned slots. The crowding results in clusters of closely spaced satellites in geostationary orbit. Table 7.1 is a partial list of such clusters (as of July 2013).

A recurrent problem for the radar and optical sensors used to collect metric and radiometric data on satellites in clusters is cross tagging of the data, particularly in clusters that contain similar satellites, like those owned by ASTRA (Atmospheric & Space Technology Research Associates). The following discriminants are used for resolving the cross tagging confusion.

1. *Location*: Information on relative east–west locations of satellites can often be obtained from the owners/operators of the clustered satellites and can be used with both optical and radar sensors for proper tagging of data. This technique is particularly useful when the cluster consists of just a few satellites. The example in figure 7.21 is from data collected by the space-based visible sensor.

2. *Orbit*: Clusters of satellites are built up one satellite at a time over months or years. Thus precise and regular maintenance of data on the orbits of the clustered satellites will enable discrimination and proper data tagging. This technique tends to be useful with radars that are able to detect and track satellites one at a time, unlike the optical sensors with large fields of view, such as the space-based visible sensor (SBV; see chapter 5), or the GEODSS sensors (see chapter 4), both of which detect the entire cluster simultaneously. This method also requires the availability of precise orbital elements (rather than the publicly available, analytically computed imprecise orbits). Hence this method is used post facto to check and correct the tagging of orbital data at an orbit-processing facility like the Joint Space Operations Center (JSpOC) of Air Force Space Command at Vandenberg Air Force Base in California or the Lincoln Space Situational Awareness Center (LSSAC) in Lexington, Massachusetts, or by satellite owners. Examples are the ASTRA clusters, whose orbital data are

Table 7.1 Partial list of geostationary satellite clusters

Satellite number	Satellite name	East longitude (deg)
28238	DTV 7S	240.95
36499	EchoStar 14	241.10
27378	EchoStar 7	241.20
31102	Anik F3	241.30
26761	XM-1 ("Rock")	244.75
26724	XM-2 ("Roll")	244.75
29520	XM-4	244.72
37843	ViaSat 1	244.90
25558	Satmex 5	245.10
23313	Solidaridad II	245.33
37748	SES-3	256.90
24315	AMC-1	257.00
28644	SPACEWAY 1	257.10
31862	DTV 10	257.20
36131	DTV 12	257.23
25004	EchoStar 3	298.20
39008	EchoStar 16	298.50
27852	Rainbow 1	298.65
39078	Amazonas 3	299.00
35942	Amazonas 2	299.00
26494	Astra 2B	19.35
26853	Astra 2C	19.2
29055	Astra 1KR	19.2
31306	Astra 1L	19.2
33436	Astra 1M	19.2
26638	Astra 2D	28.0
38778	Astra 2F	28.2
25462	Astra 2A	28.2
37775	Astra 1N	28.2

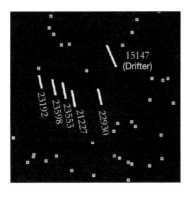 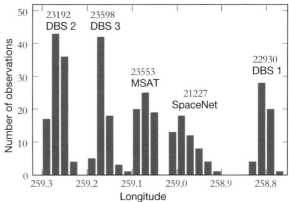

Figure 7.21   Data from Space-Based Visible (SBV) sensor showing longitudinal discrimination of geosynchronous satellites in a cluster. DBS, Direct Broadcast Satellite; MSAT, Mobile Satellite.

maintained by SES, a Luxembourg-based company providing satellite communications and broadcast services (see table 7.1).

3. *Radiometric data*: Clusters sometimes consist of nonidentical satellites whose measured radar cross sections or visual magnitudes show consistent differences. Maintaining a history of these measurements enables near-real-time discrimination and proper tagging of data using both radar and optical sensors. Such discrimination is demonstrated in the case of DSP 20 (Defense Support Program satellite 20), launched on 9 May 2000 (described in subsection 6.1.5.1 of chapter 6).

4. *Polarization*: Even identical, clustered satellites often have slight differences in the pointing of their solar panels or Earth-facing antennas. These differences result in consistently differing polarization ratios (see subsection 2.1.1.2.2 of chapter 2) as measured by radars and can be used to identify the satellites and tag the data correctly.

Obviously, a combination of these techniques is very useful. It is desirable for sensor operators to unambiguously tag the data rather than leave the task to the central processing facility, particularly because information on east–west location, radar cross section or visual magnitude, and polarization is readily available to the sensor operators in near real time.

7.1.7   Characterization of Satellites with Wideband Radars

Characterizing satellites from images generated by wideband radars is generally the best means of identifying their form and function. The process of such characterization has been extensively covered in chapter 3, hence is not discussed here.

7.2   Electro-Optical Characterization of Deep-Space Resident Space Objects

During the time period covered by this book (ca. 1970–2000), only passive optical sensors, operating at visible wavelengths, were developed and fielded for deep-space surveillance (by various laboratories or by Air Force Space Command). These sensors are described in chapters 4 and 5. Passive optical characterization of a resident space object differs from radar characterization in some fundamental ways:

1. Illumination of the target RSO is from the Sun, hence the measurement at the optical receiver (telescope) is always bistatic.

2. The illumination from the Sun is inherently broadband and noncoherent (i.e., phase uncontrolled), whereas the illumination from a radar is precisely controlled in both frequency and phase.

3. Solar energy is depolarized, and the polarization at the optical receiver is due to the surface characteristic of the RSO and the properties of the intervening atmosphere, whereas the transmitted energy of radar is controlled in polarization.

4. Optical receivers used in deep-space surveillance are broadband and do not resolve the spectral content of the received signal, whereas radar receivers can split the received signal spectrally over a narrow bandwidth.

5. Ground-based optical receivers can generally operate only at night when the Sun does not directly, or indirectly through atmospheric scattering, illuminate and interfere with the receivers, whereas radar receivers can operate 24 hours a day. Space-based optical receivers are somewhat less limited than their ground-based counterparts since there is no atmospheric scattering, and hence the sky background is always dark except if the receiver is pointed towards the direction of the Sun or the Earth.

The parameters that can be used for optical characterization of resident space objects are metric data, derived element sets, measured visual magnitude and its temporal signature in both principal and orthogonal polarizations, polarization ratio, and spectral content of the signature. Unfortunately, the last parameter is rarely available for deep-space RSOs except for very bright objects, primarily because of the lack of sensitivity of the detectors. Hence the passive optical sensors fielded for deep-space surveillance yield only metric data, bistatic visual magnitude, and temporal signatures. No polarization or spectral data are available. Thus the optical data are quite limited compared to radar data in their ability to characterize deep-space RSOs. Passive optical characterization is illustrated in chapters 4 and 5, with examples for both ground-based and space-based electro-optical systems using the optical signature and visual magnitude (see figures 4.19, 4.20, 4.21 in chapter 4; and the discussion on SBV photometry and figure 5.18 in chapter 5).

7.3   Characterization of Radar Ocean Reconnaissance Satellite Debris by Analysis and Fusion of Radar and Optical Data

The problem of resident space object characterization is, in many instances, amenable to and appropriate for data fusion using multiple phenomenologies. One such instance, described in this section, demonstrated the potential of combining radar and optical measurements with metric data and theoretical analysis to perform remote characterization of a specific population of space-debris objects in low Earth orbit.

The population in question represented space-debris objects from the Soviet RORSATs (Radar Ocean Reconnaissance Satellites). These objects were detected by the NASA Goldstone radar in California in 1989 [10]. The radar was used to observe the space-debris environment at a range of altitudes as a proof of concept for detecting small (2–5 millimeters in size) space-debris objects in low Earth orbit. It detected "swarms" of debris objects in periods of time during which the flux of debris through the radar beam increased over the average by a factor of from 6 to 7. Further observations and analysis located six such swarms and determined that they lay in the same, near-circular orbit; the objects were thus associated with one another.

The Goldstone radar observations occurred when planning for the construction of the International Space Station (ISS) was in full swing; knowledge of the space-debris environment in low Earth orbit was vital to the safety of the station and her crew. At that time, NASA's knowledge of that environment was split over two size regimes: objects larger than about 10 centimeters in diameter, which could be tracked by the Space Surveillance Network's ground-based radars; and objects smaller than 0.1 centimeter in diameter, which could be studied by analyzing impacts on returned surfaces such as those of the NASA Long Duration Exposure Facility, the European Retrievable Carrier, and the Space Shuttle. The lack of information on space-debris objects from 0.1 to 10 centimeters in size, however, prompted NASA to sponsor measurements of such objects by Lincoln Laboratory's Haystack radar in Westford, Massachusetts (see section 2.2 of chapter 2).

The data from the Haystack radar were provided to NASA for analysis to estimate statistics on the number of detections of space-debris objects, their orbit regimes, and their sizes (as characterized by the recorded radar cross sections) [11]. An example of the results is shown in figure 7.22 (ca. 1995), which illustrates the rate of object detections when the Haystack radar beam is pointed at zenith, as a function of the altitude at which the detections were made. The lower curve is a simulation of the detection rate that was expected on the basis of the number of objects in the resident space object catalog maintained by the U.S. Air Force Space Surveillance Center in 1994. The upper curve is the rate of detections observed in the processed Haystack radar data. Figure 7.22 illustrates that most of the detections of debris objects are concentrated at altitudes from 800 to 1,000 kilometers, with a smaller detection peak located at an altitude of about 1,400 kilometers. NASA combined these data with previous data sets from Goldstone radar and other information to draw the following conclusions [10, 11]:

1. Most of the debris objects were small (smaller than 2 centimeters); this conclusion was based on the observed radar cross sections of the objects.

2. The debris objects were mostly spherical in shape and were concentrated at altitudes of about 900 to 1,000 kilometers in circular orbits with orbital inclinations of about 65 degrees.

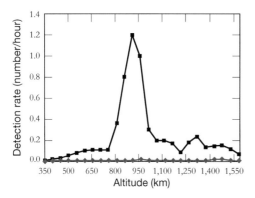

Figure 7.22 Comparison of space-debris detections rates observed by Haystack radar (*upper black line*) and in resident space object (RSO) catalog (*lower red line*) as a function of altitude.

3. There were 50,000–70,000 debris objects with a diameter greater than 8 millimeters at these altitudes.

4. The detected debris objects were most likely drops of liquid sodium-potassium (NaK) coolant that leaked from the coolant loops of the Bouk class of nuclear reactors used on Soviet RORSATs.

The conclusion that the coolant had leaked from the Soviet satellites was based on the known orbits of the parent objects (RORSATs) and on knowledge of the design of and disposal techniques for their nuclear reactors. When a RORSAT reached the end of its operational lifetime, its reactor was separated from the satellite and boosted into a disposal orbit. Thirty of these debris objects were known to exist in circular orbits at about 65 degrees inclination at altitudes of from 850 to 1,000 kilometers. When the reactor separated, the coolant loops were apparently severed, at which time the coolant was presumed to have leaked out. NASA therefore requested that Lincoln Laboratory detect, track, and characterize in detail a small sample set of these debris objects by using both radar and optical sensors. The results of the study, conducted between 1995 and 2000, showed that a representative set of the objects, termed "anomalous debris," conformed to the NASA hypothesis [12, 13].

7.3.1   Detection Strategies for Anomalous Space-Debris Objects

The characteristics of the radar and optical sensors used to detect anomalous space-debris objects are summarized in tables 7.2 and 7.3. Before any data could be collected with the sensors, however, candidate objects from the anomalous debris population had to be identified. Three strategies were used by the sensors to detect and acquire anomalous debris objects [14, 15]:

Table 7.2   Characteristics of radars used in this study

Radar	Band	Frequency	Instantaneous field of view (mrad)	Signal-to-noise ratio on 1 m² target at 1,000 km
Haystack (Westford, MA)	X	10 GHz	1	58 dB
Millstone Hill (Westford, MA)	L	1.3 GHz	8	50 dB
TRADEX (Kwajalein Atoll)	L	1.3 GHz	8	48 dB

Table 7.3 Characteristics of optical sensors used in this study

Telescope	Aperture	Sensor	Instantaneous field of view	Sensitivity (apparent magnitude)
Firepond (Westford, MA)	1.2 m	CCD photo-polarimeter	0.06 deg	~13 m_v
Experimental Test Site (ETS; Socorro, NM)	0.75 m	Vidicon Camera	1–2 deg	~15.5 m_v

1. Search the orbital planes of presumed parent objects;
2. Conduct stare-and-chase searches of low Earth orbit space with optical and radar systems;
3. Search the orbital planes of previously detected anomalous debris objects.

The results of each method are discussed below.

7.3.1.1 Search of Orbit Planes of Presumed Parent Objects

The Millstone Hill radar was used to conduct searches of the orbits of ten presumed parent objects. Due to the small size of the debris objects, the search was conducted by observing a low-range, high-elevation point in each orbit to improve the detection sensitivity of the radar, which could then detect an object about 2 centimeters in size using coherent integration or one about 3.5 centimeters in size using noncoherent integration. The searches were centered about the location of the parent objects (which are relatively large and easily tracked). No debris objects were detected with this search mode. This result suggested that, because any recent slow-velocity leak would have resulted in a relatively small separation between the parent and debris objects, none of the parent objects had released any debris objects larger than about 2 centimeters within the past 30 days or so.

7.3.1.2 Stare-and-Chase Searches

The two 75-centimeter telescopes at Lincoln Laboratory's Experimental Test Site (ETS) in Socorro, New Mexico (see chapter 4), were used to conduct stare-and-chase searches of low Earth orbit space for orbital debris in 1994 under an Air Force program [14]. The telescopes were equipped with Vidicon sensors at that time and could detect objects as faint as about 15.5 magnitude, that is, objects of about 2 centimeters in size at a range of 1,000 kilometers with reflectivity of 20 percent under good lighting conditions (solar phase angle less than 40 degrees)—or even smaller objects having higher surface reflectivity. While conducting several stare-and-chase searches that year, the ETS systems detected and characterized one "anomalous debris" object [14].

TRADEX L-Band radar located on Kwajalein Atoll also conducted stare-and-chase searches in 1995 [15]. The radar was pointed at a specific azimuth and (high) elevation point and interrogated altitudes from 500 to 1,200 kilometers. Within this range, TRADEX could detect any object larger than 3.5 centimeters; and, indeed, it detected four additional anomalous debris objects of that size in the volume searched.

7.3.1.3   Search of Orbit Planes of Debris Objects

If all the anomalous debris objects related to a particular parent object had been produced at or close to the same time (as NASA suspected), then it seemed reasonable to conclude that the orbital plane of any detected anomalous debris object would contain many other such objects. Therefore the orbital planes of the debris objects detected by the ETS and TRADEX radars were searched with the Millstone L-Band radar, which could detect objects larger than 2 centimeters in those orbital planes. The along-orbital-plane searches resulted in the detection of six additional debris objects. The Haystack radar, which could detect objects larger than 1 centimeter at the ranges searched, was also used to perform along-orbital-plane searches; it detected eight additional debris objects, but because their smaller size made them more difficult to characterize with the available sensors than the eleven objects discovered by other means, they were not studied extensively.

The orbital parameters of the eleven debris objects found as a result of these searches are summarized in table 7.4. The object numbers assigned them were local and temporary and did not represent actual object numbers from the Space Surveillance Network catalog. From table 7.4, it is evident that all eleven debris objects reside in nearly circular orbits with inclinations near 65 degrees and at altitudes of about 950 kilometers. Also of key interest is that the existence of multiple debris objects in each orbital plane implies a common parentage and time of origin. Both properties are consistent with those of drops of a leaking liquid.

After discovery of the eleven anomalous debris objects, characterization of their size, shape, and surface material properties was begun with the radar and optical sensors listed in tables 7.2 and 7.3. In addition, thermal modeling was applied to analyze whether the liquid drops were likely to evaporate or remain in orbit for an extended

Table 7.4   Orbital parameters for anomalous space-debris objects

Object number	Inclination (deg)	Eccentricity	Right ascension of ascending node (deg)	Semimajor axis in units of Earth radius
81215	64.96	0.004	304.38	1.146
33562	65.04	0.004	77.291	1.146
33609	64.96	0.005	203.35	1.146
33612	64.97	0.005	187.93	1.146
33616	64.69	0.006	207.65	1.147
39969	64.69	0.006	222.31	1.147
39970	64.65	0.006	207.86	1.147
39971	64.96	0.004	300.13	1.146
39972	64.97	0.005	202.81	1.146
39973	65.05	0.004	87.32	1.146
39974	65.06	0.004	152.82	1.145

period of time. Each of these characterization efforts is described below and serves as an illustration of how different types of data can be used to determine debris object properties.

7.3.2   Characterization of Anomalous Space-Debris Objects with Radar Data

The Millstone and Haystack radars were both used to track and characterize the eleven anomalous debris objects listed in table 7.4. The data collected include the following:

1. Radar cross-section signatures (RCS as a function of time), which were used to estimate the objects' size and to characterize their temporal variability, spin periods, or both;

2. Polarization data, specifically the ratios of the orthogonal polarization (OP) RCS to the principal polarization (PP) RCS, which were used to assess the general shape of the objects;

3. Metric data (range, range rate), which were used to determine the precise orbits of the debris objects and to calculate their area-to-mass ratios and hence their mass and density.

7.3.2.1   Radar Signature and Polarization Analysis

A number of Millstone Hill radar tracks, which typically lasted about 10 minutes, were collected on each debris object (table 7.5). Figure 7.23 shows principal and orthogonal polarization radar cross sections (PP and OP RCS; measured in dBsm units against time in seconds) during a single track of objects 81215 and 33562. Neither track showed any variation in the PP RCS at periods longer than the coherent integration interval (about 1–3 seconds) for either object. The OP RCS was smaller than the PP RCS by about 25 dB and was dominated by noise. As can be seen in table 7.5, the many tracks collected on the eleven different anomalous debris objects all show similar behavior.

The behavior shown in figure 7.23 is atypical for most space-debris objects, however. The Millstone Hill radar has collected many hundreds of tracks on debris objects that do not belong to this anomalous population. Analysis of the tracks [16] showed that the PP RCS/OP RCS ratio was typically just a few dB and often varied significantly over the course of the track. Figure 7.24 illustrates this point by presenting RCS data as a function of time for four other debris objects.

Although the observed radar cross-section signatures for anomalous debris objects were unusual for orbital debris objects, they were not unprecedented. The anomalous

Table 7.5   Millstone Hill radar cross-section and polarization results for anomalous space-debris objects

Object number	Number of tracks	PP RCS (dBsm)	Polarization ratio (dB)	Radius (cm)	Sigma (cm)
81215	51	−20.8	−24.2	2.84	0.09
33562	45	−23.6	−21.6	2.55	0.06
33609	30	−22.4	−23.3	2.68	0.08
33612	23	−22.1	−25.3	2.71	0.05
33616	25	−30.7	−25.1	1.94	0.09
39969	12	−34.3	−21.2	1.70	0.08
39970	10	−32.3	−25.2	1.83	0.02
39971	4	−21.9	−24.5	2.73	0.03
39972	5	−22.1	−24.0	2.70	0.02
39973	3	−25.4	−21.2	2.38	—

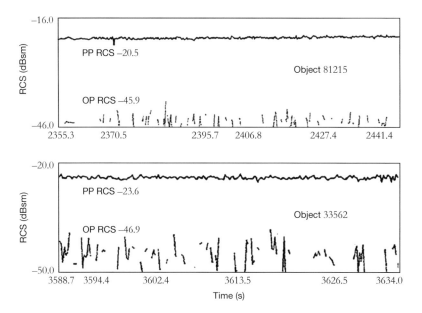

Figure 7.23   Radar cross-section signatures of anomalous space-debris objects.

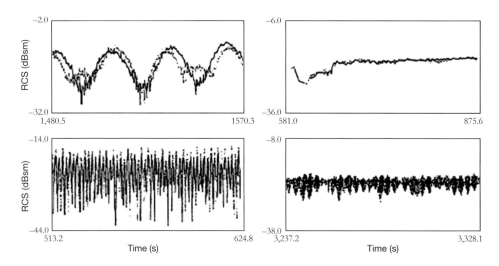

Figure 7.24   Radar cross-section signatures of typical "nonanomalous" space-debris objects.

debris objects display the same RCS behavior (invariant PP RCS and very low OP RCS) as the calibration spheres LCS-1 and LCS-4, which are 1-meter-diameter aluminum spheres in circular orbits at altitudes of 2,800 and 800 kilometers, respectively. The inference from this comparison is that the anomalous debris objects were spherical in shape.

Furthermore, the amplitude of the PP RCS allows estimation of the size of a debris object. Table 7.5 also presents the mean PP RCS, the mean polarization ratio (OP RCS divided by PP RCS), the estimated radius, and the uncertainly in the radius estimate for each of the eleven anomalous debris objects. Because the PP RCS value was consistently small enough that the target size was well within the Rayleigh scattering region of the Millstone Hill radar, size estimation could proceed without ambiguity [17]. The OP RCS of these debris objects was consistently at or near the noise level. Since we assumed the debris objects were perfectly conducting, we represented the relation between the debris object's RCS and radius as

$$\frac{\sigma}{\pi a^2} = 9\left(\frac{2\pi a}{\lambda}\right)^4,\tag{7.1}$$

where a was the radius, λ was the wavelength of the radar (23 centimeters for Millstone Hill radar), and σ was the PP RCS value [17]. The uncertainties observed were based on the variability of the mean RCS in the principal polarization channel. The RCS estimate at the Millstone Hill radar was accurate to less than 1 dBsm, based on calibration with frequent tracks of LCS-4. The same calibration uncertainty should apply to the small spheres in question because the radar attempted to track all objects at approximately the same signal-to-noise ratio (30–36 dB). The high SNR was achievable in the case of the small spheres because of the high sensitivity of and the low slant range from the radar.

7.3.3   Density Estimation

The Earth's atmosphere extends beyond the altitudes of 800–1,000 kilometers at which the anomalous debris objects resided. With accurate metric data over many orbits, it was therefore feasible to estimate an area-to-mass (A/M) ratio for the objects using the effects of atmospheric drag on them. Since the debris objects were spherical, the A/M ratio did not depend on aspect angle. The area and volume of a debris object could be estimated by using the calculated size from table 7.5, leading to estimates of its mass and density.

We represented the force on the debris object from atmospheric drag as

$$\bar{F}_{\text{drag}} = -\frac{1}{2} C_d \left(\frac{A}{M} \right) \rho V_z^2 , \tag{7.2}$$

where C_d was the ballistic coefficient, A/M was the ratio of the projected surface area to mass, ρ was the density of the atmosphere at the altitude of the debris, and V_z was the velocity of the debris object relative to the atmosphere. We assumed that C_d, the shape of the debris object, ρ, and V_s were known, and A/M was to be determined using the tracking data. The thermospheric density was provided by the MSIS83 thermosphere model [18], and the Earth's atmosphere was assumed to be corotating with the planet to facilitate computation of V_z.

The thermosphere models were known to have errors ranging from 15 to 30 percent for use in drag analysis [19, 20, 21]. These errors were corrected for by introducing a scale factor, S_o, for the thermospheric density so that $\rho_{\text{true}} = S_o \rho_{\text{model}}$. To determine S_o, contemporaneous tracking data on LCS-4, which was at a similar altitude, were utilized. LCS-4 was a well-characterized sphere, which facilitated accurate calculation of its ballistic coefficient (about 2.2 for a sphere) and the drag force. A value for S_o was determined from the tracking data and was then used to correct the thermospheric density.

Values of A/M for the anomalous debris objects were calculated with the aid of a precision orbit determination code called "DYNAMO" (see subsection 2.1.3.2 of chapter 2). In calculating the drag force, DYNAMO introduced a multiplicative drag scale factor S_d (not to be confused with S_o) into equation 7.2. S_d was solved for during the tracking data fit and was representative of the error in the thermosphere model. If S_d was greater or less than unity, the thermosphere model density would be too small or too large. For each of the anomalous debris objects, an initial value was assumed for A/M: $(A/M)_{\text{initial}} = 0.1$ square centimeter per gram. S_d was then calculated for intervals with sufficient tracking data. The scale factor derived from contemporaneous LCS-4 tracking data was introduced, and the drag force was expressed as

$$\bar{F}_{\text{drag}} = -\frac{1}{2} S_d \left(\frac{A}{M} \right)_{\text{initial}} C_d \frac{\rho_{\text{true}}}{S_o} V_z^2 . \tag{7.3}$$

The true value of A/M was then estimated from

$$\frac{A}{M} = \frac{S_d}{S_o} \left(\frac{A}{M} \right)_{\text{initial}} = 0.1 \frac{S_d}{S_o} . \tag{7.4}$$

The mass and density of the debris object then follow. Table 7.6 presents the results of this calculation for nine of the eleven anomalous debris objects (two did not have

Table 7.6   Mass and density estimates for anomalous space-debris objects from Millstone Hill radar data

Object number	Area-to-mass ratio (cm²/g)	Number of tracks	Mass (g)	Density (g/cm³)
81215	0.253 ± 0.007	31	100.2 ± 9.1	1.044 ± 0.06
33562	0.273 ± 0.066	6	74.8 ± 9.6	1.077 ± 0.11
33609	0.225 ± 0.011	7	100.2 ± 10.8	1.244 ± 0.10
33612	0.256 ± 0.006	5	90.1 ± 5.4	1.031 ± 0.04
33616	0.349 ± 0.009	6	33.8 ± 4.0	1.108 ± 0.08
39969	0.464 ± 0.22	3	19.6 ± 3.0	0.953 ± 0.09
39970	0.545 ± 0.043	2	19.3 ± 2.1	0.752 ± 0.07
39971	0.247	1	94.8	1.112
39972	0.307	1	74.6	0.905

sufficient tracking data). The mean density of the debris objects was 1.03 ± 0.05 grams per cubic centimeter. The uncertainties were computed from the variation in the individual calculations.

Don J. Kessler and colleagues [10] suggested that eutectic sodium-potassium (NaK) alloy[2] was probably used as the coolant in the nuclear reactors of the Soviet RORSATs. Figure 7.25 shows the variation of density of the eutectic sodium-potassium alloy with temperature [22]. It is clear that the expected density of eutectic sodium-potassium alloy at 300° K was about 0.87 gram per cubic centimeter, similar to the derived mean density of the anomalous debris objects.

Several caveats were warranted. In the 800-kilometer-altitude regime, the errors in the determination of C_d were on the order of ± 10 percent. In addition, the theory behind the calculation of C_d depends on the scattering mechanism, and there were no measurements of the scattering of atmospheric constituents off liquids. For example, if the liquid particles did not scatter, but instead were absorbed, C_d would be 1.0, which would increase A/M by 2.2 and decrease the anomalous debris objects' mass and density. Finally, the uncertainty in determining the radius of a debris object from

---

2. A eutectic alloy is a homogeneous solid mix of atomic or chemical species with a specific atomic or molecular ratio such that the alloy melts *as a whole* at a specific (eutectic) temperature. The eutectic temperature is the lowest possible melting temperature over all of the mixing ratios of the component atomic or chemical species. See https://en.wikipedia.org/wiki/Eutectic_system (last modified on 31 July 2016).

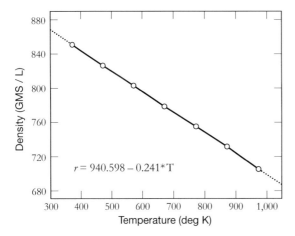

Figure 7.25   Density of eutectic sodium-potassium alloy as a function of temperature [22].

its radar cross section also affected the calculation. The estimated A/M ratios, mass, and densities are presented in table 7.6.

The results from the radar characterization of the anomalous debris objects are summarized as follows:

1. The objects exhibited, without exception, a polarization ratio at all aspect angles indicative of a spherical shape;
2. The debris objects' radii ranged from 1.6 to 2.8 centimeters; and
3. The objects had a mean density near 1 gram per cubic centimeter, consistent with the density expected for eutectic sodium-potassium alloy at temperatures of about 300° K.

7.3.4   Characterization of Anomalous Space-Debris Objects with Optical Data

Some of the physical properties of orbital debris objects can be determined by using optical sensors (telescopes) with appropriate instrumentation and examining the sunlight reflected off the objects. The gradual change of the Sun–debris object–observer angle (phase angle) during any given orbital pass provides an opportunity to characterize a wide variety of solar reflection geometries. Analysis of the phase curves generated from the high-precision photometric and polarimetric data collected during these orbital passes (brightness or percent polarization against solar phase angle) allows analysts to estimate the size of the objects and to determine whether their surfaces were metallic or dielectric. Each of these topics will be discussed below.

The telescopes at the Experimental Test Site were used together with those at the Firepond facility, collocated with the Millstone and Haystack radars in Westford, Massachusetts. The characteristics of the two telescope systems are summarized in table 7.3. Since the work at the test site involved only one of the anomalous debris objects (number 81215), the overall analysis of optical data concentrated on the Firepond measurements taken on seven of the anomalous debris objects.

Although the Firepond telescope had a small field of view (0.06 degree), which usually makes acquiring and tracking anomalous debris objects challenging, a real-time link allowed the Millstone and Haystack radars to guide the pointing of the telescope, permitting data collection over a wide range of phase angles. Firepond employed a combination charge-coupled device photometer/polarimeter mounted at the focal plane of its telescope. Three elements were in the optical path, the first of which was a rotating filter wheel containing a set of standard UVBRI (ultraviolet, blue, red, and infrared) astronomical filters. The V (visible light) filter (about 480–700 nanometers) was used for these data collections. The second element was a half-wave plate made of a birefringent material, which introduced a 90-degree phase difference in the ordinary and extraordinary wave components of the light passed through it so that the components emerged orthogonally polarized. The third element was a Savart plate, which spatially separated the two orthogonal polarizations, producing two images on the charge-coupled device. In practice, sunlight reflected by orbiting objects exhibits partial linear polarization in the plane perpendicular to the plane of incidence containing the Sun, observer, and object. Proper rotation of the half-wave plate during an observation caused the two images to correspond to the polarizations in the plane of incidence and perpendicular to that plane. The measured intensities in the two polarizations allowed analysts to determine the percentage polarization of the reflected sunlight from the brightness of the two images:

$$\%\,\text{polarization} = 100\,\frac{(I_1 - I_2)}{(I_1 + I_2)}, \tag{7.5}$$

where I_1 and I_2 are the measured intensities in the two planes, whose sum is the photometric brightness of the image, $I = (I_1 + I_2)$. The photopolarimeter measurements were calibrated by observing photometric and polarimetric standard stars during each observation session. The observed counts were then transformed to V-band magnitudes, and the resulting magnitudes were normalized to a constant range of 1,000 kilometers using the ephemeris information of the object tracked.

7.3.4.1   Photometric Measurements

The predicted brightness of a Sun-illuminated object is given by

$$m = V_{\text{Sun}} - 2.5\log\left(\frac{\rho A \Phi(\varphi)}{R^2}\right) \tag{7.6}$$

where m is the object magnitude, V_{Sun} is the V-band magnitude of the Sun (-26.74), ρ is the reflectivity (or albedo) of the object, A is the cross-sectional area of the object, R is the range to the object, and $\Phi(\phi)$ is the phase function of the object [16]. The phase function describes the directional properties of the reflected light (similar to gain for an antenna). From the photometric phase curves for two of the anomalous debris objects displayed in figure 7.26, it can be clearly seen that the phase functions were fairly flat over all phase angles, which is typical for the seven anomalous debris objects, and quite atypical of other (nonanomalous) previously cataloged debris objects observed by Firepond. Some noise in the data is due to relatively low count rates; the total throughput of the photopolarimeter was relatively low. A flat photometric phase curve is characteristic of a specularly reflecting sphere; its phase function is $\Phi = (4\pi)^{-1}$. Figure 7.27, presented for comparison, shows the photometric phase curve for an anomalous debris object and for LCS-4, a known specular sphere. The phase curves are similar in shape.

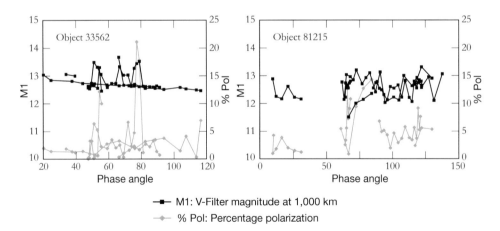

Figure 7.26   Photometric phase curves and polarimetric phase curves for two anomalous space-debris objects.

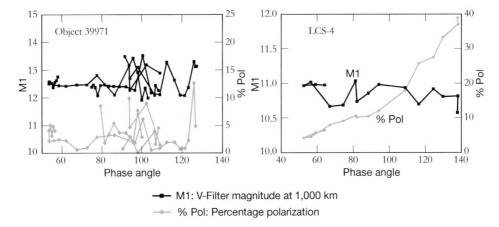

Figure 7.27   Phase curves and polarization phase curves for anomalous space-debris object 39971 (left) and the radar calibration sphere LCS-4 (right).

7.3.4.2   Polarimetric Measurements

The polarization measurements, made concomitantly with the photometric measurements described in subsection 7.3.4.1, were used to determine the reflectivity of the anomalous debris objects. Some of the reflected light exhibits partial linear polarization in the plane perpendicular to the Sun–debris object–observer plane by an amount that depends on the optical indices of the debris object's surface material. A polarimetric phase curve was therefore constructed and fit to the Fresnel equations for reflection to obtain a pair of optical indices, (n, k). It was assumed that the angles of incidence and reflection were equal to half the observed phase angle. Figure 7.27 shows examples of the polarimetric phase curves for anomalous debris object 39971 and for LCS-4. The anomalous debris objects typically show very low polarization at all phase angles (with some variations due to low count rates or variable air mass), which is characteristic of a metallic surface. The normal reflectivity, r, of the debris object may be computed from the following equation [23]:

$$r = \frac{(n-1)^2 + k^2}{(n+1)^2 + k^2} \,. \tag{7.7}$$

The values of (n, k) derived from the fits to the phase curve are not necessarily unique, but all values of (n, k) that do fit the measurements produce similar values of r (less than 1 percent scatter). A comparison of polarimetric data taken on the anomalous debris objects with polarimetric phase curves for liquid metals known to be used as nuclear reactor coolants by the Soviet Union shows that sodium, potassium, and lithium are the only metals whose polarization properties are similar to those of the anomalous debris objects [24]. We were unable to locate any measurements of the indices of refraction of solid or liquid sodium-potassium in the literature that would have allowed a direct comparison.

Once the phase function and reflectivity of the anomalous debris objects were estimated, equation 7.6 was inverted to obtain the cross-sectional area, A, and then the radius of the objects. The phase function for a specular sphere was used to perform this calculation. The results of this analysis are presented in table 7.7. The radii in table 7.7 are consistent with those calculated from the radar data of the principal polarization RCS (table 7.5) for individual debris objects.

The optical data collected on these anomalous debris objects provided a second distinct method for estimating their size (active versus passive illumination). A comparison of the sizes derived from radar (table 7.5) and optical (table 7.7) observations showed consistent estimates of the size from the two methods for the seven debris objects on which there were adequate photometric and polarimetric data. Moreover, the objects' photometric phase curves showed their brightness to be relatively constant with changing phase angle; this property is unusual for most space-debris objects and is exhibited only by specular spheres. The optical polarization was consistently low for the debris objects observed, suggesting that they have metallic surfaces. All of these inferred properties are consistent with the radar data analysis and with the hypothesis of leaked sodium-potassium coolant.

Table 7.7   Size estimates for anomalous space-debris objects from optical measurements

Object number	Magnitude	Reflectivity	Radius (cm)
81215	12.6 ± 0.4	0.89	2.9 ± 0.5
33562	12.8 ± 0.3	0.89	2.6 ± 0.4
33609	12.5 ± 0.2	0.84	3.0 ± 0.3
39969	13.6 ± 0.08	0.89	1.8 ± 0.1
39970	13.5 ± 0.2	0.89	1.8 ± 0.5
39971	12.6 ± 0.4	0.89	2.9 ± 0.5
39972	12.7 ± 0.3	0.84	2.9 ± 0.4

7.3.5   Properties of Anomalous Space-Debris Objects Inferred by Theoretical Modeling

A third approach for characterizing the anomalous space-debris objects was to apply what was learned by remote sensing in an attempt to derive additional useful information through analysis. In this case, the question of greatest interest was, given that there were a large number of these debris objects on orbit, how long would they remain there? Because all of the data collected were consistent with the leaking liquid sodium-potassium coolant hypothesis, if the objects were still in a liquid state on orbit, could their orbital lifetimes be influenced by evaporation?

To determine whether this was the case, both a thermal and an evaporation model were developed for the anomalous debris objects. Both models were relatively low fidelity and meant to provide a bound on the possibility of evaporative influence on orbital lifetime. The thermal model addressed the question of whether the objects were likely to be liquid or solid (frozen) in the orbital environment since this would greatly influence their evaporative properties. The freezing point of a eutectic sodium-potassium mixture is 260.85° K, which is not far below the expected on-orbit equilibrium temperature (about 300° K) for the anomalous debris objects [22, 25].

7.3.5.1   Time-Dependent Thermal Model

To compute the temperature of the anomalous debris objects in low Earth orbit, a simple, time-dependent thermal model was used. Following the principle of conservation of energy, the rate of change in temperature of a single debris object was written as

$$\frac{dT}{dt} = \frac{(E_{\text{absorbed}} - E_{\text{radiated}})}{Cm}, \tag{7.8}$$

where the numerator represented the difference between the energy absorbed by the object and the energy radiated by the object, and the denominator (referred to as "thermal mass") represented the product of the specific heat, C, and the mass, m. Given an initial temperature, this differential equation could be solved numerically for the thermal history of the object.

The absorbed energy originates from multiple sources, the most important of which are the Sun, Earth-reflected sunlight, and terrestrial long-wave radiation ($E_{\text{absorbed}} = E_S + E_{\text{reflec}} + E_{\text{therm}}$). The absorbed solar energy can be written as

$$E_S = F_S \alpha A_p. \tag{7.9}$$

Because the anomalous debris objects were assumed to be spherical their projected area was circular, $A_p = \pi r^2$. The absorptivity, α, was assumed not to vary with wavelength; F_S was the incident solar flux (1,360 watts per square meter). Since the anomalous debris objects were in low Earth orbit, eclipsing took place regularly. During eclipsing periods, the solar energy input was set to zero.

Earth-reflected sunlight can be described by an equation of the same form as equation 7.9. The energy flux of reflected sunlight depends on the altitude of the debris object, the albedo of the Earth, and the illuminated surface area of the Earth "visible" to the object. This last quantity is related to the direction cosine of the position vector of the object, the solar position vector, and the altitude of the object. The geometry is illustrated in figure 7.28. The quantities presented in figure 7.29 were digitized and used as a lookup table to estimate Earth-reflected solar flux [26]. This lookup table can be scaled to any globally averaged albedo.

The terrestrial long-wave radiation input was represented by

$$E_{\text{therm}} = F_{\text{therm}} \alpha A_p . \tag{7.10}$$

The thermal energy flux, F_{therm}, was taken to be 238 watts per square meter at the surface of the Earth. This value is a global average and assumes that the Earth is a diffuse radiator [26, 27]. To determine the radiated energy of the Earth, it was assumed that the debris object radiated as a graybody. The energy it radiated to free space could then be written as

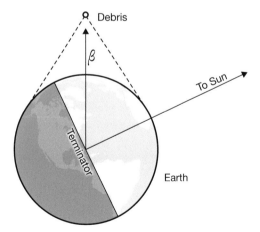

Figure 7.28   Surface area of the Earth visible to a debris object at altitude h and the geometrical relationship between the object and solar position vectors.

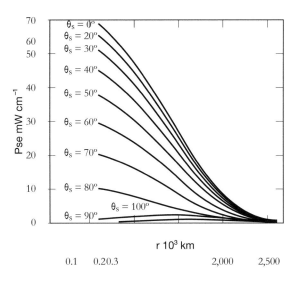

Figure 7.29   Earth-reflected solar irradiance for a spherical satellite as a function of altitude, where θs is the angle between the satellite and solar position vectors. Pse, Earth-reflected solar irradiance; r, altitude [26].

$$E_{\text{radiated}} = \sigma \varepsilon T^4 A_s .\tag{7.11}$$

The object's surface area was $4\pi r^2$. Emissivity, ε, was assumed not to vary with wavelength. The radiated energy from the Earth incident on the object was affected by the proximity of the warm Earth. The proximity reduced the area of the sphere radiating to free space by a factor of Ωr^2, where $\Omega = 2\pi(1 - \cos(\beta))$. At the altitude of the sphere, h, the Earth subtended a half angle,

$$\beta = \sin^{-1}\left(\frac{r}{r+h}\right).\tag{7.12}$$

The anomalous debris objects were assumed to consist of eutectic sodium-potassium and to be spherical in shape, as suggested by the radar and optical measurements. The specific heat and density of eutectic sodium-potassium were determined by using data from standard engineering tables. The data were fit to functions in order to extrapolate the quantities down to the freezing point of eutectic sodium-potassium (260.85° K). The specific heat was represented by

$$C(T) = 1112.19 - 0.608T + 0.000384T^2 .\tag{7.13}$$

And the density was represented by

$$\rho(T) = 940.6 - 0.241T .\qquad(7.14)$$

As the temperature of the anomalous debris drop varied in the simulation, these quantities were allowed to change. The mass was assumed to remain constant.

The optical properties of sodium-potassium were not well known. Unpublished work by Russian researchers indicates that the absorptivity and emissivity coefficients of the material are believed to be roughly 0.156 and 0.132, respectively. The value for the absorptivity of the material is reasonable, based on the radar and optical characterization work carried out at Lincoln Laboratory, the optical work of Russian researchers, and previously published work [25, 28]. Earlier, the anomalous debris objects were shown to be metallic and highly reflective, with low optical polarization ratios. Because these data indicate that the objects were unlikely to be transparent, the reflectivity and absorptivity coefficients should sum to 1.0. The reflectivity values of 0.84–0.89 then lend credibility to the absorptivity value of 0.156. Although no such "verification" exists for the emissivity value, the emissivity values for liquid sodium and liquid potassium are generally low at the temperatures expected on orbit. The normal emissivity of liquid sodium is on the order of 0.01–0.05 at about 300° K [29, 30]; that of liquid potassium is about 0.01 at about 300° K [30]. These values are significantly lower than the value reported for sodium-potassium (0.132). Nevertheless, the values $\alpha = 0.156$ and $\varepsilon = 0.132$ were assumed to apply to sodium-potassium.

7.3.5.2   Predicted Thermal Behavior of the Anomalous Debris

For the purpose of predicting the temporal history of the temperature of an anomalous debris object, the initial temperature of the object was arbitrarily chosen to be 285° K at an arbitrary initial time. A simple integration was performed for several orbits. At each time step, the instantaneous debris position vector in orbit was rotated into the Earth-centered reference frame; both the energy input from the Sun and the Earth and the energy loss due to surface emissivity were calculated. The net energy input to the debris object and the object's thermal properties were used to calculate the temperature change during the integration time step. The temperature was adjusted, and the integration continued. Other initial temperatures from about 250° K to about 320° K were checked—they had no effect on the conclusions reached.

Estimated thermal history curves are shown in figure 7.30 for the eleven anomalous debris objects in table 7.4. Individual thermal histories are difficult to discern, but the envelope of the temperature variations is evident. Because none of the anomalous

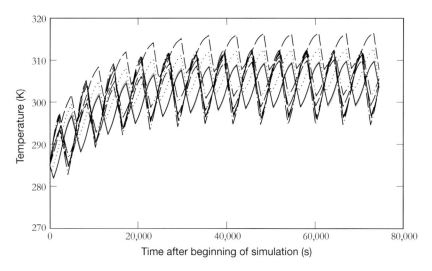

Figure 7.30   Predicted thermal behavior of eleven anomalous debris objects in their orbits.

debris object temperatures falls below 280 °K, the objects are likely to be *liquid* in orbit.

The thermal model described in the previous paragraphs has limitations. The albedo of the Earth was assumed to be a globally averaged constant, with a value of 0.34. In reality, the albedo of the Earth is dependent on the reflective nature of the surface and the amount of cloud cover. The surface albedo is, in turn, dependent on the physical nature of the surface; water, snow, vegetation, and desert all reflect light differently. Cloud albedo is also dependent on the type and thickness of the cloud cover. A second issue related to albedo is that, for low-altitude objects, the debris object temperature is dependent on the average albedo of the Earth scene that is visible to it, not the albedo of the entire globe. This number can vary from about 0.25 to about 0.70, according to the literature. To address both of these issues, a test was run with the Earth albedo set to 0.0 to determine how much the debris object temperature decreased. The largest temperature decrease observed was about 10° K, which resulted in temperatures that remained above about 280° K. Although the model was not high fidelity, the calculation was adequate to demonstrate that the temperature in orbit of the anomalous debris objects stayed far enough above the freezing point of eutectic sodium-potassium that the objects were likely to be liquid.

Finally, the computed debris object temperatures depended *strongly* on the assumed absorptivity and emissivity values. If these values were inaccurate, the conclusion that the debris objects were liquid could be incorrect. Based on the available literature on the subject, however, reasonable estimates have been made of the absorptivity and emissivity values for sodium-potassium, according to which the sodium-potassium debris objects are expected to remain in a liquid state. The evaporative lifetime of the anomalous debris objects could therefore be estimated.

7.3.6 Evaporative Lifetime of Anomalous Debris Objects and Consequences for Orbit Evolution

The evaporation of the anomalous debris objects and the timescale over which evaporation occurs can have important implications for their orbital lifetimes. The atmospheric drag force on the objects depends on the ratio of projected area to mass. Loss of mass due to evaporation would lead to a change in the A/M ratio and thus in the drag force on the objects.

The flux of evaporating particles was estimated from the general principles of kinetic theory. Particles in the liquid with kinetic energies greater than the work necessary to move them from their initial position to infinity will escape into the surrounding space. This "work function" for the liquid can also be thought of as the latent heat of evaporation for a single particle. A minimum "escape velocity" for the particle can be derived from $mv^2/2 = U_o$, where U_o is the latent heat of evaporation [31]. The escaping flux is then estimated by calculating the first moment of the liquid's velocity distribution function. Because, however, neither a value for the latent heat of evaporation for the sodium-potassium liquid nor a velocity distribution function for the sodium-potassium drop was available, the assumption was made that the rate of evaporation, W, was equal to the rate at which the vapor particles would strike the liquid if the liquid and the vapor were in equilibrium, the maximum rate of evaporation [31]:

$$W(\text{kg}/\sec) = \kappa A_s P \sqrt{\frac{M}{2\pi RT}} . \tag{7.15}$$

And because no measurements of the vapor pressure of the eutectic sodium-potassium system at the low temperatures experienced on orbit were available in the literature, an evaporation theory that applied to binary alloys was adopted. The evaporation rate, modified to account for the two different components, was expressed as

$$W_i(\text{kg}/\sec) = \kappa_i A_s P_i^o \gamma_i(x) N_i \sqrt{\frac{M_i}{2\pi RT}} , \tag{7.16}$$

where i refers to a specific component—sodium (Na) or potassium (K) in this case—κ is the evaporation coefficient (assumed to be 1), A_s is the surface area of the object, P^p is the vapor pressure for a pure element at temperature, T, γ is the activity coefficient, x is the mass of the evaporated element, N is the mole fraction, M_i is the molar weight, and R is the gas constant (8.31 Joules/mole-°K). The mole fractions were expressed as

$$N_{Na} = \frac{\left[\dfrac{(a-x)}{M_{Na}}\right]}{\left[\dfrac{(a-x)}{M_{Na}} + \dfrac{(b-y)}{M_K}\right]} \quad (7.17a)$$

$$N_K = \frac{\left[\dfrac{(b-y)}{M_K}\right]}{\left[\dfrac{(a-x)}{M_{Na}} + \dfrac{(b-y)}{M_K}\right]}. \quad (7.17b)$$

where a and b are the initial masses of the two elements in the drop, and x and y are the current masses of those element in the drop.

Note that, as time passes and more of each element evaporates from the drop, the mole fractions change (as do the atomic fractions of sodium and potassium and the activity coefficient, γ). Determination of the timescale for evaporation of the drops required a numerical solution of the following coupled differential equations:

$$\frac{dx}{dt} = W_{Na}(x,y), \quad (7.18a)$$

$$\frac{dy}{dt} = W_K(x,y). \quad (7.18b)$$

To calculate evaporation rates, the vapor pressures of sodium, P_{Na}, and potassium, P_K, and the activity coefficients must be specified. The vapor pressures of sodium and potassium at low pressures and temperatures ranging from their normal melting points (371° K for sodium; 337° K for potassium) to about 463° K are given by

$$\log P_{Na} = 8.08 - \frac{5479}{T}, \quad (7.19a)$$

$$\log P_K = 7.56 - \frac{4587}{T}, \quad (7.19b)$$

Table 7.8 Activity coefficients for liquid sodium-potassium at 298.15° K

Atomic fraction (sodium)	γ_{Na} (±0.06)	γ_K (±0.06)
0.0	3.00	1.00
0.1	2.43	1.01
0.2	2.02	1.05
0.3	1.71	1.11
0.4	1.49	1.19
0.5	1.29	1.29
0.6	1.19	1.49
0.7	1.11	1.71
0.8	1.05	2.02
0.9	1.01	2.43
1.0	1.00	3.00

where vapor pressure, P, is in torrs and temperature, T, is in degrees Kelvin [32]. The activity coefficients are given in table 7.8.

Drops with masses of 1 and 57 grams, representing the extremes of the masses measured for the eleven anomalous debris objects in table 7.6, were chosen for the simulation of evaporative lifetimes. To facilitate the calculation of evaporation rates, the drops were assumed to maintain an average temperature over an orbit, and three temperatures were considered: 300°, 330°, and 280° K. These temperatures correspond to those of debris objects in orbits that are eclipsed, orbits that are not eclipsed, and an arbitrary low temperature, respectively.

For eutectic sodium-potassium, the drops were 77.8 percent potassium by mass; therefore the 57-gram (1-gram) drop initially contained 44.346 grams (0.778 gram) of potassium and 12.654 grams (0.222 gram) of sodium. The initial atomic fraction of sodium in the drops was 0.3267 (32.67 percent of the atoms in the drops were sodium atoms); this was used to interpolate the initial activity coefficients for the drops. The initial vapor pressures were calculated using equations 7.19a and 7.19b. We assumed that the vapor pressures remained constant as the elements evaporated and the composition of the drops remained uniform. Note also that the vapor pressure formulas were being applied at temperatures below which they are deemed applicable because the sodium-potassium drop was still liquid at these temperatures. Given these initial conditions and assumptions, equations 7.18a and 7.18b were integrated using a time step of 86,400 seconds (24 hours). The initial evaporation rates were so small

(4.03348×10^{-12} kilogram per square meter per second for sodium; 4.21491×10^{-9} kilogram per square meter per second for potassium), that the large time step was acceptable. The simulation was stopped if a phase transition occurred because of the changing composition of the drop.

Figure 7.31 presents the results of the evaporation simulation for the 57-gram drop at 300° K. The simulation covered about seventy-nine years. The mass loss from the drop in total and for the individual components is shown in the figure's top panel. The sodium component evaporated very little over the course of the simulation, whereas the potassium component evaporated at a relatively rapid pace. After about seventy-nine years, the composition of the drop had changed sufficiently to raise its melting point above 300° K, and the liquid-to-solid phase transition began. The figure's middle panel shows the change in the radius of the drop; the radius decreased by about 0.75 centimeter during the simulation. The bottom panel shows the evaporation rates for each component of the alloy. The rates slow as the elements evaporate from the drop and the mole fractions decrease. The effect of the phase transition on the drop is discussed below.

Figure 7.32 shows the results for the 1-gram drop at a temperature of 300° K. Sufficient potassium evaporated from this drop for solidification to begin after only about twenty years. This counterintuitive result occurred because, even though the evaporation rate per unit surface area was slightly slower for smaller drops, over a given period of time, the mass loss per unit mass ($\Delta m/m$) was much greater ($\Delta m/m$ about r^{-1}), causing the smaller drops to evaporate more quickly overall than their larger counterparts. The radius of the 1-gram drop decreased from 7.5 millimeters to about 4.5 millimeters over the course of the simulation.

Table 7.9 shows that the sodium-potassium drops always go through a phase transition induced by the change in composition as the drops evaporate. The phase diagram for sodium-potassium is presented in figure 7.33. The eutectic mixture is marked as the starting point in the simulations. The melting point for the various mixtures is marked by the top line in the plot (liquidus). The fraction of potassium decreases relatively rapidly as the drop evaporates at constant temperature. Eventually, the melting point rises above the temperature maintained in the simulation, and solidification of the drop begins. For the temperature range considered in this note, the sodium-potassium changes to a mixture of liquid and solid when the liquidus is crossed. The composition of the liquid (solid) is given by the composition at which the constant temperature line intersects the liquidus (solidus). For example, at 300° K, when the potassium mass fraction decreases to about 40 percent, the liquid has a composition of about 40 percent potassium and about 60 percent sodium by mass/weight, whereas

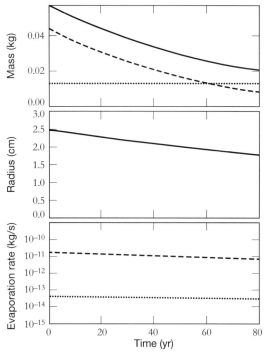

Temp = 300°K, Initial mass = 57 gm, Vapor pressure: Foust

Figure 7.31   Simulated evaporation rate for a 57-gram drop of sodium-potassium alloy at 300° K. The dashed line represents the potassium component and the dotted line the sodium component, whereas the solid line represents the sum of the two.

the forming solid has a composition of about 6 percent potassium and about 94 percent sodium by mass/weight. The relative compositions of this liquid/solid mixture *will not change unless the temperature is changed*; for additional evaporation of the potassium to occur, the temperature must be increased. The simulation was stopped at the phase transition because very little is known about how the drop solidifies or the differences in evaporation from the liquid and solid alloys. Examination of tables 7.9 and 7.10 shows that, once solidification began, only 15–50 percent of the drop's original mass remained. Complete evaporation of the drops could therefore take much longer than the timescales presented above.

Since the evaporation rates are very sensitive to the value of vapor pressures, an assessment of the accuracy of the vapor pressure values given by equation 7.19 was warranted. Vapor pressure formulas for liquid sodium and potassium had to be

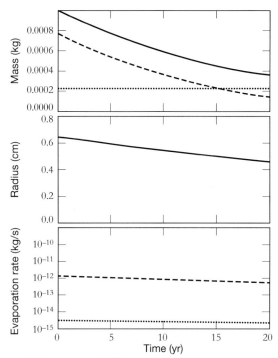

Temp = 300°K, Initial mass = 1 gm, Vapor pressure: Foust

Figure 7.32 Simulated evaporation rate for a 1-gram drop of sodium-potassium alloy at 300° K.

Table 7.9 Evaporation simulation results for sodium-potassium drops

Initial/final mass (g)	Temperature (°K)	Initial/final radius (cm)	Phase transition (years)
57.0/28.36	280.0	2.507/1.975	634.26
57.0/20.88	300.0	2.511/1.767	78.10
57.0/16.05	330.0	2.518/1.615	4.466
1.0/0.497	280.0	0.652/0.514	164.36
1.0/0.366	300.0	0.653/0.460	20.24
1.0/0.281	330.0	0.655/0.420	1.15

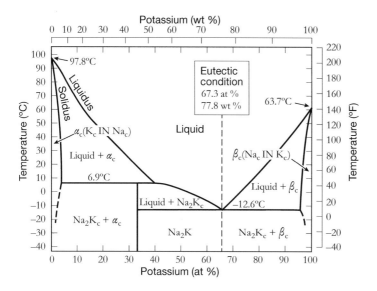

Figure 7.33 Sodium-potassium phase diagram [32]. at %, atomic percentage; wt %, weight percentage.

Table 7.10 Simulation results for equalized pressures for sodium-potassium drops

Initial/final mass (g)	Temperature (°K)	Initial/final radius (cm)	Phase transition (years)
57.0/26.54	280.0	2.507/1.933	13,354.12
57.0/18.05	300.0	2.511/1.724	12,255.03
57.0/43.79	330.0	2.518/2.333	352.55
1.0/0.465	280.0	0.652/0.503	3,460.61
1.0/0.317	300.0	0.653/0.448	3,175.79
1.0/0.166	330.0	0.655/0.365	4,610.94

extrapolated down to temperatures below the freezing points of the individual elements for the simulation. However difficult to achieve, such an extrapolation seemed to be the only course of action because the elements remained in the liquid state while in solution and there was no direct measurement of the evaporation rate for sodium-potassium. The uncertainty in the vapor pressure values was estimated by comparing extrapolations of the vapor pressure values derived by various researchers [33]. These comparisons are presented in figure 7.34. At 300° K, the sodium vapor pressure agrees

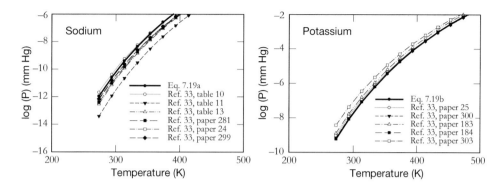

Figure 7.34   Comparison of sodium (left) and potassium (right) vapor pressure values (in millimeters of mercury) extrapolated from the results of various researchers.

to within one to two orders of magnitude and the potassium vapor pressure values to within one order of magnitude with the values quoted in Foust [32], whose vapor pressure values are the lowest for potassium and among the highest for sodium in the literature. Thus the original calculation may prove to be a lower limit on the evaporation lifetimes.

Two additional assumptions used in the simulation of evaporation might have affected the calculated evaporation rates. The first assumption was that of a constant vapor pressure value in equilibrium with the drop. For potassium, the equilibrium vapor pressure at all temperatures considered is significantly larger than the ambient atmospheric pressure at the altitude of the drop (P_{atm} is about 10^{-8} pascal at 925 kilometers) [34]. It is unlikely that the potassium vapor pressure around the drop would attain levels about 20 to 6,000 times that of the neutral atmosphere; the potassium atoms would therefore diffuse away from the drop. Except at 330° K, sodium vapor pressures remain well below atmospheric pressure. If it were assumed that the potassium atoms diffused away from the drop, the evaporation of the drop would be slowed. To estimate the difference in evaporation rates, a series of model runs was conducted in which vapor pressures above the ambient atmospheric pressure were reduced to 10^{-8} pascal to simulate the outflow induced by the pressure gradient. Although the sodium vapor pressures at the lower temperatures were *not* increased to remove the pressure gradient (since an increase would imply a second source of sodium atoms), at 330° K, the sodium vapor pressure *was* decreased, to 10^{-8} pascal. The results are presented in table 7.10 and should be considered crude estimates.

The second assumption—the uniform composition of the drop—implied that there was a process on orbit that continually mixed the drop as its components evaporated. Without mixing, only the potassium near the surface of the drop would evaporate quickly, and after it did, the drop could be surrounded by a thin sodium "shell," which would evaporate at a much slower rate. This effect could also slow the evaporation of the drop, but the effect might be minimal since the drop was mostly potassium in the early stages of evaporation.

For drops in certain orbits, the temperature varied as the drop passed through sunlight and shadow instead of remaining constant, as might be assumed for computational ease. The preceding analysis has shown that the evaporation rate of the drops was sensitive to their temperature. The effect on the evaporation rate of allowing the temperature of the drops to vary as they evaporated was estimated by calculating the amount of material that evaporated from the drops over the course of one orbit. In one case, the temperature was allowed to vary, whereas, in the other, the temperature was held fixed at the average temperature over the orbital cycle (about 303° K for both 57-gram and 1-gram drops). It was found that for the 57-gram drop, only 5.32×10^{-10} kilogram evaporated when the temperature varied in a realistic fashion during the orbit, whereas 4.64×10^{-8} kilogram evaporated when the temperature was held constant. The 1-gram drop behaved in the opposite manner; 9.86×10^{-11} kilogram evaporated when its temperature was allowed to vary, compared with 3.16×10^{-11} kilogram when the temperature was held constant. The reason for this behavior was that the temperature variation for the smaller drop was larger over the course of an orbit, as shown earlier. The 1-gram drop reached higher temperatures than a 57-gram drop in the same orbit, hence evaporated more quickly.

The simulation was stopped when the drops began to solidify since there was no information available to model the evaporation for a mixed liquid-solid state. Evaporation of a liquid at the same temperature requires an amount of heat (per mole of substance) equivalent to the latent heat of vaporization at that temperature, whereas evaporation of a solid (sublimation) at constant temperature requires an amount of heat (per mole of substance) equivalent to the sum of the latent heats of fusion and vaporization at that temperature. A solid will thus evaporate at a slower rate since the amount of energy required to liberate its atoms is greater.

Finally, the calculated evaporation lifetimes for the anomalous debris drops were compared with the estimated orbital lifetimes of debris objects before decay into the atmosphere, calculated using the atmospheric models and lifetime charts provided in King-Hele [35]. The estimated object lifetimes assume a ballistic coefficient, C_D, of 2.2 for the spheres, an atmospheric model representing an average amount of solar activity, and the initial A/M ratios derived from radar and optical analysis. The estimates were accurate only to about ±20 percent and are presented in table 7.11.

Table 7.11   Approximate orbital lifetimes of sodium-potassium drops

Altitude (km)	Lifetime (m = 57 g) (years)	Lifetime (m = 1 g) (years)
850	172.8	45.2
925	288.0	75.3
1,000	576.0	150.6

Comparison of the estimated orbital lifetimes with the calculated evaporation lifetimes presented in table 7.10 suggests that it will take much longer for the drops to evaporate significantly than it will for their orbits to terminally decay. Previous work indicates that 1-centimeter-diameter sodium-potassium spheres at about 273° K (m = 0.46 gram) and an altitude of 950 kilometers would decrease in diameter due to evaporation at a linear rate over "hundreds of years," but would reenter the atmosphere in about eighty years [36]. This orbital lifetime is in approximate agreement with our orbital lifetime estimate for a 1-gram drop. We therefore concluded that the drops were not likely to evaporate prior to terminally decaying because of atmospheric drag.

7.3.7   Conclusions

The following conclusions on the anomalous space-debris objects can be derived from the fusion of radar and optical data and analytical derivations:

1. The objects were small and spherical in shape.
2. The objects' density was consistent with that of eutectic sodium-potassium alloy.
3. The objects' surfaces were metallic and highly reflective, and the measured reflectivity was consistent with the other known optical properties of sodium-potassium.
4. The temperatures of the objects on orbit were likely to stay high enough that the objects remained in the liquid state (hence they are often referred to as "drops" or "droplets").
5. Evaporation of the objects was not likely to be the limiting factor to their orbital lifetimes.

The work presented here helped promote general acceptance of the hypothesis advanced by the Orbital Program Office at NASA's Johnson Space Center, that the source of the detected anomalous space-debris objects was leaked reactor coolant.

References

1. Raup, R. C. "Bayesian Characterization and Detection of Rare Binary Events," MIT Lincoln Laboratory Technical Report 876, 15 March 1990.
2. Abbott, R., and T. Wallace. 2007. Decision Support in Space Situational Awareness. *MIT Lincoln Laboratory Journal* 16 (2): 297–335.
3. Ruck, G. T. 1970. *Radar Cross Section Handbook*. New York: Plenum Press.
4. Lambour, R., N. Rajan, T. Morgan, I. Kupiec, and E. Stansbery. 2004. Assessment of Orbital Debris Size Estimation from Radar Cross-Section Measurements. *Advances in Space Research* 34:1013–1020.
5. Baker, D. N. (December 2000). The Occurrence of Operational Anomalies in Spacecraft and Their Relationship to Space Weather. *IEEE Transactions on Plasma Science* 28 (6): 2007–2016.
6. Earl, M. A., and G. A. Wade. (May–June 2015). Observations of the Spin-Period Variations of Inactive Box-Wing Geosynchronous Satellites. *Journal of Spacecraft and Rockets* 52 (3): 968–977.
7. Spence, L. B., and R. Sridharan. "A Radar Observable of Three-Axis Stable Geosynchronous Satellites," MIT Lincoln Laboratory Project Report STK-108, 18 February 1981.
8. Hunt, S. M. "Radar Observations of Sun Tracking Three-Axis-Stabilized Spacecraft," MIT Lincoln Laboratory Project Report STK-166, 9 April 1990.
9. Seniw, W. P., and L. B. Spence. "FLTSATCOM V: Millstone Radar Support," MIT Lincoln Laboratory Project Report STK-115, 2 December 1981.
10. Kessler, D. J., M. J. Matney, R. C. Reynolds, R. P. Bernhard, E. G. Stansbery, N. L. Johnson, et al. "The Search for a Previously Unknown Source of Orbital Debris: The Possibility of a Coolant Leak in Radar Ocean Reconnaissance Satellites," NASA Johnson Space Center Report JSC-27737, 21 February, 1997.
11. Stansbery, E. G., T. J. Settecerri, M. J. Matney, J. Zhang, and R. Reynolds. "Haystack Radar Measurements of the Orbital Debris Environment: 1990–1994," NASA Johnson Space Center Technical Report TR-27436, April 1996.
12. Sridharan, R., W. I. Beavers, E. M. Gaposchkin, R. L. Lambour, and J. E. Kansky. "A Case Study of Debris Characterization by Remote Sensing," MIT Lincoln Laboratory Technical Report 1045, 18 December 1998.
13. Sridharan, R., E. M. Gaposchkin, T. Moore, and L. Swezey. "Debris Characterization: An Interesting Example," in *Proceedings of the 1996 Space Surveillance Workshop*, pp. 57–67, Lexington, MA: MIT Lincoln Laboratory, 1996.

14. Pearce, E. C., M. Blythe, D. Gibson, and P. Trujillo. "Space Debris Measurements: Phase One Final Report," in *Proceedings of the 1994 Space Surveillance Workshop*, pp. 125–134, Lexington, MA: MIT Lincoln Laboratory, 1994.

15. Izatt, D. L. 1995. Stare-and-chase at TRADEX. *Proceedings of the 1995 Space Surveillance Workshop*, pp. 145–153. Lexington, MA: MIT Lincoln Laboratory.

16. Sridharan, R. "Characteristics of Debris and Implications for Detection and Tracking," *Proceedings of the American Astronautical Society*, pp. 96–117, February 1996.

17. Ruck, G. T., ed. 1970. *Radar Cross-Section Handbook*. New York: Plenum Press.

18. Hedin, A. E. 1983. A Revised Thermospheric Model Based on Mass Spectrometer and Incoherent Scatter Data: MSIS83. *Journal of Geophysical Research. Atmospheres* 88:10170.

19. Gaposchkin, E. M., and A. J. Coster. 1988. Analysis of Satellite Drag. *Lincoln Laboratory Journal* 1:203.

20. Gaposchkin, E. M., and A. J. Coster. 1986. Evaluation of Recent Atmospheric Density Models. *Advances in Space Research* 6:157.

21. Marcos, F. A. 1990. Accuracy of Atmospheric Drag Models at Low Satellite Altitude. *Advances in Space Research* 10:417.

22. Miller, R. R. 1955. *Liquid Metals Handbook: Sodium (NaK) Supplement*. Ed. C. B. Jackson. Washington, DC: Atomic Energy Commission and Department. of the Navy.

23. Hecht, E. 1988. *Optics*. Reading, MA: Addison-Wesley.

24. Russian Institute of Physics and Power Engineering. (http://www.rssi.ru/general/spacer.html). Accessed in 1999.

25. Meshcheryakov, S. A. "Physical Characteristics of Alkaline Metals and Behavior of Liquid Metal Coolant Droplets in Near-Earth Orbits," in *Proceedings of the Second European Conference on Orbital Debris*, ESA SP-393, pp. 257–259, Noordjwik, Netherlands: ESA Publications Division, 1997.

26. Cunningham, F. G. 1962. Earth Reflected Solar Radiation Input to Spherical Satellites. *ARS Journal* 32 (7): 1033–1036.

27. Wolfe, W. L., and G. J. Zissis. 1978. *The Infrared Handbook*. 1st ed., 15–55. Washington, DC: Office of Naval Research.

28. Morgan, R. 1922. The Optical Constants of Sodium-Potassium Alloys. *Physical Review* 20 (3): 203–213.

29. Jackson, J. D., D. K. W. Tong, P. G. Barnett, and P. Gentry. 1987. Measurement of Liquid Sodium Emissivity. *Nuclear Energy* 26 (6): 387–392.

30. Kamiuto, K. 1986. Radiative properties of liquid potassium. *Journal of Nuclear Science and Technology* 23:372.

31. Ohno, R. 1972. Kinetics of Evaporation of Various Elements from Liquid Iron Alloys under Vacuum. In *Liquid Metals*. 1st ed., ed. S. Z. Beer, 37–80. New York: Marcel Dekker.

32. Foust, O. J., ed. 1972. *Sodium NaK Engineering Handbook*. 1st ed. vol. 1., 1–59. New York: Gordon and Breach.

33. Nesmeyanov, A. N. 1963. *Vapor Pressure of the Chemical Elements*. 1st ed., 128–143. New York: Elsevier.

34. Jursa, A. 1985. *Handbook of Geophysics and the Space Environment*. 2nd ed., 14–30. Bedford, MA: Air Force Geophysics Laboratory, Hanscom Air Force Base.

35. King-Hele, D. 1987. *Satellite Orbits in an Atmosphere: Theory and Applications*. 1st ed., 258–263. Glasgow: Blackie & Son.

36. Kessler, D. J., M. J. Matney, R. C. Reynolds, R. P. Bernhard, E. G. Stansbery, N. L. Johnson, et al. "A Search for a Previously Unknown Source of Orbital Debris: The Possibility of a Coolant Leak in Radar Ocean Reconnaissance Satellites," IAA 97-IAA.6.3.03, presented at 48th International Astronautical Congress, 6–10 October, Turin, 1997.

8

Conjunctions and Collisions of Resident Space Objects

Richard I. Abbot

Because of the seemingly unlimited volume of space available for resident space objects (RSOs) above the Earth, collisions of RSOs with one another have not been considered a problem until recent years (see figures 6.1 and 6.2 in chapter 6). But because of the ever-increasing number of RSOs launched into orbit, the popularity of certain orbit regimes, such as low Earth orbit (LEO; less than about 3,000 kilometers above the Earth) and geosynchronous orbit (GEO; about 36,000 kilometers above the Earth), and, more significant, the ever-growing number of RSOs produced by the fragmentation of satellites and rocket bodies in orbit, conjunctions (close approaches) and even collisions between RSOs are increasing in frequency, the latter resulting in hundreds more RSO fragments (space debris). NASA's Don Kessler was the first scientist to conduct detailed research into the likelihood of satellite collisions in low Earth orbits [1, 2]. He showed that, absent effective measures for controlling the production of space debris from fragmentation of satellites and rocket bodies, there was a significant possibility of a cascading series of collisions and the runaway generation of space debris, as is known to have happened in the asteroid belt of our solar system. The Orbital Debris Program Office at NASA's Johnson Space Center, the leader in studies of space debris, regularly publishes its research work and the *Orbital Debris Quarterly News* [3, 4, 5]. Space is now a risky place for functioning satellites: many tens of thousands of objects as small as 1 centimeter pose potentially catastrophic threats to mission-critical payloads.

Significant work has been done at MIT Lincoln Laboratory on the topic of space debris. As described in section 2.2 of chapter 2, the laboratory's Haystack Long-Range Imaging Radar (LRIR) and Haystack auxiliary radar (HAX) have been the leading sensors collecting radar detection data on space debris, data that have been used by NASA's Orbital Debris Program Office to model the numbers, sizes, and distribution of space-debris objects [6]. The laboratory has also done significant analytical and numerical work on collision probabilities in both low Earth and geosynchronous orbits.

8.1   The General Problem of Conjunctions and Collisions

Fragmentation events occur in orbital space with some frequency due not only to collisions between resident space objects but also to spontaneous breakups of RSOs, resulting in a significant increase in the number of space-debris objects. Thus a critical problem is emerging in orbital space, particularly in the denser parts of low Earth orbit space. Given the large and increasing number of space-debris objects (about 100,000 RSOs larger than 1 centimeter, as estimated by the Orbital Debris Program Office in 2000), calculations indicate that a collision between a large RSO and a space-debris object may occur every 400 days on average.[1] Among key avenues pursued to address the problem of systematically identifying where such collisions are most likely to occur are

1. Developing a computational scheme that can rapidly estimate conjunctions for large catalogs of objects (100,000 or more);

2. Improving covariance analysis to effectively cull the number of conjunctions of critical interest;

3. Devising new covariance-based statistics such that, regardless of orbit-estimation quality, all true conjunctions of critical interest are successfully tagged; and

4. Introducing an automated, adaptive tasking scheme to ascertain all potential collisions with a low false-alarm rate, while providing quantification that ties conjunction-prediction performance to sensor-network performance and tasking parameters.

As a specific example of assessing the collision probability, using a 50,000-object catalog in the most dangerous space-debris regime, a protocol can be established whereby sensors take a series of observations over 10 days to predict the estimated 1.4 conjunctions of less than 10 meters for the 11th day. Only 9 apparent conjunctions per day need to be monitored to identify those 1.4 true conjunctions. The sensor-network load necessary is 3.8 times the loading necessary to maintain the latent day-to-day catalog, driven by the need for additional data to ensure accuracy on the order of 10 meters for all the objects in the catalog.

---

1. Arthur Lue, MIT Lincoln Laboratory, personal communication, March 2011.

Predicting when and where these catastrophic events will most likely occur is now an urgent task for the space surveillance community. The task is composed of many technical challenges to achieve needed improvements in three key areas:

1. Computational speed: developing faster, more efficient algorithms to assess and cull conjunction events of critical interest in a complex space environment;

2. Accuracy: developing and honing our understanding of the evolution of orbit confidence, including the crucial role that non-Gaussian uncertainty dynamics plays in detailed prediction of future events in the orbital space environment; and

3. Action: developing an automated protocol for reacting to future risks and collision threats in the context of expected improvements of the Space Surveillance Network (SSN).

Innovations in all three areas have been achieved by the laboratory, though more recently than the time frame covered by this book (1970–2000),[2] hence they will be left to other authors. But, from 1997 onwards, in partnership with private satellite-owning companies, the laboratory conducted an investigation of conjunctions and collision probabilities in geosynchronous orbits and developed an automated system for monitoring close approaches and warning about them.

8.2   Conjunctions and Collisions in Geosynchronous and Geostationary Orbits

Next to low Earth orbit space, geosynchronous and geostationary orbit space is the most crowded orbit regime. Almost every space-faring country or private multinational organization launches and maintains satellites in this regime for applications related to communications, monitoring, intelligence, and weather forecasting. Locations in the geosynchronous orbit regime are eagerly sought from and granted by the International Telecommunication Union (ITU). Generally speaking, the individual satellites are expected to be maintained within tight latitude and longitude bounds: 0.05×0.5 to 0.1×0.1 degree for most commercial satellites and larger bounds for other satellites [7]. As of 2000, some 750 satellites along with rocket bodies had been launched into near-geosynchronous orbit; approximately 260 of the satellites were actively maintained [8]. Several hundred other satellites are known to be drifting

---

2. Ibid.

there, along with hundreds of smaller resident space objects, some of which are lost (not recently tracked) or undetected [8].

The increasing collision threat in the geosynchronous belt arises from the following facts:

1. Satellites have grown in size to satisfy their owners' requirements for ever greater capabilities in data collection and transmission. It is no longer unusual for satellites weighing up to 3,000 kilograms to be launched into geosynchronous orbit. These satellites often measure 25 meters from tip to tip of their solar arrays and about 5 meters in other dimensions.

2. The shortage of desirable locations in geosynchronous orbit space (those with direct communications access to populated areas) has led to multiple satellites from the same or different agencies being clustered within, say, 1 degree of longitude. More than twenty clusters contain from two to eight satellites (see table 6.1 for a sample).

3. Unless actively maintained, geostationary satellites tend to drift in longitude and therefore toward one another. The principal longitudinal variation of the geopotential leads to a tangential acceleration of a satellite in the geostationary belt that is approximately sinusoidal, with four nodes [9]. Near two of these nodes, at 75.1 degrees east and 105.3 degrees west sublongitudes, geostationary satellites will drift back to a node and oscillate around it; near the other two nodes, at 11.5 degrees west and 161.9 degrees east sublongitudes, they will drift away from a node. Probably ten or more defunct satellites are parked at the stable node at 105.3 degrees west sublongitude so as not to drift toward other satellites, not including an equal number of actively maintained satellites stationed there to minimize station-keeping maneuvers.

4. Although the probability of a collision is still low, the consequences are quite severe: the loss of the satellite(s), which can cost more than $200 million each, the loss of comparable revenue from commercial applications of the satellite(s) during the long time needed to replace it (them), and, most significant, the long-term threat to other satellites from space debris created by the collision.

Unlike collisions between RSOs in low Earth orbit, those between RSOs in geosynchronous orbit happen at relatively low velocities, on the order of tens of meters per second; but such collisions are still strong enough to render satellites defunct. Concern over possible collisions in the geosynchronous belt was addressed by Lincoln Laboratory in response to requests from commercial satellite owners/operators.

8.2.1   The Geostationary Orbit

The concept of a geostationary satellite orbit is believed to have originated with the Russian theorist Konstantin Tsiolkovsky, who wrote articles on space travel in the late nineteenth and early twentieth centuries. A vision of how satellites in such an orbit could be used for communications is widely credited to Arthur C. Clark [10], who worked on many details of communications satellites, including their orbit characteristics, frequency needs, and use of solar energy for power. A geosynchronous satellite has an angular velocity matching that of the Earth, which theoretically requires a near-circular orbit with a semimajor axis of 42,164.2 kilometers. Figure 8.1 summarizes its

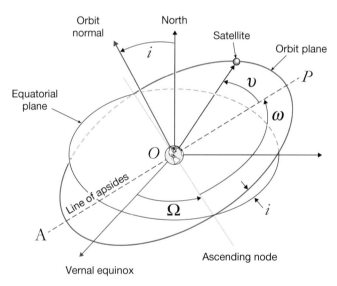

Figure 8.1   Parameters for an artificial satellite in orbit around the Earth. The orbital ellipse (shown in red) is described by its semimajor axis, a, and eccentricity, e. Other parameters are i, the inclination of the orbital plane to the equatorial plane of the Earth; A and P, the apogee and perigee of the orbit (points on the orbit farthest from and closest to the Earth); Ω, the right ascension of the ascending node, measured from the vernal equinox to the intersection of the north ascending orbit with the equatorial plane of the Earth; ω, the argument of perigee measured from the ascending node to the perigee; v, the true anomaly measured from the perigee to the instantaneous satellite location; and r_p and r_a, the perigee and apogee distances given by $a(1 - e)$ and $a(1 + e)$. The line of apsides connects the perigee and apogee.

important orbital parameters. A geostationary satellite remains over a given location above the equator. A geosynchronous orbit does not necessarily make a satellite geostationary. If the orbit is slightly inclined to the equator, during the course of one day, a satellite's latitude will increase and decrease through zero degrees, tracing a small figure eight over the surface of the Earth. In addition, if the geosynchronous orbit is not circular, the satellite will, on average, rotate at the same rate as the Earth, but when it is at perigee (the closest point to the Earth on its orbit), it will move faster and at apogee (the farthest point), it will move slower. This change in velocity will add a slant to the small figure-eight shape. Without a zero inclination and eccentricity, the geosynchronous satellite will thus not be geostationary.

The first geosynchronous satellite was Syncom 2, which NASA launched into orbit in July 1963. Although geosynchronous, it was not stationary over one location, as announced by NASA after the launch. The first truly geostationary satellite was Syncom 3, which NASA launched in August 1964 [11]. As of 2000, several hundred geosynchronous satellites orbited the Earth in a narrow belt near the required Earth distance of 42,164.2 kilometers, or at altitudes of about 36,000 kilometers above the Earth. There is now an international agreement that inactive satellites should be lofted to an orbit about 300 kilometers above the geosynchronous belt, but many old, defunct geosynchronous and geostationary satellites predate this agreement. In addition, satellites still fail in situ and cannot be lofted. It was such a failure that galvanized the work discussed here.

A geostationary satellite position is inherently unstable—natural forces acting on the satellite will cause the satellite to drift in position. The Earth is not a perfect sphere, and the flattening near the poles caused by Earth's rotation is well known. There is also an ellipticity along the Earth's equator. Although the difference between the longest and shortest radius of the equator does not exceed 192 meters, this difference can have a significant effect on a geostationary satellite, giving it a tangential acceleration. The following three natural forces perturb the orbital position of a geostationary satellite:

1. The tangential forces produced by the ellipticity of the Earth's equator, which cause the satellite to drift in longitude;

2. The torques due to the gravity of the Sun and Moon, which cause a long-term evolution of the satellite's orbital inclination from 0 degrees to about 15 degrees and back in a 54-year cycle; and

3. The solar radiation pressure, which causes an annual periodic change in the satellite's orbital eccentricity.

These three natural perturbing forces, which all require counteracting maneuvers by the satellite operator to keep the satellite geostationary, are illustrated in figure 8.2, panels a, b, and c, respectively, and explained in greater detail below.

Mathematically, the nonsymmetrical gravity field potential of the Earth is described in terms of a spherical harmonics expansion. The zonal terms of this expansion are rotationally symmetric (i.e., longitude independent) and quantify the rotational flattening of the Earth. The unsymmetrical mass distribution inside the Earth is quantified by the tesseral terms of the expansion (i.e., longitude dependent). The dominant two tesserals give a longitude dependence that is approximately sinusoidal with four nodes. At these nodes, or equilibrium points, the acceleration is zero; a satellite will therefore stay at the node if it is stopped there. This situation is highly unlikely, however, because of the natural perturbing forces. Two of these equilibrium points are stable because a small deviation from the node's longitude point will cause the satellite to drift back to the node and oscillate about it. The other two equilibrium points are unstable because a satellite will drift away from the node, given any deviation in longitude. One can think of the stable points as gravity "wells" and the unstable points as gravity "peaks." The deeper of the two wells (stable points) is at 75.1 degrees east longitude over the equator and is associated with Asia and Africa, whereas the shallower of the two wells is at 105.3 degrees west longitude (over the equator south of Denver) and is associated with North and South America. The higher of the unstable geopotential peaks is in the western Pacific at 161.9 degrees east longitude, and the lower peak is at 11.5 degrees west longitude in the eastern Atlantic [12]. An interesting aspect of a satellite left to drift near the western Pacific peak is that it will move down the peak and have enough energy to climb the eastern Atlantic peak and drift to the other side of the Pacific peak, visiting both geopotential wells in the process.

The second perturbing force acting on a geostationary satellite is the gravitational attraction of the Sun and the Moon, which do not lie in the equatorial plane of the geostationary orbit. The out-of-plane force of the Sun is at its maximum at midsummer and midwinter, and at zero in spring and fall. A similar attraction occurs for the Moon during its monthly cycle, with the acceleration at its maximum twice per lunar period and passing through zero in between. The lunar and solar perturbations are predominantly out of plane and thus cause a change in the inclination of a geostationary satellite's orbit that is both periodic and secular in nature. These perturbations along with an influence of the zonal (latitude-dependent) terms of the Earth potential cause the satellite's orbital inclination to increase to 15 degrees in a period of 27 years and then return to 0 degrees in the next 27 years.

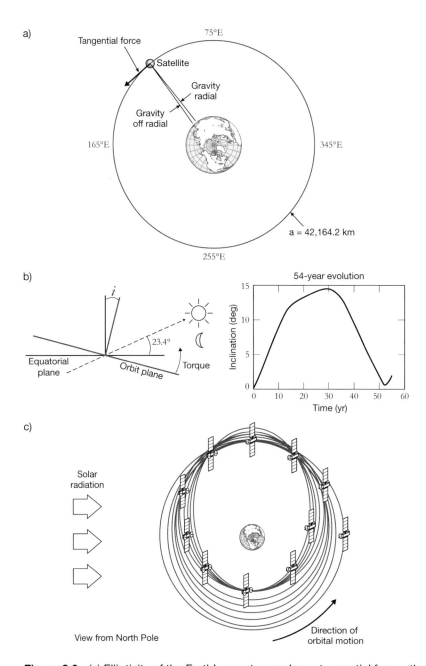

Figure 8.2 (a) Ellipticity of the Earth's equator produces tangential forces that cause a drift in a geostationary satellite's longitude. (b) Torques of the Sun and Moon cause a long-term evolution of the geostationary satellite's orbital inclination from 0 degrees to 15 degrees and back in a 54-year cycle. (c) Solar radiation pressure causes an annual periodic change in the geostationary satellite's orbital eccentricity.

The third important perturbing force on geostationary satellites is caused by electromagnetic solar radiation pressure (SRP). This force has become more significant as the satellites and their solar panels have grown ever larger, exposing ever greater effective area to the Sun. The SRP force, which is always perpendicular to the Sun-oriented solar panels, when integrated over one-half of the orbit, causes a small velocity increment that tends to raise the apogee of the satellite's orbit. Over the other half of the orbit, the small velocity increment opposes the orbit velocity, which tends to lower the perigee. During the year, as the Earth moves around the sun, the solar radiation pressure force changes direction, causing the eccentricity of the satellite's orbit to increase for half the year and decrease for the other half with a magnitude on the order of 0.0005 for typical large satellites in the geosynchronous belt.

Figure 8.3 shows the changes in semimajor axis, eccentricity, and longitude resulting from natural perturbing forces acting on a geosynchronous satellite that started to drift near the western Pacific peak of the Earth's gravity field potential. The drifts and oscillations caused by natural perturbing forces require the satellite operator to perform counteracting maneuvers to keep the satellite within its latitude and longitude boundaries.

Study of conjunctions of active satellites in geostationary orbit and the manner in which these satellites can avoid collisions requires knowledge both of the modes of orbit control exercised by the satellite owners/operators and of the data available to determine the satellites' orbits.

8.2.2 Orbit Control

Each geostationary satellite is assigned a longitude slot in which the satellite must be kept. The primary limitation in spacing satellites along the geostationary belt is avoiding electromagnetic interference between them in the communications frequencies allocated for up- or downlinks. The longitude slots are assigned by the International Telecomunication Union with coordination by regional agencies, such as the Federal Communications Commission (FCC) in the United States. For commercial satellites, the slots range in longitude width from ±0.05 degrees to ±0.1 degrees [7]. Some satellites (e.g., meteorological and military communications satellites, as well as satellites for some mobile telephone systems) often have larger longitude station-keeping or control boxes because they have a wider coverage beam or use a tracking rather than a fixed antenna on the ground.

The primary orbital parameters of concern that change because of natural perturbing forces are orbital longitude, inclination, and eccentricity. Drift in orbital longitude must be counteracted; otherwise, the geostationary satellite will move out of its slot

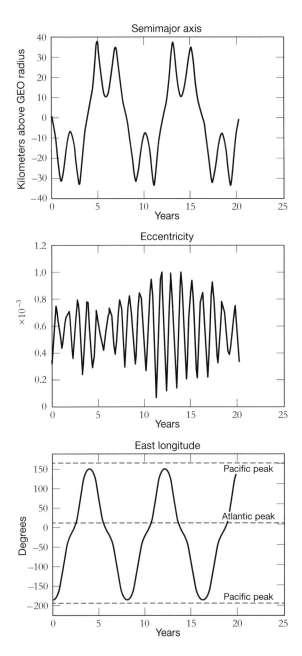

Figure 8.3 Evolution under natural forces of the semimajor axis, eccentricity, and longitude of a geosynchronous satellite that started to drift near 173 degrees east longitude, that is, near the western Pacific peak of the Earth's gravity field potential. One orbital position is plotted per day. (Top) Over an eight-year period, the semimajor axis of the satellite's orbit varies from −35 to +35 kilometers from the geosynchronous radius. (Middle) The satellite's orbital eccentricity varies yearly. (Bottom) Over an eight-year period, the satellite's longitude moves east over the Atlantic peak on to the other side of the Pacific peak until it turns near 150 degrees east and then moves westward back to the initial longitude.

within 7–14 days. Orbital inclination must be maintained or the satellite will describe an ever larger figure eight (i.e., its latitude excursion will increase) and thus may require a tracking antenna on the ground. Generally speaking, bounds of inclination of ±0.1 degree are maintained, although if control in inclination is not as critical (because of wide coverage beams or tracking antennas), inclination can be allowed to drift for some years. Orbital eccentricity must also be maintained close to zero lest the satellite drift out of the allowable longitude coverage region during an orbit. The maintenance of a geostationary satellite in its assigned slot is called "station keeping." Two types of maneuvers are done for this station keeping: in the east or west direction for longitude control and in the north or south direction for inclination control. The longitude maneuvers are often slightly modified to control eccentricity. The lifetime of a satellite, typically from 10 to 20 years, is limited by the availability of station-keeping fuel. The control of inclination (north or south maneuvers) over the lifetime consumes more than 90 percent of the station-keeping fuel. When inclination control is not as stringent and when a ground antenna can continuously track, the operator can let the satellite drift to save fuel.[3] For example, a 3-degree inclination bound can be maintained for about 7.5 years if the right ascension of the ascending node starts at 270 degrees. If the maximum possible inclination bound is only 0.5 degree, then at least one maneuver is required per year [12].

In theory, one should be able to precisely predict when an operator should initiate a maneuver from knowledge of the orbital mechanics described above and of the station-keeping bounds. In practice, however, the time when a satellite can undergo a station-keeping maneuver depends on (1) how well the operator knows the true position of the satellite; (2) how well the previous maneuver met its objectives; (3) how much "wiggle room" the operator likes to maintain for the satellite in the longitude slot; and (4) how much equipment or manpower is available.

An additional effect on longitude drift must be considered for orbit control. Satellites must maintain attitude control for proper orientation to the Earth and the Sun. One method of doing this is with momentum wheels, which use gyroscopic stiffness to provide three-axis stabilization by absorbing external torque disturbances with a gradual spin-up or spin-down [13]. For the momentum wheels to function properly, their stored momentum must be kept within allowable limits. When critical limits are exceeded, the wheels must be adjusted, which involves a thruster firing of suitable magnitude and orientation—from several times a day in extreme cases to several times

---

3. Geosynchronous satellites are controlled by ground-based operators who monitor their status and take corrective actions as necessary to ensure mission performance.

a month. The resultant change in satellite velocity is small (less than 0.01 meter per second) but can still produce a noticeable drift. Momentum wheels can also be used to advantage to provide a small contribution to the east or west station keeping.

8.2.3   Tracking Data and Orbit Estimation for Geosynchronous and Geostationary Satellites

Satellite operators collect tracking data in order to determine the orbits for their satellites and to compute when station-keeping maneuvers are required. They may do so less frequently (e.g., once per hour) or much more frequently for a limited period following a maneuver in order to check performance of the maneuver and to derive a new orbit. The tracking data consist of range, azimuth, and elevation measurements. The range measurements can be obtained from time delays of a signal sent to and returned by the satellite through a transponder or, much less commonly, by using the satellite beacons with ranging stations (usually two) with the largest possible separation, or baseline.

Independent assessment of the range data by Lincoln Laboratory analysts indicates that the data, although precise to within a few meters, can be poorly calibrated and have large bias errors, primarily due to not having an independent method of calibration [14]. (The necessity to independently calibrate metric data collection by radar and optical tracking systems is demonstrated in subsection 2.1.3.2 of chapter 2.) The angle measurements from satellite ranging stations may have errors on the order of tens of millidegrees and are therefore only marginally useful for high-accuracy orbit determination. Range rate is not usually measured. The consequence of poorly calibrated range data can be severe. Large biases in these data will cause an apparent shift of the satellite in longitude and, to a lesser extent, in inclination. These shifts can lead to a satellite drifting outside its allocated station-keeping box in violation of ITU regulations.

There are four primary components in determining and maintaining a satellite's orbit: tracking data, force models, an estimation theory (to tie these components together), and covariance (i.e., error analysis). Each of these components is explained below.

1. The tracking data consist of the range data collected by the satellite's operator through a transponder system (see above) and the metric tracking data collected by sensors of the Space Surveillance Network—mainly from deep-space radars, ground-based optical systems (such as GEODSS), and space-based optical systems like the SBV (Space-Based Visible) sensor (see chapters 2, 4, and 5).

2. The force models take into account the Earth's geopotential, solar and lunar gravitational forces, solar radiation pressure, and, if available, the timing and size of the operator-controlled maneuvers of the satellite.

3. An estimation theory uses the tracking data and force models to determine the satellite's orbit and to ascertain unknown parameters like the satellite's area-to-mass (A/M) ratio for computing the effect of solar radiation pressure. The theory most commonly used, known as "special perturbations," entails numerical integration of the orbit motion of the satellite, and can be either least squares or a sequential method based on a Kalman filter or a similar algorithm (see subsection 2.1.3.2 of chapter 2).

4. Covariance (i.e., error analysis) is needed to set a threshold level for close approaches when the satellite's operator should take avoidance action.

The U.S. Air Force Space Command has deployed a combination of ground-based deep-space radars, and ground- and space-based optical systems for the purpose of finding and tracking geosynchronous and geostationary satellites (see chapters 2, 4, and 5). The optical systems provide precise directional information to a satellite with respect to the sensor location (either an azimuth-elevation pair or a right ascension–declination pair of observations). Radar systems provide directional information (usually azimuth and elevation measurements, which are less precise than the information from optical systems), as well as range and range rate, the latter two measurement types typically being the most precise. Metric observations from both radar and optical sensors are fused in orbit determination, making the most of the precision of the two systems.

8.2.4   Collision Threat in the Geostationary Belt

The realm of geostationary satellites is a bustling belt-like region of space. Satellites are regularly launched into this belt, older satellites are retired, and other satellites have become prematurely defunct and are left to drift through the active satellite population. Satellites occasionally suffer catastrophic failure caused by strong solar activity, malfunction of a component or software, or, occasionally, indeterminate factors. Since 1977, by international agreement, aging satellites must be guided into graveyard orbits 300 kilometers above their geosynchronous or geostationary orbits before they run out of station-keeping fuel [8]. Satellites often share common regions of space in clusters. Vigilance is therefore required on the part of satellite operators to avoid close approaches and collisions, as is more accurate satellite tracking plus improved orbit determination methods and strategies for quick and accurate decision making.

By 2000, the number of actively controlled satellites in the geostationary belt was more than 400. The total number of drifting uncontrolled geosynchronous satellites (with a mean motion of 0.9–1.1 revolutions per day or with semimajor axes of 39,664–45,314 kilometers, respectively, and orbital eccentricity of less than 0.1) was nearly 750. Of these drifters, 141 were in librating orbits whose longitude extent is limited, and hence they impact only a limited region of the entire population in the geostationary belt.[4] Of these librators, about 36 oscillate in the geopotential well centered at 105.3 degrees west with periods of 2.5–6 years, about 90 oscillate in the other geopotential well centered at 75.1 degrees east with periods of 2.5–5.5 years, and about 15 oscillate about the unstable points passing through both wells with periods of 8–10 years. The remaining uncontrolled drifter satellites are circulators and though they are far enough from geostationary orbit space not to be captured in oscillation, the eccentricity of their orbits is large enough that that they can cross the active geostationary population at perigee or apogee. They drift around the Earth with periods proportional to their semimajor axis. Figure 8.4 shows a one-day snapshot of the region within 200 kilometers of the geostationary belt, illustrating the potential threat of the uncontrolled inactive satellite population to the controlled active population, based on common radial distances from the Earth.

8.3   The Geosynchronous Monitoring and Warning System

Lincoln Laboratory became actively involved in the study of potential collisions between geosynchronous satellites in early 1997, when Telstar 401 failed on orbit in situ because of a geomagnetic storm brought on by a mass ejection from the solar corona [15]. Because it failed at 97 degrees west longitude, Telstar 401 repeatedly oscillates between 113 and 97 degrees west longitude with a period of 2.5 years. Unfortunately, this oscillation causes the satellite to pass through a dense population of geostationary satellites serving the Americas [16]. It was feared that a malfunction or failure of any of these satellites might be attributed to a collision with Telstar 401, with attendant legal consequences. Figure 8.5 shows the first cycle of Telstar 401's drift through the geopotential well centered at 105 degrees west longitude. The first predicted crossing came with Galaxy IV in June 1997. The estimated separation distance between the two satellites was less than 1 kilometer, so an avoidance maneuver was suggested for Galaxy IV, even though such a maneuver can require additional

---

4. The geosynchronous and geostationary satellite numbers are derived from the publicly available satellite catalogs of U.S. Air Force Space Command, as is figure 8.4.

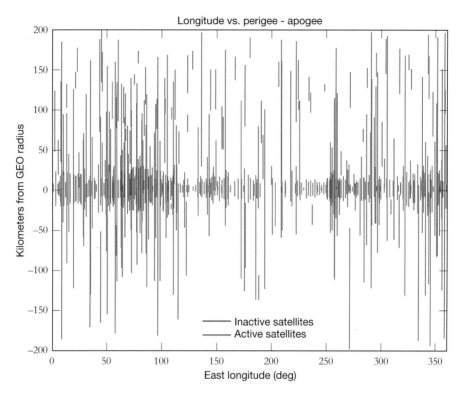

Figure 8.4 Snapshot on a given day of the radial distances from the Earth (determined by the perigee and apogee) plotted against the longitude of all active and inactive geostationary satellites within 200 kilometers of the geostationary radius. The active satellites (shown in blue) stay nearly at the same longitude, whereas the inactive satellites (shown in red) drift in longitude at a rate that depends on how far they are above or below the geostationary radius. An animation of these data would show how the inactive population drifts by the active satellites and thus could pose a potential threat if they have common radial distances.

station-keeping propellant and therefore shorten the useful life of the satellite. The maneuver for Galaxy IV resulted in a new predicted crossing distance of 6 kilometers. To increase the separation distance, some type of avoidance maneuver strategy, whether an unscheduled maneuver or a modified planned one, was implemented on eight of the fourteen crossings that occurred with Telstar 401 that year [16].

In 1997, Lincoln Laboratory initiated a cooperative research and development agreements (CRDA) with four commercial satellite-owning companies, all of which

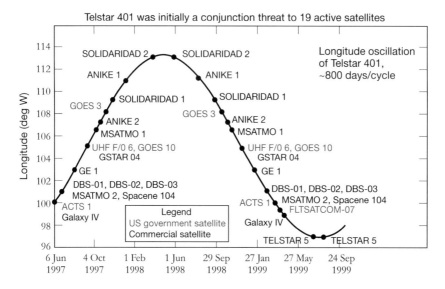

Figure 8.5   After Telstar 401 failed in January 1997, it began to drift in the geopotential well centered at 105 degrees west longitude. This figure, from Martin [11], shows the oscillation of Telstar 401 and the commercial and U.S. government satellites it crossed during the first two-year period, beginning in June 1997 with Galaxy IV (since retired). Telstar 5 replaced the failed Telstar 401 during this period.

had many assets at risk because of the drifting Telstar 401 [16]. Besides monitoring Telstar 401 satellite crossings, the CRDA concentrated on

1. Studying the orbit accuracy of geostationary satellites as a function of tracking type (i.e., radar, optical or both) and tracking density;

2. Understanding the threat posed to the active satellite population by the entire drifting satellite population (the active versus inactive problem);

3. Monitoring the calibration of CRDA company range data and using these data in orbit estimation;

4. Understanding how to model the satellite-specific station-keeping maneuvers in the orbit-estimation process; and

5. Developing an automated monitoring and warning system for conjunctions so as to avoid collisions.

Figure 8.6 illustrates the distribution of the number of close approaches versus separation distance between all active and inactive (drifter) satellites in

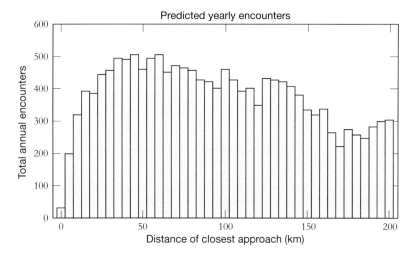

Figure 8.6 Histogram of the predicted number of encounters of all active satellites with the inactive satellites over one year. The peak depends on the radial distribution of the drifter population. This histogram will stay nearly the same in shape but will scale as both populations increase.

geosynchronous orbit for 1997, when Telstar 401 failed [16]. The peak of this distribution depends primarily on the variance of the radial distribution of the drifter population [17]. The question is invariably asked about the probability that a collision will occur between geosynchronous satellites. Assuming a collision radius of 50 meters (i.e., a cross-sectional area of about 8,000 square meters), the cataloged geosynchronous satellite population at the time of the Telstar 401 failure in 1997 was predicted to experience about one collision every thousand years [18]. This rate of collision will continue to increase, however. And an order-of-magnitude calculation like this one does not consider that there are longitude regions in which satellites are clustered. It also assumes that active satellites cannot collide with each other because they are maintained at their assigned position. A collision between geosynchronous satellites and the subsequent loss of an active geostationary satellite would have an enormous impact. Besides the monetary loss of the satellite, valuable communications would be disrupted or possibly lost completely over the affected area on the Earth until the satellite could be replaced. A collision could also create a debris population that would make that longitude region of space unusable with no easy means to clear it.

Although the local threat from Telstar 401 could be handled by analysts, the overall collision threat in the geosynchronous belt called for creation of the automated

Geosynchronous Monitoring and Warning System (GMWS) [19]. Figure 8.7 illustrates the components of this system, which was designed to

1. Maintain a list of the CRDA company active satellites that need to be protected from collision;

2. Form a threat list using the most recent historical orbit information for all the inactive drifters to determine which of them can cross the active geosynchronous belt;

3. Supplement the threat list with a list of other active satellites that can occasionally pose significant threats;

4. Combine tracking data from the Space Surveillance Network with the CRDA company ranging data into an orbit-determination process to update the orbit state for all the satellites; and

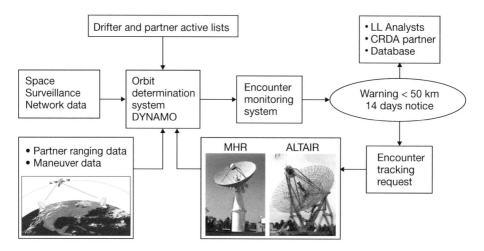

Figure 8.7   Geosynchronous Monitoring and Warning System (GMWS). This automated system, which monitors the active satellites owned by cooperative research and development agreement (CRDA) companies and potentially threatening inactive drifting satellites or other active satellites, computes high-precision orbits for all geosynchronous satellites by using a Lincoln Laboratory orbit-determination system known as "DYNAMO." Fusing Space Surveillance Network data with commercial tracking data, which usually are collected from two widely spaced ground stations, GMWS determines a long-term watch list of potential close encounters and a two-week warning list of close encounters that may require some precautionary action. CRDA, cooperative research and development agreement; MHR, Millstone Hill radar.

5. Incorporate station-keeping maneuver information obtained two weeks in advance from all CRDA companies.

Several tools were developed to help the GMWS visualize satellite crossings. In one known as the "box-crossing tool," drifter and active satellite trajectories were shown relative to the station-keeping box and to each other. Figure 8.8 shows an example of this tool [20]. The trajectory of the drifting satellite is shown in two and three dimensions as it crosses through the station-keeping box of the active satellite. Relative

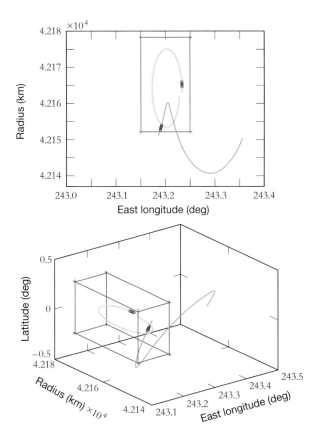

Figure 8.8   (Top) Crossing of a drifter satellite through the control box of an active satellite as a function of radius and longitude. The drifter trajectory is in blue, the active trajectory is in green, and the control box is in red. Positions along each trajectory are marked by the dots when the two satellites were at their closest. (Bottom) Crossing in three dimensions.

positions shown are at the predicted time of closest crossing. A quick look clearly reveals that this crossing does not pose any significant threat of collision.

The use of the CRDA company ranging data was an important aspect of the Geosynchronous Monitoring and Warning System. The company satellites had transponders that returned a time-marked transmitted signal back to the ground station, thus enabling calculation of the distance to the satellite. Range data were collected daily, e.g., hourly, or after a station-keeping maneuver to confirm the performance of the thrusts and the new orbit. Range data after a maneuver were very important for orbit processing because Space Surveillance Network tracking would most often not be immediately available. These data helped laboratory analysts in their estimation of the maneuver so that a continuous and current orbit could be maintained. The company data also supplemented the SSN data for satellites that were not visible to the Millstone Hill or Reagan Test Site radars. Most operators calculated the range to their satellites from two ground stations, which significantly improved the observability of satellite orbits. Finally, the range data were accurate to within 2–6 meters when properly calibrated using independent data (see subsection 2.1.3.2 of chapter 2 for details).

The calibration of the commercial range data is vital to ensure their accuracy. Lacking an independent "truth" source for their satellites' orbital positions, most operators calibrate their data by using one or more range stations and an orbit-estimation program. Because the SSN data have high accuracy and provide an independent estimate, analysts at the laboratory have regularly conducted joint calibrations of the CRDA company transponder range data, which they have found to exhibit sometimes large unmodeled biases, biases that vitiate computation of satellite locations using range transponder data alone. In one egregious case, computation from transponder data falsely determined a satellite to be outside its station-keeping box. The calibration process is iterative, with a bias and a sigma estimated and applied for each tracking station in combination with orbit determination with those data and the SSN data until the bias estimates converge. Best results are obtained when the calibration process is carried out during maneuver-free periods.

Figure 8.9 shows CRDA company range data before and after the calibration process. Figure 8.9 (top) shows that the range data had a bias error of −75 meters, which then changed to a bias error of −50 meters (generally such a jump could be explained by some change in the ranging hardware). As figure 8.9 (bottom) shows, the laboratory's calibration process corrected the range data to bring it to a zero mean error with a sigma (or noise) level of about 7 meters. The remaining error in the data is due to orbit error being reflected in the range data error.

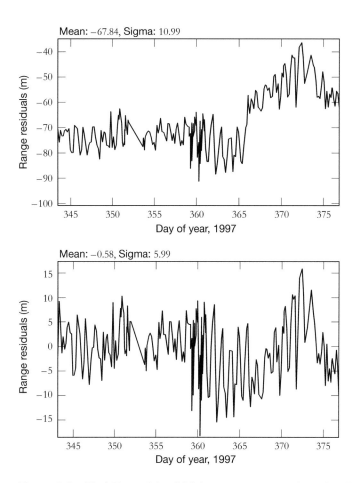

Figure 8.9 (Top) Error of the CRDA company range data showing a range bias in the data as well as a jump in the bias. (Bottom) Error after the bias was determined and removed from the range data through the calibration process.

Returning to the operation of the Geosynchronous Monitoring and Warning System, the next step was to determine the orbits for all near-geosynchronous drifting satellites and then to assess which of these would actually come close to one of the CRDA company active satellites over a prediction period of 60 days [19]. For the drifters in orbits with nonzero inclination, the close approach would occur near the intersection of the two orbital planes. The specified close-approach volume was conservatively made slightly larger than the size of the station-keeping box. Satellite pairs that pass this criterion were then put onto a watch list, where the term "watch" indicated that a close approach might occur. The watch could be visualized in various ways [20]. Figure 8.10 (top) shows the longitude of a drifting geosynchronous satellite for nearly a year (from 15 February to 10 December 2006). It also shows when the drifter could have a radial distance from the center of the Earth such that it could cross the station-keeping box of an active satellite and become a threat. Initially, the radial distance did not meet the criterion for intersecting the active satellite station-keeping boxes. But the orbital semimajor axis evolved so that (after August 2006) the drifter finally met the radial distance criterion for intersecting several active satellite station-keeping boxes, although it crossed through the plane of the boxes at the critical radial distance for only three active satellites. Figure 8.10 (bottom) shows a three-dimensional view of one of these three crossings (in September 2006). The crossing distance was 50 kilometers, thus of no concern. As discussed later, the orbit accuracy even 60 days ahead of any given time is generally on the order of 3–6 kilometers, so the point of crossing through the station-keeping box was well determined. The visualization allowed the operators to see where their active satellites should not be.

Thirty days before a close approach, the accuracy of the orbit and the amount of tracking data available for the drifter were assessed. If the data were insufficient, extra tracking was requested from the Millstone Hill radar and sometimes also from the ALTAIR or TRADEX radar at the Reagan Test Site if either had coverage and was available. Finally, within two weeks of a close approach, a more urgent alert—called a "warning"—was given so that the operators could adjust their station-keeping maneuvers if the close approach distance was of concern. The threshold of concern was based on orbit accuracy assessments, which are discussed below. A close approach in total distance greater than or equal to 6 kilometers was considered safe, but this information was reviewed by satellite operators and the analysts at the laboratory for the impact of any potential maneuvers in the two weeks prior to the time of the calculated close approach. The following factors were examined in this review:

1. The actual performance of recent maneuvers for the active satellite relative to the predicted performance of the maneuvers;

2. The effect of scheduled maneuvers prior to the close approach on the approach;

3. The quantity and quality of metric tracking data, the adequacy of the modeling of solar radiation pressure, and the resulting accuracies of the orbits, for the drifter and the active satellite; and

4. The geometry of the crossing in terms of the components along the radial, along-track (or velocity), and cross-track (or out-of-plane) directions for the satellite being protected.

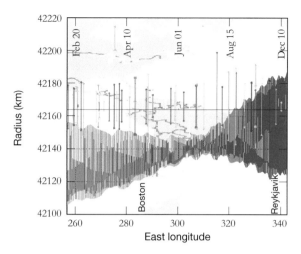

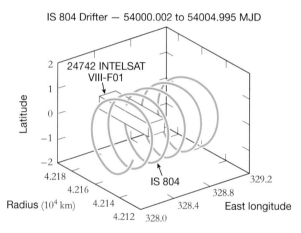

Figure 8.10 (Top) Plot of radial distance against longitude for an inactive drifting geosynchronous satellite (in 2006), shown at four-hour spacing for its propagated orbit. Each colored line represents an active satellite with its longitude and radial range computed from its perigee and apogee. The orbital evolution characterized earlier in figures 8.2a–c and 8.3 has the semimajor axis lower than the geosynchronous radius during the start of this period. As a consequence, the orbital radial distances were below the geosynchronous radius, and the drifter could not cross through the active station-keeping boxes, making a close encounter impossible. Initially (in the summer), the satellite's eccentricity decreased, keeping the radial spread smaller. As the orbit evolved (toward the end of 2006), the semimajor axis and eccentricity increased, and the drifter crossed radially into other active satellite boxes. The determining factor as to how close the drifting satellite will get to the active satellite in its box depends on where the drifter crosses through the station-keeping box of the active satellite as it passes through the active satellite's orbital plane. (Bottom) Three-dimensional view of a single box crossing. This figure shows the projected crossing of the drifter through an active satellite's station-keeping box (in September 2006).

Generally, when a close approach was calculated to be within 4 kilometers with a particular geometry, a modification to the active satellite's planned station-keeping maneuver, to include an avoidance strategy, was suggested, although it was the satellite operator's decision whether or not to implement the suggestion. Advancing or delaying the maneuver by a day was the most common strategy since it provided adequate separation (at least 6 kilometers). Another strategy, which required expenditure of additional fuel, was to change the eccentricity of the active satellite's orbit to increase the radial separation before the crossing, and then to change it back afterward. The operators had relatively rigid constraints on their station keeping, but they always seemed to find a strategy that accomplished their goals and gained a satisfactory increase in separation.

Compatibility of the laboratory's orbit calculations with the CRDA company orbit determinations (issues such as reference frames and solar radiation pressure modeling) was one of the first things checked with the companies, which would make the final decision about the safety of their satellites. Examples of close crossings, where avoidance maneuver strategies were performed, are presented later.

The most difficult aspects of monitoring the conjunction of a drifting and an active satellite were properly modeling a maneuver near a close approach and validating that the maneuver resulted in the expected performance. This validation, though not difficult for the primary component of the maneuver (either north or south or east or west), could be difficult for the coupled components of the maneuver that depended on the active satellite's attitude, hence the direction in which the satellite's thrusters fired. For an east or west maneuver, there could be coupling in the radial direction (causing a change in the orbital eccentricity), and for a north or south maneuver, the coupling could be in both the east or west and the radial directions. Often these coupling components were not given in advance, but their determination was critical for close crossings of certain geometries. As soon as the laboratory's analysts received sufficient metric tracking data, they could estimate the relevant maneuver component values and compare these with the active satellite operator's estimates. Generally, this comparison was done within a day after the maneuver, given the CRDA company's ranging data, with at least two tracks of optical or radar measurements or both.

For very close approaches, the decision to modify a maneuver or specifically perform an avoidance maneuver depended on the orbit accuracy and the encounter geometry. Orbit accuracy was the first issue addressed after the Telstar 401 failure and drift [21]. If the tracking were from a well-calibrated and accurate radar, orbit accuracy was estimated to be on the order of 0.5–2 kilometers (1 σ). If this orbit error is split into components in the along-track (or velocity) direction, the cross-track (or out-of-plane) direction, and the radial direction for the active satellite, the error was on the order of 0.3–0.5 kilometer, 0.5–1.5 kilometers, and 0.05–0.1 kilometer, respectively. If the solar radiation pressure was sufficiently well modeled, the error usually remained at this acceptable level for many weeks into the future.

With only optical tracking data—before the Deep Stare upgrades[5]—and with at least ten tracks of five to ten measurements per track, the error components in the in-track, cross-track, and radial directions were approximately 1 kilometer, 0.2 kilometer, and 0.2 kilometer, respectively. The cross-track and radial components were comparable and better determined than the in-track component. Metric tracking data from both radar and optical sensors are complementary, and total errors achieved were on the order of 0.5 kilometer, with directional components of 0.2–0.4 kilometer, 0.1–0.2 kilometer, and 0.025–0.05 kilometer.[6] With the addition of calibrated CRDA company two-station ranging, orbit accuracy on the order of 50 meters was achievable [21] The limiting factor controlling improved geostationary accuracy seemed to be the simplified average modeling of solar radiation pressure by a simple surface normal to the Sun, whereas the geostationary satellite was actually a combination of surfaces at different angles to the Sun. For a drifting satellite, its orientation to the Sun would have a seasonal variation. The momentum-wheel adjustment thrusts or attitude thrusts for dumping excess momentum could also complicate the orbit modeling for the active satellite at this level, particularly if the dumping of momentum was achieved with a few large, discrete thrusts and not with numerous daily small thrusts spread over the orbit.

Given the ability to determine orbits to within an estimated error of 0.2 kilometer, the obvious question was, why use a separation distance of 6 kilometers for a watch list and one of 4 kilometers for a possible avoidance maneuver strategy? First, these values (for estimated close approaches two weeks prior to the approach) were set in the early days of the cooperative research and development agreement, based on the conservative approach of the CRDA companies. Second, high orbit accuracy could be achieved on the company satellites by using radar, optical, and ranging data in combination, but accuracies tended to be often worse for the satellite drifting in

---

5. Surveillance systems change over time. The GEODSS had an upgrade, called "Deep Stare," in its coupled-charge device sensor and processing software about 2000, which resulted in about a fourfold improvement in tracking data accuracy, down to about 1 arc second in both angular dimensions. Note that 1 arc second at geosynchronous distances is about 175 meters.

6. With Deep Stare GEODSS and radar data or even with adequate Deep Stare data alone, 0.2-kilometer errors (1 σ) are typically achievable.

close to the active satellite because data types with similar accuracies were not routinely available.[7] Since the cost of a collision was extremely high, the distances chosen provided a comfortable margin of safety and could be achieved by appropriate timing of the normal maneuvers (thus incurring no fuel penalty). Of the geometric components, the radial separation between the active and the drifting satellite was much better determined, leading some CRDA companies to be comfortable with a lower distance threshold for an avoidance strategy. For sixty company satellites, a decision had to be made about five times per month regarding crossings within 4 kilometers—a relatively low workload for the satellite operators.

8.3.1   The Xenon Ion Propulsion System: Some Caveats

The XIPS thrusting technology is ten times more efficient than conventional liquid fuel systems for station keeping. Satellites equipped with XIPS could have bigger payloads and longer lifetimes for the same or lower cost. Although there were different strategies for using XIPS, thrusting was generally done a few times per day at irregular intervals of up to a few hours. With XIPS, the impulsive maneuver a few times per month was no longer needed, although a traditional bipropellant fuel was still sometimes used in conjunction with the system. That said, there were decided drawbacks to using XIPS. Its frequent thrusting each day made finding suitably long maneuver-free periods for metric data collection impossible. Without reliable XIPS maneuver information, accurately estimating separation distances from crossing satellites was difficult, and perfect modeling of the active satellite orbit even more so. And when the XIPS operated autonomously, as it did on many satellites (with operator intervention only when absolutely necessary), significant advance planning was needed if an avoidance maneuver strategy was required. In the absence of reliable XIPS maneuver and operational information, the Geosynchronous Monitoring and Warning System provided the best estimates possible on the active and drifter satellite orbits to the satellite operator for analysis and action.

8.3.2   Avoidance Maneuvers

Since the initial work with Telstar 401, the Geosynchronous Monitoring and Warning System has monitored more than 1,250 close approaches within 10 kilometers

---

7.   Most rocket bodies and space-debris objects, as well as tumbling defunct satellites, present a lower radar cross section to the deep-space radars, hence will not be tracked with as high an accuracy as large, actively stabilized satellites. A similar situation obtains for optical sensors.

between CRDA company satellites and about 135 unique drifters, as well as many active satellites. For as many as sixty active company satellites, the GMWS found on the order of 250 crossings per year that were closer than 10 kilometers, and it has suggested about thirty avoidance maneuvers since its monitoring began with Telstar 401 in 1997. Unscheduled maneuvers to avoid close approaches are down to about two or three per year for these sixty active company satellites, with the rest of the maneuvers being accomplished by modifying the time of the scheduled station-keeping maneuver.

The following four strategies can be used by active satellite operators to avoid a predicted close approach of either a drifting or another active satellite:

1. Delay a scheduled station-keeping maneuver;
2. Modify a station-keeping maneuver;
3. Change the satellite orbit's semimajor axis, eccentricity, or both;
4. Rotate the satellite's orbit ellipse.

Although strategies 3 and 4 could require some additional expenditures of fuel, strategies 1 and 2 should require none.

Two of these avoidance strategies are discussed for approaches that were predicted to be very close (less than 3 kilometers) to demonstrate how the strategies increased the separation distance [22]. The first example (October 2005) illustrates avoidance strategy 1, in which a scheduled station-keeping maneuver was delayed (figure 8.11). Here the GMWS predicted a 1.5-kilometer crossing distance between an active CRDA company satellite and the drifter ASC-1 (which was deployed by the U.S. Space Shuttle *Discovery* in 1985 and retired in situ in 1994). The company satellite operator was to make a scheduled east or west maneuver the day before the predicted close approach (which would have *resulted* in the close approach); delaying this maneuver by one day until after the close crossing, however, as suggested by the GMWS, was predicted to increase the separation distance to 17.5 kilometers. The operator was able to effect the suggested maneuver delay while keeping the satellite within its station-keeping box. Postencounter tracking and orbit determination, essential for the GMWS validation process, determined the actual crossing separation distance to be slightly greater than predicted, 17.9 kilometers.

A second example (May 2006) illustrates avoidance strategy 3, in which a small eccentricity change was made for the active satellite's orbit in such a way as to increase the radial separation with the drifter satellite during the closest crossing. The encounter involved the drifting Telstar 401 and an active CRDA company satellite. A west station-keeping maneuver had been performed for the active satellite seven days prior

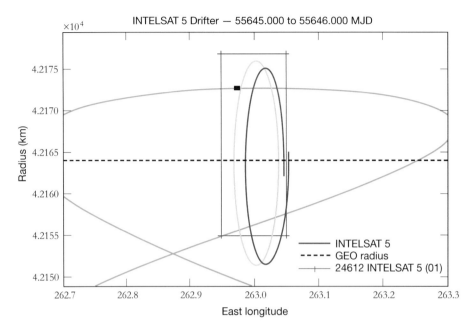

Figure 8.11   Two-dimensional view of the effects of avoiding a close crossing, estimated to be less than 1.5 kilometers. Given the value of the active satellite, and the risk of losing it, an avoidance strategy was performed: delaying by one day an east or west station-keeping maneuver. The red box is the active satellite station-keeping region in the radial and longitudinal projections. The encountering drifter satellite is ASC-1; its trajectory is shown in blue. ASC-1 is a high-inclination drifter (nearly 9 degrees) passing through the station-keeping box of the active satellite with a relative velocity of 0.5 kilometer per second. The black "x" along the ASC-1 trajectory represents where ASC-1 passed through the orbital plane of the active satellite. The green trajectory inside the station-keeping box is the trajectory predicted for the active satellite during the day of the encounter, but before the maneuver was changed. With a change of maneuver, the active satellite trajectory became the magenta path, and the separation distance was increased to 17.5 kilometers. MJD, Modified Julian Date.

to the predicted encounter. As a result of that maneuver, the GMWS estimated the crossing distance to be 2.4 kilometers, with 0 kilometers in the cross-track component, 0.2 kilometer in the radial component, and the rest in the along-track direction. There were no scheduled maneuvers before the encounter, so the operator decided to schedule an avoidance maneuver that involved a two-maneuver change of orbital eccentricity.[8] The first eccentricity maneuver resulted in a total separation of 4.6 kilometers, but, more important, increased the radial separation to 2.4 kilometers, which was considered a safe distance, given the smallness of errors in that component, as discussed in section 8.3 and also in Vigil and Abbot [21].

Maneuvers of a satellite can create close approaches just as quickly as they can eliminate them, especially if maneuver information is not available to the GMWS in a timely manner. An example of this phenomenon is a close approach of an active CRDA company satellite and a rocket body, with a separation distance of 5 kilometers, 4 kilometers of which was in the radial component. This crossing had been on the watch list. Up to nine days before the crossing, the separation distance was estimated to be 57 kilometers, which was generally considered to be safe. The next calculated orbit for the active satellite indicated that a maneuver needed for it to stay in its station-keeping box had actually occurred. When the GMWS finally received the latest maneuver information requested from the operator, it discovered there had actually been two east or west maneuvers over two days. After the maneuvers were included in the orbit fit, the separation distance dropped to 5 kilometers—safe but still a drastic change from the prior estimate.

8.3.3 Collision Threat between Active Satellites in Geosynchronous Belt

A number of geostationary satellites require station-keeping strategies that are subject to additional constraints. The ring-shaped geostationary region of the geosynchronous belt has just one dimension (i.e., longitude) in which to allocate different satellites. With increasing demand for geostationary satellite services over certain regions of the world, many geosynchronous satellites today exist in clusters, which consist of active satellites in neighboring boxes plus others that are collocated or share common "deadband" regions. A cluster can provide connected or individual satellite services from a number of satellites. A well-known example of collocated satellites is the Astra cluster at 19.2 degrees east longitude ±0.10 degree with six satellites (as of about 2000),

---

8. The operator made this avoidance maneuver one of the yearly sets of required eccentricity control maneuvers. As a result, there was no additional fuel expenditure beyond what had been already planned; thus the predicted lifetime of the satellite, which is entirely determined by the available fuel, was not affected.

which are kept separated by the eccentricity and inclination of their orbits. Two satellites may also be collocated for a short time when one replaces another. From the surveillance perspective, a cluster is defined as two or more satellites that have either bordering or shared control boxes. Such satellites can routinely come so close to one another (in a controlled manner) that tracking sensors can mistag them (erroneously assign the tracking data on one satellite to another in that cluster).

Satellites collocated in clusters can be owned or operated by a single owner or operator or by several. For some configurations, the single operator of collocated satellites must keep them within the beamwidth of a fixed ground-station antenna, hence must satisfy their station-keeping requirements while also ensuring that they stay sufficiently separated to avoid colliding with one another. Different operators of collocated satellites must pay strict attention to their own station keeping to avoid interference or a possible collision. Because it is in the best interest of the different operators to share orbit information, they routinely do so.

There are four approaches to collocation of geosynchronous satellites. In the first, different operators are involved, and the risk of a collision is ignored (the probability of collision is considered insignificant by the operators). Signal interference can, of course, be monitored by each operator. In the second approach, the satellites are controlled independently, but a safe separation distance is agreed upon and checked both before and after maneuvers. In the third approach, the collocated satellites are kept separated by longitude, eccentricity, or eccentricity and inclination in combination. This approach can still involve different operators who either exchange information or assume that they are keeping to individual allocated orbital regions (control boxes). In the fourth and final approach, collocated satellites are kept separated by longitude, eccentricity, or eccentricity and inclination, but with offsets so that station keeping for all satellites is done on a predefined schedule. With the same station keeping, they all move in their respective control boxes in the same manner. Clearly, routine proximity checks should be made when using any of these approaches. Figure 8.12 illustrates both longitude station keeping and a collocation of two satellites.

Satellites that have larger longitude control for their station-keeping boxes may overlap satellite clusters. Generally, there is no coordinated effort by operators during these longitude crossings, although proximity analysis must be maintained by the surveillance community. From time to time, a geostationary satellite operator performs a relocation—whether simply to move the satellite to a different longitude, to replace an older satellite with a newer and more capable one over a given service area, or to move an older satellite closer to a gravitational stable point over the equator to conserve fuel and lengthen its lifetime. The rate at which this relocation is effected

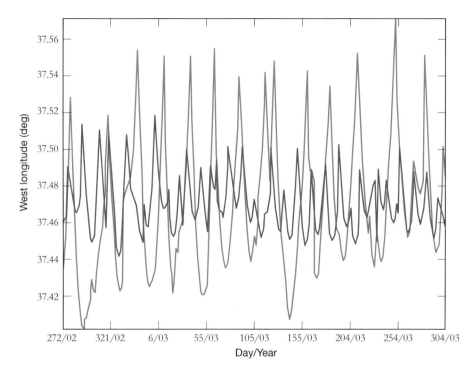

Figure 8.12   Example of longitude station keeping and collocating two satellites, which shared the same longitude slot for 3.5 years. This figure illustrates slightly more than a year of this collocation. One satellite was Telstar 11 (red), which had a longitude box size of ±0.05 degree, and the other was Satcom C1 (blue), with a longitude box size of ±0.1 degree. Collocating these two satellites at the same longitude forced operators from different companies to develop a strategy to keep the satellites separated. The maneuvers of each satellite occur at the greater longitude as they drift in that direction toward the limit of their control box. Lincoln Laboratory played a role in monitoring this collocation and occasionally suggested avoidance strategies to keep the satellites at safe distances. Satcom C1 has since been retired by being boosted into a safe supersynchronous orbit above the geostationary radius.

depends on how much fuel and time the operator wishes to allocate. Because the relocating satellite crosses other active satellites during this move and is more exposed to the population of defunct satellites, careful monitoring is required to avoid a possible collision.

8.4   Summary

The sudden failure of Telstar 401 in situ and its resultant oscillatory drift (with a period of 2.5 years) through a fairly dense part of the geostationary belt was the prime cause for significant research at Lincoln Laboratory in collaboration with the CRDA companies that owned the satellites in that part of space on possible collisions in the geostationary belt. The first problem addressed was the repeated close approaches of Telstar 401 to some seventeen active company satellites. The research was generalized to the problem of the close approaches and collision threat between approximately 400 active satellites in the entire geostationary belt and the several hundred drifting defunct satellites, rocket bodies, and other space-debris objects (down to the smallest size trackable by the Space Surveillance Network). The laboratory developed an automated software system called the "Geosynchronous Monitoring and Warning System" to compute orbits using SSN metric tracking data and CDRA company ranging data to predict close approaches two weeks beforehand and to help satellite operators decide when and how to maneuver their satellites to safely avoid collisions. Along the way, laboratory analysts helped the CRDA companies calibrate their ranging systems and develop collision avoidance strategies that required little or no additional expenditure of fuel—a vital commodity on board satellites. The laboratory then addressed the problem of possible collisions between clustered satellites. The sharing of orbit and maneuver data between operators of satellites collocated in a cluster was an immense help in avoiding close approaches. Finally, the laboratory turned its attention to the residual, smaller problem of possible collisions between geostationary satellites and those in high-eccentricity orbits that happen to pass through the geostationary belt—a problem still under study.

The success of Lincoln Laboratory thus far in addressing these problems can be summarized as follows:

1. A tremendous level of trust was established between the private satellite-owning companies of the CRDA and the laboratory;

2. The laboratory's independent calibration of the companies' ranging data was a vital contribution to their joint efforts;

3. The laboratory's Geosynchronous Monitoring and Warning System has pre-dicted and monitored more than 1,250 close approaches and helped the com-panies plan avoidance strategies for some ten close approaches per year with little or no additional expenditures of fuel.

References

1. National Research Council. 1995. *Orbital Debris: A Technical Assessment.* Washington, DC: National Academy Press.
2. Kessler, D. J., G. C. P. Burton, and G. Cour-Palais. 1978. Collision Frequency of Artificial Satellites: The Creation of a Debris Belt. *Journal of Geophysical Research* 83:2637–2646.
3. *Orbital Debris Quarterly News (ODQN)* http://orbitaldebris.jsc.nasa.gov/newsletter/newsletter.html.
4. Stansbery, E. G., D. J. Kessler, T. E. Tracy, M. J. Matney, and J. F. Stanley. 1996. Characterization of the Orbital Debris Environment from Haystack Radar Measurements. *Advances in Space Research* 16 (11): 5–16.
5. Xu, Y. L., M. Horstman, P. H. Krisko, J. C. Liou, M. Matney, E. G. Stansbery, et al. 2009. Modeling of LEO Orbital Debris Population for ORDEM2008. *Advances in Space Research* 43 (5): 769–782.
6. Settecerri, T. J., E. G. Stansbery, and T. J. Hebert. "Radar Measurements of the Orbital Debris Environment: Haystack and HAX Radars, October 1990–1998," NASA Johnson Space Center Report JSC-28744, 1999.
7. W. Flury, "Collision Probability and Spacecraft Disposition in the Geostation-ary Orbit," *Advances in Space Research*. vol. 11, no. 12, pp. (12)67–(12)79, 1991.
8. Johnson, N. L. (August 1999). Protecting the GEO Environment: Policies and Practices. *Space Policy* 15 (3): 127–135.
9. Allan, R. R. 1963. Perturbations of a Geostationary Satellite by the Longitude-Dependent Terms in the Earth's Gravitational Field. *Planetary and Space Science* 11:1325–1334.
10. Clarke, A. C. "Peacetime Uses for V2" (letter to the editor), *Wireless World*, vol. L1, no. 3, p. 58, February 1945.
11. Martin, D. H. 2000. *Communication Satellites.* 4th ed. El Segundo, CA: Aerospace Press.
12. Soop, E. M. 1994. *Handbook of Geostationary Orbits, Space Technology Library.* vol. 3. Dordrecht: Kluwer Academic.

13. Berlin, P. 1988. *Geostationary Applications Satellite*. Cambridge: Cambridge University Press.

14. Abbot, R. I., and L. E. Thornton. "GEA CRDA Range Data Analysis," MIT Lincoln Laboratory Project Report CESA-2, 28 July 1999.

15. Baker, D. N. (December 2000). "The Occurrence of Operational Anomalies in Spacecraft and Their Relationship to Space Weather" (invited paper). *IEEE Transactions on Plasma Science* 28 (6): 2007–2016.

16. Abbot, R. I., L. E. Thornton, and D. E. Whited. "Close Encounter Analyses of Telstar 401 with Satellites in the Geopotential Well 97-113 Degrees West Longitude," MIT Lincoln Laboratory Project Report CESA–1, 28 July 1999.

17. Conner, J. M., and C. A. Toews. "Estimating Upper Bounds on the Probability of Collision Between Objects in the Geostationary Environment," MIT Lincoln Laboratory Project Report CESA–10, 16 June 2004.

18. Leclair, R. A., and R. Sridharan. "Probability of Collision in the Geostationary Orbit," in *Proceedings of Third European Conference on Space Debris*, pp. 463–470, European Space Operations Centre (ESOC), Darmstadt, Germany, 19–21 March 2001 (ESA SP-473, October 2001).

19. Abbot, R. I., R. Clouser, E. W. Evans, J. Sharma, R. Sridharan, and L. E. Thornton. "Close Geostationary Satellite Encounter Analysis: 1997–2001," in *Proceedings of the American Astronomical Society* 02–115, pp. 219–237, 2002.

20. Abbot, R. I., and J. Sharma. "Determination of Accurate Orbits for Close Encounters between Geosynchronous Satellites," in *Proceedings of the 1999 Space Control Conference*," MIT Lincoln Laboratory Project Report STK-254, 13–15 April 1999.

21. Abbot, R. I., and T. P. Wallace. "Decision Support in Space Situational Awareness," *MIT Lincoln Laboratory Journal*, vol. 16, no. 2, 2007.

22. Vigil, M. C., and R. I. Abbot. "A System for Predicting Close Approaches and Potential Collisions in Geosynchronous Orbit," *Advances in the Astronautical Sciences*, vol. 131, pp. 685–702, AAS Paper 08–086, 2008.

9

Resulting Technology Applications

Joseph Scott Stuart, E. M. Gaposchkin, and Robert Bergemann

In any technological endeavor lasting many decades, inevitably there are spin-off applications that go beyond the primary thrust of the core research. Two such applications of space surveillance research are described in this chapter. It should be noted that, though work on both applications began before 2000, a substantial amount of the work reported here was accomplished in years since.

MIT Lincoln Laboratory conducted a study for the Office of the Chief Scientist of the Air Force Space Command in the mid-1980s on space surveillance, space debris, and near-Earth asteroids (NEAs). The study recommended that the optical surveillance technology developed for deep-space surveillance of resident space objects (RSOs) be adapted to search for near-Earth asteroids. In particular, this technology could be used to help NASA comply with the mandate from Congress to catalog all NEAs above a certain size in approximately ten years. Called "Lincoln Near-Earth Asteroid Research" (LINEAR), this application has been, and continues to be, an extraordinarily successful scientific endeavor.

Several optical sensors that work in both visible and infrared wavelengths, such as those of the Hubble Space Telescope and the European Infrared Astronomical Satellite (IRAS), have been launched into space for studies in astronomy. Detections of resident space objects in the focal planes of such sensors are generally considered part of the noise background and ignored. Occasionally, however, as in the case of the IRAS, these detections are archived, though not analyzed. The IRAS science team very generously shared their archive of suspected RSO detections with Lincoln Laboratory.[1] Analysis of these data has yielded important insights into the detectability of deep-space RSOs in the infrared wavelengths.

---

1. The IRAS archive of science data is located at the Space Research Institute in Groningen, Netherlands.

9.1   A Nontraditional Application of Space Surveillance Technology: Catalog Discovery and Catalog Maintenance of Near-Earth Asteroids

Small, Sun-orbiting celestial bodies whose orbits bring them near to the Earth's orbit, NEAs and other asteroids sparsely fill the inner solar system inside Jupiter's orbit, as shown in figure 9.1. Study of near-Earth asteroids is important for several reasons. NEAs provide an important link between the meteorites and the main-belt asteroids, placing the geochemical information obtained from the meteorites into a spatial context to help us better understand the formation of our solar system. Millions of years ago, NEAs collided with Earth, leaving deep geological imprints on its surface and producing extreme devastation that may have substantially altered the evolution of life on Earth. Such collisions remain a long-term hazard for Earth's inhabitants. In the early 2000s, near-Earth asteroids and near-Earth comets became important destinations for exploration by unmanned spacecraft.

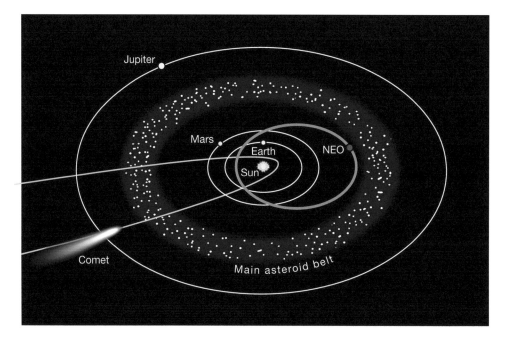

Figure 9.1   Inner solar system showing the orbits of asteroids and comets in the main asteroid belt between Mars and Jupiter (relative orbits drawn to scale). Most asteroids are in near-circular orbits and do not pose a threat to Earth. Asteroids with highly elliptical orbits that cross or approach the Earth's orbit are called "near-Earth asteroids" (NEAs).

When the first near-Earth asteroid was discovered (asteroid 433 Eros in 1898), it was not initially called a "near-Earth asteroid." It was, however, recognized as an unusual object, being the first known asteroid not confined to the main asteroid belt, and the first that crossed the orbit of a planet (Mars) [1]. Several decades passed before astronomers found asteroids that closely approached or crossed the Earth's orbit. Then, in 1932, Eugène Delporte discovered asteroid 1221 Amor, and Karl Reinmuth, asteroid 1862 Apollo. Apollo and Amor quickly became the archetypes and namesakes of two groupings of NEAs [2]. The Amor asteroids have orbits that are entirely beyond, but close to, the Earth's orbit, whereas the Apollo asteroids have orbits that cross the Earth's orbit. In 1936, Delporte found asteroid 2101 Adonis, which passes 2.4 million kilometers from the Earth's orbit [3]; the next year, Reinmuth found asteroid 69230 Hermes, which passed to within 800,000 kilometers of the Earth. After that passage, Hermes was lost until 2003, when it was rediscovered [4].

The growing number of asteroid discoveries prompted the astronomical community to establish the Minor Planet Center (MPC) in 1947 to serve as a central repository for asteroid observations, to maintain a catalog of asteroid orbits, to quickly notify astronomers of important observational opportunities, and to coordinate follow-up observations to pin down the orbits of newly discovered NEAs and other interesting asteroids.[2] The workload of the MPC was much increased with the first dedicated survey to find asteroids, the Yerkes-MacDonald Survey, started in 1950 by Gerard Kuiper [5], which ran until 1952 and discovered 1,550 asteroids. The Palomar-Leiden Survey [6] followed in 1960, finding more than 2,000 asteroids in two months of observations. Both surveys were designed to find main-belt asteroids. The first search specifically designed to find near-Earth asteroids was the Palomar Planet-Crossing Asteroid Survey [3], started in 1973 by Eugene Shoemaker and Eleanor Helin, who discovered five new near-Earth asteroids, seven new Mars crossers, and many main-belt asteroids [7].

Efforts to discover asteroids have always sought the latest technology. The switch from direct visual to photographic observation through the telescope in the 1890s led to the photographic discovery of asteroid 323 Brucia [8]. Photography dominated asteroid searches (as well as all optical astronomy) until the advent of electro-optical techniques in the 1960s and 1970s. And, in the 1980s, the advent of charge-coupled devices (CCDs) as astronomical instruments revolutionized the search for and study of asteroids of all types [9]. The first group to make use of CCDs to search for asteroids was Spacewatch, which discovered its first asteroid in 1985, although it did not

---

2. http://www.minorplanetcenter.net/iau/mpc.html.

become fully operational as a large search effort until 1990 [10]. Complementing Spacewatch, the Lowell Observatory Near Earth Object Search (LONEOS) began in 1993 with the principal goal of discovering near-Earth asteroids and comets. In 1995, Helin began a successor to the Palomar survey to find NEAs, the Near-Earth Asteroid Tracking program [11]; in 1996, the laboratory began the LINEAR program [12]; and in 2003, the Catalina Sky Survey began operations [13]. Of course, the search for asteroids, and near-Earth asteroids in particular, continues with larger telescopes, wider fields of view, and ever-advancing detector technology.

9.1.1   Early Development of Asteroid Search Capabilities at the Laboratory

Chapter 4 provides a detailed description of the development of electro-optical space surveillance technology by the laboratory, which began applying the equipment and techniques of the Air Force space surveillance mission to the job of asteroid search in 1979.

The application used the prototype Ground-Based Electro-Optical Deep-Space Surveillance (GEODSS) telescope, equipped with television camera tubes of the intensified silicon diode–array type. The video output from the GEODSS camera was integrated in an image processor and stored as a 512×512, 12-bit image on a DS-20 digital integration and storage unit. The image was then transferred to an Arvin/Echo EFS-1A Discassette video recorder, which could store up to 200 images. After enough time had passed to allow asteroids with typical apparent (angular) velocities of about 30 arc seconds per hour to move by a few resolution elements in a typical time span of about 20 minutes, the same field was repeated, and the DS-20 unit computed the difference between the new image and the stored reference image. Stationary objects (stars) would mostly subtract away in the difference image, whereas slow-moving asteroids would produce a black-white image pair that could be easily discerned by a human analyst. The system required 10–15 seconds to accumulate an image; this included moving and settling the telescope, image quality checks, and data acquisition. The time between the two images of a field was spent using the automated telescope control system to image other nearby fields. After the second image was acquired, a trained analyst spent 20–40 seconds examining the difference image to find the characteristic black-white double image of a moving object [14].

After centering each detected asteroid image on a fiducial mark etched onto the front of the camera's video tube, the telescope operator would record the telescope pointing angles and, with assistance from the telescope control software, measure the positions of four or five nearby stars from the Smithsonian Astrophysical Observatory Star Catalog, using the data to compute a local pointing model for the telescope and the astronomical coordinates of the target asteroid. This astrometry process typically took less than two minutes and produced results with a precision of 2–3 arc seconds, not including systematic errors in the star catalog. Detected asteroids were compared with those in the Minor Planet Center catalog. For known asteroids, a single observation was reported to the MPC. For previously unknown asteroids, the field would be revisited several more times during the night to provide more data for initial orbit determination [15].

Used in combination, the GEODSS telescope and the intensified video tube produced a useful field of view of 0.84 × 0.66 degree, and the digitization process resulted in a resolution of about 5 arc seconds per pixel. With a dark sky background of 20.5 m_v per square arc second, typical at the laboratory's Experimental Test Site (ETS) in Socorro, New Mexico, the system was capable of detecting asteroids down to limiting magnitudes of about 16.5 m_v with 2 seconds of integration [16]. A wide-area search was constructed by first acquiring reference images covering a 7 × 7 box totaling 27.2 square degrees and requiring about fifteen minutes of elapsed time, including image integration and telescope step and settle. A second pass of the forty-nine fields of view was then completed in 30–50 minutes with the extra time necessary to allow the operator to visually scan the difference images for objects and to measure the positions of any objects found. The system produced a potential search rate of almost 30 square degrees per hour [14].

The laboratory's Earth-crossing asteroid search was operated with funding from NASA for five to ten nights per month from January 1981 through March 1984, typically covering 300–400 square degrees of sky per month. The search provided 110 observations of asteroids to the Minor Planet Center during the development period and an additional 1,273 observations of asteroids during regular operation; it resulted in the discovery of 73 new asteroids, none of which were near-Earth asteroids.[3]

As early as 1981, the laboratory recognized that the new solid-state CCD imagers being manufactured there and elsewhere would provide greater capabilities for asteroid search than the video tube technology of the 1970s. It began planning for an upgraded asteroid survey that would use CCDs [17] and have a goal of increasing the limiting magnitude of detection to about 17.8 m_v and the area coverage rate to over 250 square degrees per hour. In addition to providing greater sensitivity and coverage rates, the CCDs were expected to enable the search and detection process to be fully automated [17], as indeed it was after the Air Force upgraded the GEODSS system to CCD cameras in 1996 (see chapter 4).

---

3. Observations and provisional discovery credits are documented in Minor Planet Circulars 5921 through 9037.

9.1.2   Initial Field Tests with Charge-Coupled Device Imagers and the Lincoln Near-Earth Asteroid Research System

Soon after the initial GEODSS upgrades, the new CCD cameras were used with the 31-inch telescopes at the laboratory's Experimental Test Site (ETS, see chapter 4) in experiments for asteroid search. Tests were conducted first with prototype cameras having 420×420–pixel CCD sensors in 1992 and then with cameras having $1,000 \times 1,000$–pixel sensors in 1993; the larger cameras successfully detected dozens of near-Earth asteroids, including some as faint as about magnitude 21 m_v, in a few nights [17].

For these experiments, the laboratory constructed a new image acquisition, storage, and processing system based on Silicon Graphics computers and hard-disk storage. The technique of imaging the same field twice, with a time delay, and subtracting the images of stars to reveal those of moving asteroids was extended to the images taken with the new cameras and to the use of more than two frames to provide more refined characterization of the static-sky background and better clutter rejection. With the larger field of view of the new cameras and new star catalogs, astrometric calibration stars and target asteroids could be imaged simultaneously, thereby producing an astrometric solution for the detected asteroids. This work also led to the development and validation of a detailed, analytical sensor performance model that could predict the performance of electro-optical sensors by accounting for all the relevant signals and noise sources [17].

After the 1993 tests, the laboratory constructed another generation of CCD cameras for the GEODSS upgrade program. These new cameras contained advanced CCD imagers—Charge-Coupled Imaging Device 16s, or CCID-16s—manufactured at the laboratory. Being larger ($2,560 \times 1,960$ pixels or 61×47 millimeters) than the previous CCDs, the CCID-16s provided a larger field of view on the sky; they were also back-illuminated devices with much greater sensitivity [17]. These features combined to make the new cameras superior to any other instrument then available for asteroid search. The combined GEODSS telescopes and CCID-16 cameras, together with advanced image-processing hardware and technology, became the heart of the laboratory's next foray into asteroid search. This effort eventually gave rise to the Lincoln Near-Earth Asteroid Research program, located at the Experimental Test Site, which led the world in discovering near-Earth asteroids for many years.

The LINEAR system comprised two GEODSS-type telescopes, with nearly identical cameras (based on CCID-16s) and identical data-processing systems, working in tandem to cover as much sky as possible. The telescopes had 1-meter primary mirrors in a folded prime-focus configuration with very stable, agile, equatorial mounts able to routinely achieve step-and-settle times of 2 seconds. The telescope's 2.19-meter focal length, combined with the CCID-16, provided a field of view of 2 square degrees. The large field of view, quick step-and-settle times, and fast readout of the CCDs allowed a single LINEAR telescope to easily cover 1,200 square degrees of sky each night of observing, with each field being visited five times to provide data for the moving-target-detection algorithm [12].

9.1.3 Data-Processing Techniques

The operational Lincoln Near-Earth Asteroid Research system included two complete computer systems for processing the prodigious amounts of data from the two telescopes (56 gigabytes per night). Figure 9.2 displays the LINEAR software processing flow. The input data typically consist of five CCD images of the same location on the sky collected with a delay of 15–30 minutes between successive images. The central box highlights the core functionality of the moving-target-detection algorithm, which has five major steps. The first step, image registration, corrects for small pointing differences between the images, which are deliberately introduced to reduce false detections from bad pixels. The images are registered by shifting the frames as necessary to align the stellar backgrounds of subsequent images with that of the first image. In background suppression, the second step, an estimate of the local background mean and standard deviation is computed at each pixel, averaging over the five images. Each pixel is then normalized by subtracting the local mean and dividing by the local standard deviation. In this way, high noise near bright stars is suppressed without destroying information in the low-noise areas of the image. In binary quantization, the third step, a simple threshold is applied to each image to produce a constant number of above-threshold pixels. The pixels whose normalized values are above the threshold are considered to be "lit" for purposes of clustering and moving-target detection [12].

In clustering & velocity, the fourth step, the binary quantized data are clustered by grouping contiguous lit pixels into clusters, and the centroids and sizes of the clusters are computed. Each cluster in the first frame is then paired with each cluster in the last frame to fall within a specified radius, selected as an upper limit on asteroid rates of motion. These pairs form the list of candidate detections or streaks. The velocity of each streak is computed by dividing the displacement from the beginning to the end of the streak by the time interval that it spans. For each candidate streak, intermediate frames are searched for clusters with the appropriate displacement to match the streak's velocity. The matching clusters are added to the candidate streak. Once all of the candidates have been gathered, the fifth and final step, detection,

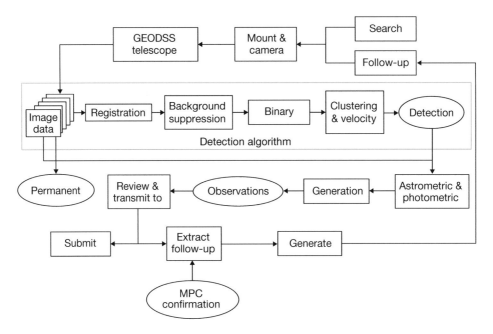

Figure 9.2   Process flow at Lincoln Near-Earth Asteroid Research (LINEAR). MPC, Minor Planet Center.

begins by rejecting streaks that have too few clusters. The remaining streaks are checked to ensure that the motion of the object matches a constant velocity profile and that the shape of the object remains reasonably consistent from frame to frame. Conflicts in which a pixel cluster is claimed by multiple streaks are resolved by assigning the popular cluster to the streak that it fits the best and rejecting the others. If the minimum number of frames necessary to allow a streak to be declared a valid detection is lower than the number of frames collected, then additional rounds of streak detection are performed. For example, LINEAR typically takes five images and requires three detections per streak. The first round of streak detection creates candidate streaks with endpoints in frames 1 and 5, the next round in frames 1 and 4. A third round of streak detection would select streaks with endpoints in frames 2 and 5, another round with endpoints in frames 1 and 3, and finally a round with endpoints in frames 3 and 5. This process ensures that all possible streaks with at least three detections are found, with preference given to longer streaks [12], and completes the five steps of the detection process.

After the final list of streaks is generated, the software returns to the original images for further processing. First, star images are found and matched to a star catalog to generate a precise plate model for converting image pixel coordinates to equatorial reference coordinates and CCD intensities to calibrated magnitudes. The center and intensity of each detection are recalculated in each input image and converted to equatorial coordinates and stellar magnitudes. The list of streaks is thus converted to a list of observations of detected asteroids in standard astronomical coordinates. The velocities of all detections are checked to flag those which are inconsistent with main-belt asteroid motions and thus likely to be near-Earth objects (NEOs) or some other category of asteroid. Up to this point, the processing of a night's worth of data is completely automated; this requires approximately five hours of computer time for a night in which no fields are lost to poor weather. The detections that are likely to be near-Earth objects are then checked by an analyst to weed out a number of common artifacts that often fool the automated detection software, such as a sporadic hot pixel or conjunction with a faint star combining with normal background noise fluctuations to produce a false detection. Although the list of candidate NEOs is checked for false detections, the larger list of probable main-belt asteroids is not checked for two reasons. First, the process of selecting detections with unusual velocities is likely to sweep up many false detections, which, not being constrained by the laws of orbital mechanics, can exhibit any velocity. Second, because potential near-Earth objects are given preferred treatment by the rest of the astronomical community, considerable telescope resources can be wasted attempting to verify false detections with NEO-like motions [12].

After the potential NEOs are checked by an analyst, the data are transferred to LINEAR computers at the laboratory's main facility in Lexington, Massachusetts, where the detections are divided into several groups. Detections of slow-moving objects whose motions are consistent with those of main-belt asteroids are reformatted and sent to the Minor Planet Center. Detections of objects whose motions mark them as potential NEOs are also reformatted and sent to the MPC, but these are sent separately so that the MPC can process them more quickly and place any potential new near-Earth objects on the NEO Confirmation Page on the center's website. A third category of detections, those of objects whose orbits are potentially consistent with NEO orbits, but whose identification as near-Earth objects is uncertain, are filtered out from the data to be sent to the MPC. Initial orbit determination is performed on these "maybe" objects, ephemerides are generated for the next night, and the objects are targeted for follow-up observations. In addition to the list of uncertain NEO detections from the previous night, follow-up fields are added for objects from the

NEO Confirmation Page regardless of which survey submitted those observations to the Minor Planet Center. The list of objects targeted for follow-up on any given night is pared down to ensure that it uses no more than 10 percent of telescope resources per night, so the telescopes can concentrate on LINEAR's biggest strength: wide-area search [12].

9.1.4   Operational Approach

To accomplish a program of efficient wide-area search, the Lincoln Near-Earth Asteroid Research program sought to cover the entire sky that was available in each lunar month with the following two additional constraints:

1. The telescopes were pointed no closer than 20 degrees from the Moon to limit degradation of LINEAR sensors' sensitivity; and
2. The region around solar opposition was covered at least twice each lunar dark period.

During the first few days of each observing run, while the Moon was nearly full and fairly close to opposition, the telescopes were scheduled to observe either far to the north or far to the south, depending on the declination of the Moon. The regions away from opposition were searched in long strips of constant declination from as far west to as far east as possible, given the limits of sunset and sunrise times. As the waning Moon moved away from opposition, one of the LINEAR telescopes pointed to the opposition region and covered a long strip of constant ecliptic latitude (within ±10 degrees) each night while the other telescope continued covering the sky away from the ecliptic in strips of constant declination. The processing software automatically sent the Minor Planet Center a summary of the fields observed by LINEAR telescopes every night. The MPC facilitated cooperation among the several near-Earth object search programs to minimize duplication of effort [12].

9.1.5   Lincoln Near-Earth Asteroid Research Program Results

Operating with GEODSS-type telescopes from 1998 to 2013, the NASA-funded LINEAR system proved to be a reliable supplier of discoveries of near-Earth objects, comets, unusual asteroids, and main-belt asteroids. Indeed, by deploying its telescopes for as many nights as possible during each lunation and by covering the entire available sky at least once each lunation, the system proved to be a key contributor to the discoveries of large near-Earth objects (diameter greater than about 1 kilometer) for every year of its operation.

9.1.5.1   Nights Observed and Sky Coverage

Initially, the LINEAR program proposed to schedule a minimum of twenty nights' observing per lunar dark period, with the expectation that some nights would be lost to poor weather. In actuality, the program was able to schedule twenty-two nights' observing per lunar month, although poor weather and equipment outages reduced that number to an average of fifteen nights (see figure 9.3).

During each night's observing, each telescope was scheduled for 500 to 600 fields, or about 1,000–1,200 square degrees of coverage. Because of poor weather, the two LINEAR telescope systems, L1 and L2, actually observed an average of 525 and 506 fields per night of observing, respectively, from January 2000 through May 2013. The L1 system averaged 1,038 square degrees per night for a total of 2,084,780 square degrees, whereas the L2 system averaged 1,002 square degrees per night for a total of 1,629,463 square degrees (see figure 9.4).

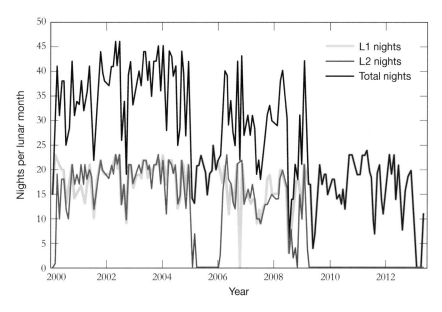

Figure 9.3   LINEAR program nights of observing. LINEAR scheduled 20 to 25 nights of observing in every lunar dark period. After accounting for outages caused by inclement weather and equipment failure, the L1 and L2 systems observed for a total of 2,795 and 1,625 nights, respectively, during the period January 2000 to May 2013.

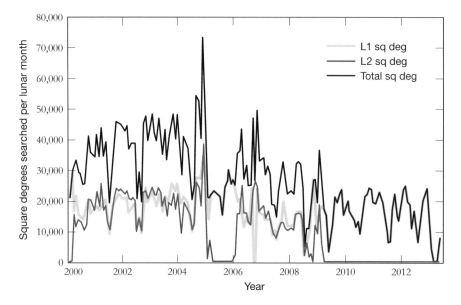

Figure 9.4   LINEAR sky coverage per lunar month. From January 2000 through May 2013, the L1 system averaged about 18,000 square degrees per operational month, while the L2 system averaged about 14,000 square degrees. Thus the LINEAR program searched about 4.8 million square degrees of sky for asteroids. The moving target detection algorithm uses five visits to each field in a night, so that represents over 22 million square degrees of sky imaging. In 2007, LINEAR increased the integration time by about 25 percent to detect fainter limiting magnitudes than previously, resulting in a concomitant decrease in the areal coverage.

9.1.5.2   Depth of Search

Using integration times that range from 5 seconds on short summer nights up to 15 seconds on long winter nights, the depth of search for the LINEAR system was generally limited to a magnitude of about 19 m_v. The limiting magnitude was calculated for each field every night. Figure 9.5 shows the sky coverage for one dark period in 2008. Figure 9.6 shows one full year, and figure 9.7 shows the 13-year period between January 2000 and May 2013. In each case, the limiting magnitude calculated for each field is converted to the integration time necessary to achieve that limiting magnitude in good weather. The good-weather-equivalent integration times for all the visits to each field are summed and then converted back to a limiting magnitude to give an impression of how the cumulative depth of search varies over the sky.

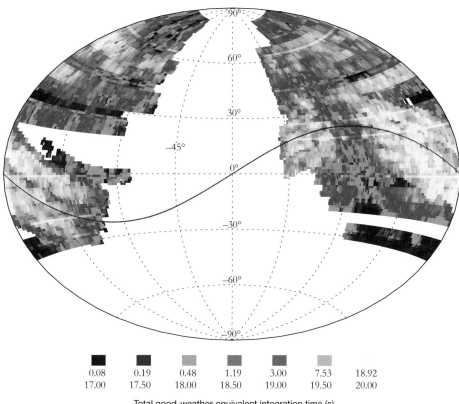

0.08	0.19	0.48	1.19	3.00	7.53	18.92
17.00	17.50	18.00	18.50	19.00	19.50	20.00

Total good-weather-equivalent integration time (s)
Equivalent coverage depth (m_v of SNR = 6.0 star)

Figure 9.5   LINEAR sky coverage for one dark period (26 March to 17 April) in 2008. SNR, signal-to-noise ratio; m_v, apparent visual magnitude.

9.1.5.3 Detection Statistics

The LINEAR program has discovered over 2,400 near-Earth objects and is responsible for 30 percent of the observations in the Minor Planet Center's database. The program has consistently maintained a highly productive search every year since beginning full-time operations in 1998, discovering more than one-quarter of all known near-Earth asteroids to date, nearly one-half of all large NEAs (diameter greater than 1 kilometer), and more than 40 percent of all known potentially hazardous asteroids (diameter greater than about 100 meters and minimum orbit intersection distance less

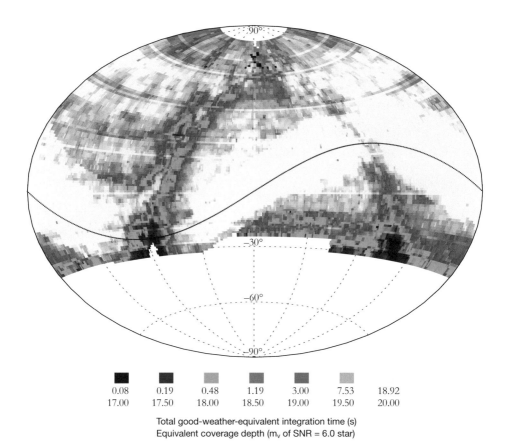

Total good-weather-equivalent integration time (s)
Equivalent coverage depth (m$_v$ of SNR = 6.0 star)

Figure 9.6   Total sky coverage by LINEAR for 2008.

than 0.05 astronomical units). Table 9.1 summarizes LINEAR observations and discoveries from pre-1998 to 2013.[4]

9.1.5.4   Archival Data

The archival data from the LINEAR survey offer a unique window into the temporal variability of the night sky that is not covered by other surveys. There have been

---

4.   LINEAR's near-Earth object discoveries and observation submissions are detailed in Minor Planet Circulars 29698 through 86285.

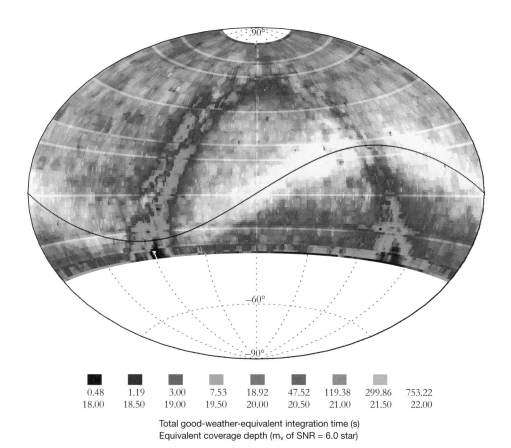

Figure 9.7 LINEAR sky coverage for 28 December 1999 to 13 May 2013. Note change in color scale relative to figures 9.5 and 9.6.

no other surveys with limiting magnitudes fainter than about 15 m_v that repeatedly imaged a large area of the sky (thousands of square degrees per night), with revisits spanning days to years. In 2008 and 2009, the laboratory's researchers teamed with astronomers from the University of Washington to extract photometric measurements for the portion of the LINEAR survey that overlaps spatially with the Sloan Digital Sky Survey (SDSS) [20, 22]. The SDSS provided excellent photometric calibration for the LINEAR images and accurate multicolor absolute photometry for the stars. The archival LINEAR data provided a temporal history of the changes in brightness of all the stars in the field of regard.

Table 9.1   Yearly summaries of Lincoln Near-Earth Asteroid Research observations and discoveries

Year	Observations accepted by Minor Planet Center	Near-Earth object discoveries	Comet discoveries	Other discoveries
2013	38,380	3	0	1
2012	1,480,069	46	9	164
2011	1,717,351	70	7	368
2010	2,193,495	104	11	892
2009	1,861,072	108	16	903
2008	2,511,332	140	8	2,183
2007	2,492,493	112	13	3,569
2006	2,234,995	96	11	1,765
2005	2,056,466	137	12	4,621
2004	3,219,318	304	25	17,444
2003	3,005,532	235	27	15,622
2002	3,076,259	286	34	31,545
2001	3,096,728	279	33	48,108
2000	2,094,399	258	29	52,523
1999	1,107,675	161	36	29,119
1998	585,480	136	17	18,152
Pre-1998	79,379	18	0	2,126
Total	32,850,423	2,493	288	229,105

The LINEAR Survey Photometric Database comprised several million individual images acquired from January 2003 (when archiving began) to late 2009. Approximately five billion brightness measurements were extracted from this imagery. These measurements were positionally clustered into about 25 million objects that had more than fifteen observations each. Objects in the database that are within ±10 degrees of the ecliptic plane have a mean of about 460 brightness measurements, whereas those away from the ecliptic plane have an average of about 200 brightness measurements. Comparison with the Sloan Digital Sky Survey reveals the LINEAR database to be more than 90 percent complete down to a limiting magnitude of m_{LINEAR} less than 19. The LINEAR Survey Photometric Database has been recently released for public use and is hosted at the website for the Los Alamos National Laboratory Sky Database for

Objects in Time-Domain (SkyDOT: skydot.lanl.gov) [21]. In addition to its use to characterize RR Lyrae variable stars and thereby map out the structure of the Milky Way galaxy [22], the data have also been used to characterize the optical variability of blazars[5] identified by the Fermi Gamma-ray Space Telescope [23].

9.1.6 Contributions to Comet Science

In addition to being the world's most productive asteroid search program, LINEAR has significantly advanced the field of comet science. Based on algorithms used to detect Earth-orbiting satellites, the LINEAR detection algorithm is fundamentally a moving-object detector. Any object in motion across the fixed-star pattern within the dynamic range of the algorithm (about 0.1–10 degrees per day) is duly recorded. Because these rates of motion are characteristic of comets as they enter the inner solar system, the LINEAR system has discovered more than 200 of them, making it the most prolific ground-based discoverer of comets in history.

Most of the comets discovered by LINEAR are found on their inbound trajectories as they pass the orbits of Saturn or Jupiter. At this point, a comet starts to brighten as volatile materials are sublimated by solar heating, and the comet becomes detectable by the LINEAR system, although typically the system does not notice any comet-like trailing feature that would clearly identify the object as a comet at this time. Thus the LINEAR object-detection observation is routinely passed to the Minor Planet Center along with hundreds of thousands of asteroid observations generated each month. One of two possible actions results in the object being identified as a comet:

1. The orbit of the object is calculated and determined to be comet-like, as opposed to asteroid-like, and the Minor Planet Center requests an observer with a large telescope to check the object for a tail; or

2. If the object is posted on the MPC's NEO Confirmation Page because of its interesting rate of motion, a follow-up observer may detect a tail.

This process of comet discovery is fundamentally different from what it was in the pre-LINEAR era, when amateur observers usually discovered comets by scanning regions close to the Sun. By the time comets are near the Sun, they have heated up and formed the characteristic tail that makes them detectable. There are, however, two deficiencies in this earlier approach:

---

5. https://en.wikipedia.org/wiki/Blazar (last modified on 23 July 2016): "A blazar is a very compact *quasar* (quasi-stellar radio source) associated with a presumed *supermassive black hole* at the center of an active, giant elliptical galaxy. Blazars are among the most energetic phenomena in the universe."

1. Only comets that travel close enough to the Sun and are active enough to develop a large tail are discovered; and

2. The comets are discovered only after they have substantially completed their inbound trajectories. Thus the heating and tail formation process are not observed or recorded.

By finding comets far from the Sun, LINEAR overcomes both these deficiencies, discovering comets that never form tails large enough to be visible, and, more important, discovering comets early in their trajectories. This early detection enables comet scientists to gather observations covering the interval in the comets' orbits from when they become active and evolve a tail to when they break into pieces. In addition, finding them far from the Sun allows scientists time to schedule additional observations by other assets, such as the Hubble Space Telescope and the Keck Observatory. These observation opportunities have led to some striking discoveries and have resulted in the dedication of a special section of *Science* magazine to comets [19], with a special focus on comet C/1999 S4 LINEAR, which was discovered on 27 September 1999, just inside the orbit of Jupiter (figure 9.9).

By June 2000, the comet had a well-developed tail, as shown in the CCD image in figure 9.8, and was expected to be visible to the naked eye at a closer approach to the Earth (the dark-adapted eye at a dark site is sensitive to objects of magnitude 5–6). In reality, C/1999 S4 LINEAR peaked with a magnitude of about 6.5 m_v in late July (visible through binoculars), before disintegrating from 21 to 24 July 2000. Because of the long time between discovery of the comet and its closest approach to Earth, the Hubble Space Telescope was scheduled to observe it in July 2000 and recorded the comet's activity and residual cometesimals. These images of comet C/1999 S4 LINEAR provided a wealth of insights into both comet evolution and comet function (figure 9.8) [19].

9.1.7   Other Contributions

9.1.7.1   Unusual Asteroids

Not surprisingly, among LINEAR's large number of discoveries of asteroids and comets are some interesting and unique objects. A notable discovery, made in February 2003, is that of an asteroid, provisionally designated "2003 CP20," but now numbered and named "asteroid 163693 Atira," whose orbit is entirely inside the Earth's orbit—the first known member of a new class of inner-Earth-orbit asteroids. The existence of such asteroids, though theorized for years, was not proven until the

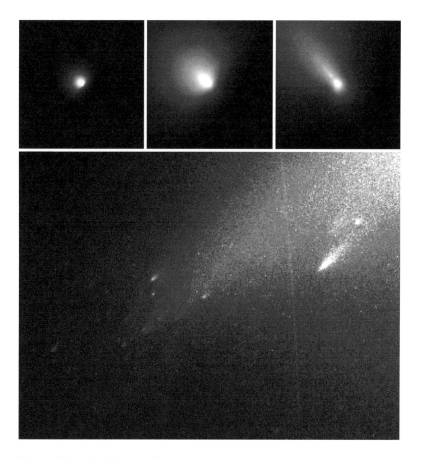

Figure 9.8   (Top) Hubble Space Telescope observations show comet C/1999 S4 LINEAR flaring up and beginning to disintegrate in early July 2000. (Bottom) Later observations show the cometesimals remaining a few days after the breakup of S4 LINEAR on 24 July 2000 (image credits: NASA; Space Telescope Science Institute (STScI); H. Weaver and P. Feldman, Johns Hopkins University; M. A'Hearn, University of Maryland; C. Arpigny, Liège University; M. Combi, University of Michigan; M. Festou, Midi-Pyrénées Observatory; and G.-P. Tozzi, Arcetri Observatory).

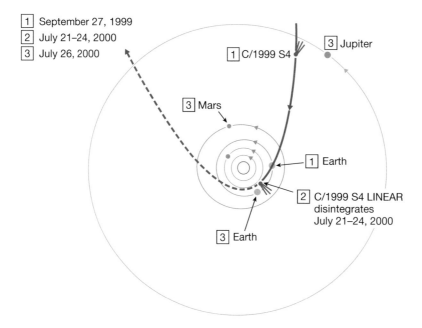

1 September 27, 1999
2 July 21–24, 2000
3 July 26, 2000

3 Jupiter

1 C/1999 S4

3 Mars

1 Earth

2 C/1999 S4 LINEAR
disintegrates
July 21–24, 2000

3 Earth

Figure 9.9   Path of comet C/1999 S4 LINEAR through the solar system. It was discovered at point 1 and disintegrated in late July, 2000, near its closest approach to Earth. The position of Earth is shown as a green circle for the discovery epoch (1) and the disintegration epoch (3).

discovery of Atira [24]. LINEAR has also discovered two near-Earth objects in resonance with the Earth, both with unique horseshoe-type orbits, asteroids 54509 YORP and 2002 AA29. Although asteroid YORP maintained its horseshoe-type orbit only through the year 2006, asteroid 2002 AA29 will likely be the Earth's companion for at least another hundred years. Asteroid YORP is also significant because measurements of its rotation rate over several years provided the first observational evidence for the Yarkovsky-O'Keefe-Radzievskii-Paddack (YORP) effect, in which differential solar illumination and thermal radiation alters the rotational state of an asteroid.[6] In January 2000, LINEAR discovered a Sun-grazing asteroid—asteroid 2000 BD19 (137924)—one with the closest known approach to the Sun [26]. Even though no cometary activity has been spotted in association with the asteroid, some astronomers have suggested that it is, in fact, an extinct comet [27]. In June 1999, LINEAR

6. See note 4.

discovered the first known retrograde asteroid, which is now numbered 20461 and, in honor of its backward orbit, named "Dioretsa" [28].

Besides discovering asteroids with unique orbits, LINEAR has also found a number of asteroids with unique light curves. Radar observations have shown asteroids 1999 KW4 (66391) and 2000 DP107 (185851) to be binary, orbiting each other while orbiting the Sun [29, 30]. Finally, an early LINEAR discovery—asteroid 25143 Itokawa, named, at the laboratory's suggestion, in honor of the early Japanese rocket developer Hideo Itokawa—was chosen by the Japanese as the target destination for the *Hayabusa* mission to an asteroid. The *Hayabusa* spacecraft rendezvoused with asteroid Itokawa in 2005 and returned to Earth with samples in June 2010 [31, 32].

9.1.7.2   Near-Earth Asteroid Population Estimates

Because NEAs have a range of orbits that make some easier to discover than others, there is an observational bias that makes the distribution of orbital parameters for the known near-Earth asteroids unrepresentative of the distribution for those as yet undiscovered. This bias depends in complicated ways both on the orbits of the NEAs and on the surveying strategy used. An important aspect of understanding the distribution of near-Earth asteroids is to correct for this observational bias to obtain the true, unbiased distribution of the asteroids' orbital parameters.

After only a few years, LINEAR's asteroid survey amassed far more asteroid observations than any other single survey before it. LINEAR gathered these observations while operating in a consistent way night after night, year after year. The huge number of asteroid observations and the consistency of operations became an unmatched resource for estimating the total number of near-Earth asteroids, including those not yet discovered, while accounting for the observational biases that are inherent in any telescopic survey.

In 2001, LINEAR's asteroid observations were used to estimate the number of near-Earth asteroids as a function of their absolute magnitude and orbital parameters [33]. This population estimate was significantly more precise than previous estimates and provided a basis upon which the NEA research community could plan the next ten years of NEA survey efforts. In 2003, the NEA population estimate based on LINEAR's observations was combined with asteroid albedo measurements from a variety of sources and with visible and near-Infrared reflectance spectroscopy to obtain a debiased estimate of the number of near-Earth asteroids in terms of their diameter, albedo, and spectral classification [34]. These results combined the largest set of existing NEA discovery statistics from a single survey with the largest existing set of physical data on NEAs and of corrections for observational bias. The result provided a comprehensive estimate of the total near-Earth asteroid population in terms of orbital parameters, absolute magnitudes, albedos, and sizes. This improved description helped researchers plan surveys to find and study the remaining undiscovered near-Earth asteroids, to connect such asteroids both to their origins in the main belt and to meteorite samples, to compare the lunar and terrestrial cratering record to the current population of potential impactors, and to understand the magnitude of the NEA impact hazard to the Earth's biosphere.

9.1.8   The Ceres Connection

Under the rules of the International Astronomical Union, the discoverer of an asteroid has the right to name it. In order for an asteroid to be formally numbered, and thus eligible for naming, its orbit must be well determined so that the asteroid will not be lost in the future. Determining a good orbit normally takes a few apparitions, or perhaps five years for a main-belt asteroid. Observing since March 1998, the LINEAR program has accrued discovery credit for approximately 231,000 objects, of which more than 127,000 have been numbered and are available to be named. Each month several hundred more LINEAR discoveries are numbered, thus continuously adding to the total. By 2001, LINEAR had accrued enough naming rights to precipitate serious thought on how to employ these rights to greatest benefit (the International Astronomical Union forbids the use of naming rights for financial gain).[7] After careful consideration, the laboratory decided that the highest and best use of the rights was to promote science education in the international community. To that end, it named LINEAR-discovered asteroids in honor of middle and secondary school students—and their teachers—who demonstrated excellence in select science competitions. The name chosen for the asteroid-naming program was the Ceres Connection because Ceres, discovered by Italian astronomer Giuseppe Piazzi in 1801, was the first known asteroid in our solar system, a dwarf planet and the largest of the main-belt asteroids. Developed in cooperation with Science Service, Inc.—now known as the "Society for Science and the Public (SSP), an organizer and administrator of several national and international competitions—the Ceres Connection fits in well with the educational outreach objectives of Lincoln Laboratory, its parent, MIT, and NASA. The program was inaugurated on 23 October 2001, with an awards presentation (by Grant Stokes of the laboratory) in Washington, D.C., to the forty finalists and their teachers in the

---

7. See note 4.

Discovery Science Challenge Competition (now known as the "SSP Middle School program"). The asteroid name is either the award recipient's last name or, if an asteroid already had that or a similar-sounding name, a name is derived from a combination of the recipient's first and last names.

During the 2001–2002 academic year, the Ceres Connection awarded additional naming honors to the forty finalists and their teachers in the Intel Science Talent Search and to 105 student winners at the International Sciences Fair held in Louisville, Kentucky. In addition to rewarding the specific achievements of these students, the Ceres Connection is intended to promote popular interest in science education. Since the inauguration of the Ceres Connection awards, more than 2,800 top-ranking students and their teachers have returned to the classroom knowing that excellence in science can result in a part of the solar system being officially named in their honor.

9.2   Serendipity in Space Surveillance: Resident Space Objects Detected from the Infrared Astronomy Satellite

This book has described techniques and technology developed at the laboratory for focused, directed deep-space surveillance and for space situational awareness in both near-Earth and deep-space regimes. Occasionally, serendipity played a part in these developments (see, for example, subsection 7.1.3 of chapter 7). Over the years, a large number of optical systems, both ground based and space based, have been developed primarily for astronomical studies. These systems have produced entirely serendipitous detections of resident space objects, which were either ignored or archived depending on the scientific objectives of the system developers. The discussion in this section pertains to the analysis of data from one such system by Robert Bergemann and E. M. Gaposchkin (both then at the laboratory) [35].

The European Infrared Astronomy Satellite (IRAS) was launched in 1983 to perform an all-sky survey in the infrared portion of the spectrum. Although stellar objects—stars, nebulae, comets, and asteroids—were the primary objects of interest, the IRAS also serendipitously detected resident space objects. The RSO data were archived at the IRAS Processing and Analysis Center (IPAC). A number of attempts have been made to retrieve these data and demonstrate that they could be used to characterize infrared emission of orbital debris. A. R. W. de Jonge, P. Wesselius, and R. M. van Hees [36] and Phillip Anz-Meador and colleagues [37] used unprocessed IRAS detector data to search for space debris. Kimberly Lynn Dow [38] used objects found in the Sky Brightness Images database over the entire 10-month IRAS mission. All studies concluded that the IRAS sensors had the capability to observe quite small space-debris objects—Anz-Meador claimed sizes down to 1 millimeter—but none of the studies' researchers systematically attempted to isolate data on known satellites. The Astronomical Group at Groningen, Netherlands, reprocessed all the raw IRAS data tapes to find all satellite and space-debris detections. This effort produced a space-debris database containing more than 190,000 detections, generously made available to the laboratory prior to the launch and deployment of the Space-Based Visible (SBV) space surveillance sensor (see chapter 5). Scientists at the laboratory analyzed this database to achieve the following two objectives, which were helpful in designing the processing system for the SBV sensor:

1. Early insight into the capability of a space-based optical sensor to detect resident space objects; and

2. Modeling the radiometric and metric data of RSOs. The fact that the radiometric data spanned multiple windows of the infrared spectrum was a distinct plus.

Because the RSO detection data were collected in the six months of IRAS operations in 1983, they were analyzed using the Space Surveillance Catalog of resident space objects from 1983, made available to the laboratory by the U.S. Naval Space Surveillance System analysts. The first step was to identify the objects that were detected using positional data. Once the identifications were confirmed, radiometric analysis of the data could proceed.

9.2.1   The Infrared Astronomy Satellite Instrument

The IRAS operated from March 1983 until November 1983 in a Sun-synchronous orbit. Its orbital plane (800 kilometers altitude, 99.2 degrees inclination) was nearly perpendicular to the Sun line, as illustrated in figure 9.10, and its telescope was always pointed away from the Earth center, that is, toward local zenith. The satellite's attitude control system allowed pole-to-pole scans in ecliptic longitude from 60 to 120 degrees, most of them between 84 and 96 degrees. The IRAS detectors covered four long-wavelength infrared bands with flux density reported at 12, 25, 60, and 100 microns.

9.2.2   Metric Data Analysis

Some calibration uncertainties were associated with the reported IRAS data made available to the laboratory. These have been recognized by other scientists; and various

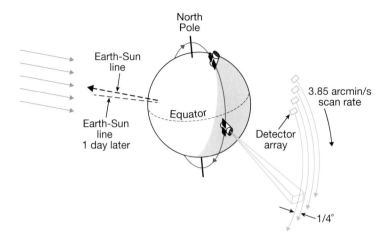

Figure 9.10   IRAS orbit geometry.

corrections or zero corrections have been advocated. Since the laboratory did not have all the primary resources, it decided to take the "zero corrections" path.

The next analysis issue was identification of a potential resident space object with a set of detections on the focal plane of the IRAS instrument [41, 42]. Given the altitude of the Infrared Astronomy Satellite above the Earth and its pointing away from the Earth, it was more likely to detect deep-space than near-Earth RSOs. The estimate was that a deep-space resident space object should move in the focal plane at between about 0.25 and 10 arc minutes per second and could take up to 10 seconds to cross the focal plane, crossing each detector at a different time. It could be identified by correlating the detection time and position, as well as the relative velocity as determined by the time history of the detections, to the satellite catalog of date; thus radiometric data could be associated with the RSO. The correlation process required knowledge of the inertial pointing of the boresite of the focal plane, the angular offset of the detectors showing a detection, and the ephemeris of the IRAS. The first two pieces of information were available with high precision (less than 1 arc minute) in the IPAC database, but a precise ephemeris was not available since it was not needed for the satellite's primary mission, observing stellar objects. But because correlation of the detection data to an RSO demanded accurate knowledge of the position of the platform, laboratory analysts calculated a precise ephemeris by retrieving all the tracking data of the IRAS from the archives of the Space Surveillance Network and processing them with the numerical orbit estimator

DYNAMO (see subsection 2.1.3.2 of chapter 2). In addition to accounting for the normal geopotential, the gravitational potential of the Sun and the Moon, the solar radiation pressure, the earthshine, and the atmospheric drag on the IRAS, determining the satellite's precise orbit also required modeling the low thrust caused by the constantly escaping cryogen.[8] The cryogen thrust was modeled as a constant along-track thrust. The area-to-mass ratios of the satellite were held constant for estimating the effects of the solar radiation pressure, the earthshine, and the atmospheric drag; the thrust resulting from the cryogen was a free parameter estimated daily in the orbit fit.

Although the average estimated error for the IRAS ephemeris for an arbitrary time during the mission was 40 meters, the maximum error was an order of magnitude larger. Readers may have noticed that the ephemeris error for the Space-Based Visible sensor on the Midcourse Surveillance Experiment (MSX) satellite (with sublimating solid hydrogen as cryogen) was kept to less than 10 meters (see chapter 5). The difference was primarily due to the following two factors:

1. Whereas accurate and frequent ranging and tracking data were required for the MSX satellite and available from the S-band ranging station of the Air Force Space Command and from the Millstone Hill and other radars, such data were neither required nor available for the Infrared Astronomy Satellite.

2. Complete knowledge of the physical structure and orientation of the MSX satellite throughout its mission, particularly with respect to atmospheric drag and cryogen venting, made modeling its orbit much easier than modeling the orbit of the IRAS, for which such knowledge was unavailable.

The results of processing the IRAS data, drawn from Lane, Baldassini, and Gaposchkin [39], are shown in tables 9.2 and 9.3. The conclusions were the following:

1. The Infrared Astronomy Satellite in a zenith-pointing stare mode detected at least four resident space objects per day in deep-space orbits.

2. Since the IRAS instrument was not meant for precision metrics, its data errors were rather large, although comparable to the errors of the GEODSS sensors of that era.

3. Clearly, other such astronomical instruments will also be able to detect deep-space RSOs and could provide a supplemental data stream for the Space Surveillance Network.

---

8. The focal plane of the IRAS instrument was cooled to 3–4° K with a liquid helium cryogen, the depletion of which determined the useful lifetime of the satellite.

Table 9.2   Results from all Infrared Astronomy Satellite tracks on known resident space objects

Total number of tracks	1,168
Number of days processed	277
Number of distinct satellites observed	399
Average number of tracks per day	4.2
Standard deviation of number of tracks per day	2.3
Average right ascension bias (over number of tracks)	–5.9 mdeg
Average declination bias (over number of tracks)	–0.6 mdeg
Standard deviation of right ascension biases (over number of tracks)	33.0 mdeg
Standard deviation of declination biases (over number of tracks)	27.6 mdeg
Average right ascension standard deviation (over number of tracks)	10.9 mdeg
Average declination standard deviation (over number of tracks)	7.0 mdeg
Standard deviation of right ascension standard deviations (over number of tracks)	13.1 mdeg
Standard deviation of declination standard deviations (over number of tracks)	12.2 mdeg

Table 9.3   Results from all Infrared Astronomy Satellite tracks on known geosynchronous resident space objects

Total number of geosynchronous tracks	137
Average right ascension bias (over number of tracks)	–4.7 mdeg
Average declination bias (over number of tracks)	–1.2 mdeg
Standard deviation of right ascension standard deviations (over number of tracks)	42.8 mdeg
Standard deviation of declination biases (over number of tracks)	26.4 mdeg
Average right ascension standard deviation (over number of tracks)	9.5 mdeg
Average declination standard deviation (over number of tracks)	3.6 mdeg
Standard deviation of right ascension standard deviations (over number of tracks)	11.4 mdeg
Standard deviation of declination standard deviations (over number of tracks)	7.0 mdeg

9.2.3   Radiometric Data Analysis

The objective of analyzing the IRAS radiometric data on detected and identified resident space objects was to derive information about infrared observations of artificial satellites [40]. The process was as follows:

1. Screen the individual detections for resident space object images (assuming point sources) that did not completely cross the detector since partial "hits" would give biased estimates;
2. Collate the detections for each RSO; and
3. Color-correct and analyze the flux-density measurements to derive temperature and emissivity and to obtain other physical properties of each RSO by using information such as range and physical size.

Figures 9.11 and 9.12 display some of the results of the radiometric analysis. The temperature of a detected RSO can be calculated by the ratios of the flux at 12 microns and the flux at 25 microns, and also by similarly using the flux ratio at 25 microns and 60 microns. The calculated temperatures should be similar, assuming that the emissivity does not change with wavelength (a reasonable assumption over this wavelength range for satellite materials). The temperatures calculated by using the two flux ratios before screening (figure 9.11) are clearly not similar. But, just as clearly, the temperatures calculated after screening was introduced (figure 9.12) are much better clustered at about 300° K, which is consistent with typical satellites in space.

The IRAS collected good-quality radiometric data on a number of spin-stabilized cylindrical satellites of various sizes in geosynchronous orbit. Table 9.4 shows the

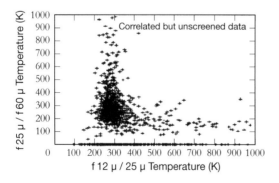

Figure 9.11   Resident space object temperatures from correlated but unscreened data.

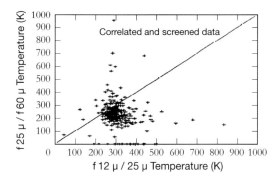

Figure 9.12   Resident space object temperatures from correlated and screened data.

Table 9.4   Infrared Astronomy Satellite detections of Hughes HS-376 cylinders

SID	Satellite name	12 microns Jansky(s)	#	25 microns Janskys	#	60 microns Jansky(s)	#	T12/25 °K	T25/60 °K
12065	SBS-1	1.86 ± 0.04	13	2.42 ± 0.11	12	1.20 ± 0.19	14	297	223
13069	Westar 4	1.59 ± 0.06	16	2.30 ± 0.08	16	1.01 ± 0.12	16	284	247
13269	Westar 5	1.78 ± 0.03	6	2.48 ± 0.22	5	1.02 ± 0.19	5	288	263
13651	SBS-3	2.03 ± 0.10	2	2.68 ± 0.10	2	1.25 ± 0.10	2	296	235
13652	Anik C3	1.87 ± 0.10	12	2.67 ± 0.11	12	1.00 ± 0.34	10	286	287
13431	Anik D1	1.83 ± 0.11	10	2.74 ± 0.08	10	1.14 ± 0.08	10	279	261
14158	Galaxy I	1.61 ± 0.19	12	2.39 ± 0.16	12	1.04 ± 0.39	12	280	250
14134	Palapa B-1	1.52 ± 0.10	2	2.64	1	0.98 ± 0.10	2	260	291

results of the analysis of these data; figure 9.13 shows the detected fluxes against time. Because these are spin-stabilized cylinders whose spin axis is perpendicular to the equator (and nearly perpendicular to the boresite of the IRAS), there is no temporal variation, thus demonstrating the stability of the IRAS detectors and constancy of their calibration.

The IRAS also collected radiometric data on a large number of rocket bodies in deep space, most of them in high-eccentricity orbits with apogees at near-geosynchronous altitudes and perigees at altitudes of a few hundred kilometers. Because the rockets carry cooled fuel, their bodies are generally covered in white paint

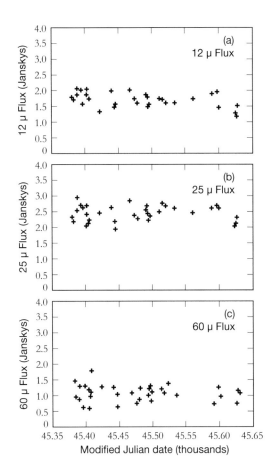

Figure 9.13 Detected flux on spin-stabilized cylindrical satellites in three infrared bands: (a) 12-micron flux; (b) 25-micron flux; (c) 60-micron flux.

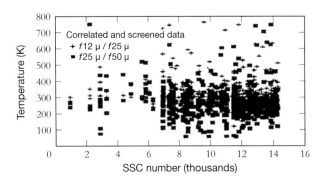

Figure 9.14 Plot of detected rocket body temperatures against Space Surveillance Catalog (SSC) numbers from correlated and screened data.

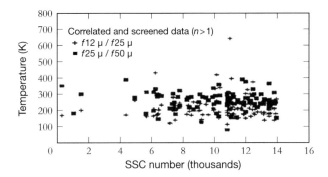

Figure 9.15 Detected rocket body temperatures with more than two detections from correlated and screened data.

(titanium dioxide) to prevent absorption of heat while they are on the launch pad. In orbit, rocket bodies tend to tumble every 10 to 30 or more seconds. A question arose as to whether the white paint would deteriorate with time, causing an increase in temperature in older rocket bodies. But the calculated temperatures plotted against the serial catalog numbers assigned the objects in the Air Force Space Command's Space Surveillance Catalog (figures 9.14 and 9.15) clearly show this not to be the case: there is no distinct trend in the temperatures of the rocket bodies launched over several years.

The IRAS also detected a few three-axis-stabilized satellites in geostationary orbits (see Gaposchkin and Bergemann [35] for more details).

In summary, the IRAS serendipitously collected a large number of stable, reasonably calibrated multispectral infrared data on resident space objects in the process of observing stellar objects. Luckily, these data were not dismissed as noise but were calibrated and archived and made available to the laboratory. They show that, given the physical parameters of the RSOs, a considerable amount of information can be derived on the temperature and other parameters that characterize the RSOs' infrared emissions. These infrared data on RSOs were particularly valuable because they were acquired from a space-based platform with no atmospheric windows or attenuation to be concerned with.

References

1. Scholl, H., and L. D. Schmadel. 2002. Discovery Circumstances of the First Near-Earth Asteroid (433) Eros. *Acta Historica Astronomiae* 15:210–220.

2. Leverington, D. 2012. *A History of Astronomy: From 1890 to the Present*, 107–108. New York: Springer Science & Business Media.

3. Hodgson, R. G. (June 1977). Long-Lost Planet 1936 CA ("Adonis") Recovered. *Minor Planet Bulletin* 4:35.

4. B.G. Marsden, "Minor Planet Electronic Circular 2003-T74: 1937 UB (Hermes)," Minor Planet Center, 15 October 2003.

5. Kuiper, G. P., Y. Fugita, A. M. J. Gehrels, M. J. Anton, I. Groeneveld, J. T. Kent, et al. 1958. Survey of Asteroids. *Astrophysical Journal* 3 (Supplement): 289.

6. van Houten, C. J., I. van Houten-Groeneveld, P. Herget, and A. M. J. Gehrels. 1970. The Palomar-Leiden Survey of Faint Minor Planets. *Astronomy & Astrophysics. Supplement Series* 2 (5): 339.

7. Helin, E. F., and E. M. Shoemaker. (December 1979). The Palomar Planet-Crossing Asteroid Survey, 1973–1978. *Icarus* 40 (3): 321–328.

8. Campbell, W. W. 1892. Discovery of Asteroids by Photography. *Publications of the Astronomical Society of the Pacific* 4 (26): 264.

9. Janesick, J., and T. Elliot. "History and Advancements of Large Area Scientific CCD Imagers," in *Astronomical CCD Observing and Reduction Techniques*, S.B. Howell, Ed., p. 345, Astronomical Society of the Pacific (ASP) Conference Series, vol. 23, San Francisco: Astronomical Society of the Pacific, 1992.

10. McMillan, R. S. "Spacewatch Survey of the Solar System," University of Arizona Lunar and Planetary Laboratory Technical Report FRS-305430, January 2000.

11. Pravdo, S. H., D. L. Rabinowitz, E. F. Helin, K. J. Lawrence, R. J. Bambery, C. C. Clark, et al. 1999. The Near-Earth Asteroid Tracking (NEAT) Program: An Automated System for Telescope Control, Wide-Field Imaging, and Object Detection. *Astronomical Journal* 117 (3): 1616–1633.

12. Stokes, G. H., J. B. Evans, H. E. M. Viggh, F. C. Shelly, and E. C. Pearce. 2000. Lincoln Near-Earth Asteroid Program (LINEAR). *Icarus* 148 (1): 21–28.

13. Larson, S. M., E. C. Beshore, R. Hill, E. Christensen, D. McLean, S. Kolar, et al. (May 2003). The CSS and SSS NEO Surveys. *Bulletin of the American Astronomical Society* 35:982.

14. Taff, L. G. "Asteroid Searches—1981," MIT Lincoln Laboratory Project Report ETS-61, 25 February 1982.

15. Beatty, D. E., J. M. Sorvari, and L. G. Taff. "Artificial Satellites, Minor Planets, and the ETS," MIT Lincoln Laboratory Project Report ETS-53, 1980.

16. Taff, L. G. 1981. A New Asteroid Observation and Search Technique. *Publications of the Astronomical Society of the Pacific* 93:658–660.

17. Tennyson, P. D., E. W. Rork, and D. F. Kostishack. "Applying Electro-Optical Space Surveillance Technology to the Detection of Near-Earth Asteroids," *Proceedings of SPIE*, vol. 2198, *Instrumentation in Astronomy VIII*, pp. 1286–1297, David L. Crawford and Eric R. Craine, Eds., June 1994.

18. York, D. G., J. Adelman, J. E. Anderson, S. F. Anderson, J. Annis, N. A. Bahcall, et al. (September 2000). The Sloan Digital Sky Survey: Technical Summary. *Astronomical Journal* 120 (3): 1579–1587.

19. Boehnhardt, H. (May 2001). The Death of a Comet and the Birth of Our Solar System. *Science* 292 (5520): 18.

20. Fukugita, M., T. Ichikawa, J. E. Gunn, M. Doi, K. Shimasaku, and D. P. Schneider. (April 1996). The Sloan Digital Sky Survey Photometric System. *Astronomical Journal* 111:1748.

21. Veyette, M., A. C. Becker, H. Bozic, P. Carroll, P. Champey, Z. Draper, et al. "The LINEAR Photometric Database: Time Domain Information for SDSS Objects," *AAS Meeting 219*, 348.19, January 2012.

22. B. Sesar, S. Stuart, Z, Ivezic, D.P. Morgan, A.C. Becker, and P. Wozniak, "Exploring the Variable Sky with LINEAR: 1. Photometric Recalibration with the Sloan Digital Sky Survey," *Astronomical Journal*, vol. 142, no. 6, article id. 190, 13 pp. December 2011.

23. Ruan, J. R., S. F. Anderson, C. L. MacLeod, A. C. Becker, T. H. Burnett, J. R. A. Davenport, et al. "Characterizing the Optical Variability of Bright Blazars: Variability-Based Selection of Fermi Active Galactic Nuclei," *Astrophysical Journal*, vol. 760, no. 1, article id. 51, 11 pp., 2012.

24. "Lincoln Laboratory Discovers Inner Earth Orbit Asteroid," MIT Lincoln Laboratory Press Release, https://www.ll.mit.edu/news/atira.html, February 2009.

25. Taylor, P. A., J. L. Margot, D. Vokrouhlicky, D. J. Scheeres, P. Pravec, S. C. Lowry, et al. 2007. Spin Rate of Asteroid (54509) 2000 PH5 Increasing Due to the YORP Effect. *Science* 316 (5822): 274–277.

26. Williams, G. V. "Minor Planet Electronic Circular 2000-C09: 2000 BD19," Minor Planet Center, http://www.minorplanetcenter.net/mpec/K00/K00C09.html, 3 February 2000.

27. Williams, G. V. "Minor Planet Electronic Circular 2000-C49: Revision to MPEC 2000-C09," Minor Planet Center, http://www.minorplanetcenter.net/iau/mpec/K00/K00C49.html, 12 February 2000.

28. D.W.E. Green, "International Astronomical Union Circular no. 913 (20461) Dioretsa," Central Bureau for Astronomical Telegrams, April 2010.

29. Steven, J. O., J. L. Margot, L. A. M. Benner, J. D. Giogini, D. J. Scheeres, E. G. Fahnestock, et al. 2006. Radar Imaging of Binary Near-Earth Asteroid (66391) 1999 KW4. *Science* 314 (5803): 1276–1280.

30. Margot, J. L., M. C. Nolan, L. A. M. Benner, S. J. Ostro, R. F. Jurgens, J. D. Giorgine, et al. 2002. Binary Asteroids in the Near-Earth Object Population. *Science* 296 (5572): 1445–1448.

31. Baker, J. 2006. The Falcon Has Landed. *Science* 312 (5578): 1327.

32. Krot, A. N. 2011. Bringing Part of an Asteroid Back Home. *Science* 333 (6046).

33. Stuart, J. S. 2001. A Near-Earth Asteroid Population Estimate from the LINEAR Survey. *Science* 294 (5547): 1691–1693.

34. Stuart, J. S., and R. P. Binzel. (August 2004). Bias-Corrected Population, Size Distribution, and Impact Hazard for the Near-Earth Objects. *Icarus* 170 (2): 295–311.

35. Gaposchkin, E. M., and R. J. Bergemann. "Infra-Red Detections of Satellites with IRAS," MIT Lincoln Laboratory Technical Report 1018, 26 September 1995.

36. de Jonge, A. R. W., P. Wesselius, and R. M. van Hees. 1992. Detecting Space Debris Using IRAS: Report on Work Package D. In *DOT ROD-DEB-92-14*. Groningen, Netherlands: Laboratory for Space Research.

37. Anz-Meador, P. D., D. M. Oro, D. J. Kessler, and D. E. Pitts. 1986. Analysis of IRAS Data for Orbital Debris. *Advances in Space Research* 6:139–144.

38. Dow, K. L. "Earth Orbiting Objects Observed by the Infrared Astronomical Satellite," M.S. thesis, University of Arizona, 1992.

39. Lane, M. T., J. F. Baldassini, and E. M. Gaposchkin. A Metric Analysis of IRAS Resident Space Object Detections," MIT Lincoln Laboratory Technical Report 1022, 1 November 1995.

40. Lane, M. T., J. F. Baldassini, and E. M. Gaposchkin. "The Calibration and Characterization of IRAS Metric Resident Space Object Detections," in *Proceedings of the AAS/AIAA Astrodynamics Specialist Conference*, pp. 1–14, 14–17, August 1995.

Afterword

A lasting legacy of innovative developments in the enterprise of space situational awareness has been built since the late 1960s at MIT Lincoln Laboratory in support of national security. The innovations would not have occurred but for the atmosphere that fostered creativity at the laboratory and the funding provided by the government. This book has attempted to capture the highlights of the first thirty years of that enterprise (with some extensions beyond). The editors hope that readers have enjoyed reading the text as much as the editors enjoyed participating in the work it describes.

Challenges in space situational awareness have metamorphosed from a two-sided competition to a multisided enterprise, both competitive and collaborative. The editors expect that the ethos of Lincoln Laboratory will continue to foster innovative solutions, to be reported in a future book.

Index

AFSSS, 207
Albedo, 281
ALCOR, 47–48, 75–78,82
ALTAIR, 13, 41–42, 208, 215, 221, 242, 245–246, 249, 316
A/M ratio, 269–270, 272, 282, 307
Analysis
  metric data, 352–355
  radiometric data, 356–359
ANODE, 33
Anomalous
  behavior, 243–244
  space debris, 263, 266–269, 272, 276–277, 279–280, 282
Anz-Meador, Phillip, 351–352
Arecibo, 42
ARPA, 28, 77, 208
ASC-1, 321–322
Asteroid, 346, 350
ASTRA, 257–258
Atmospheric
  extinction, 131–133
  refraction, 133
  seeing, 130–133
  transmittance, 138
Autofocus, 93–96

Baker-Nunn, 2, 6, 9, 12, 15, 30, 119, 209
Bandwidth extrapolation, 103–109
Bergemann, Robert, 5, 7, 19, 351, 359

BMEWS, 3
Bowling, Stephen, 105
Burg algorithm, 105, 108

Calibration systems
  metric, 35–41
  radiometric, 34–35
Camera, intensified image Ebsicon, 134–136
CAST, 5
Cataline Sky Survey, 332
Catalog, 203–234
  launch activity, 204–205
  maintenance, 232–234
  technology development, 203–204
CCD, 120–124, 163–172, 183–184, 194–195, 331, 333–335, 346
CCID, 334
CCID-16, 169–170, 334–335
Ceres Connection, 350–351
Cheyenne Mountain Complex, 12, 17
Clark, Arthur C., 299
Clark, Tom, 11
COBE, 225
COLA, 195–196
Collision threat, 307–308, 323–326
Comet, 345–346
Conical target, 110–114
Conjunctions and collisions, 295–326
  general problem, 296
  geosynchronous and geostationary orbits, 297

Cooley, James, 19
CRDA, 309–315, 318–321, 326

DARPA, 47–48
DAYSAT, 157–161
Debris detection, 43–47
Deep space search strategies, 212–223
Deep Stare, 319
DEFSMAC, 10–11
Delporte, Eugene, 331
Density estimation, 269–272
Difference beam, 60
Doppler
  imaging, 84–87
  resolution cell, 86
  shift, 53
  time-intensity plot, 84, 217
  tracking, 29–30
  walk, 86, 88–89, 99, 101
DSCS, 8, 127
DSP, 1, 7–9, 127, 214–218, 259
DYNAMO, 30, 33, 40, 270

Ebsicon, 121, 123, 133–136, 164–166
ECP, 87, 97–100
Electro-optical, 119
ETS, 45, 120–124,144–147, 152–153,
  264–265, 333–334
Extended range return, 88, 92

F-111, 82–83
FCC, 303
FEDS, 224–230, 233–234
Fengyun-1C, 71
FFRDC, 3
FFT, 16, 22–23, 30, 52
Firepond, 273
FLTSATCOM-V, 250–251, 255–256
Fobos-Grunt, 92–93
FPS-79, 13, 42

FPS-85, 2, 9, 17, 38, 40, 42, 208–209,
  224–225, 255
Freedom, 43–44, 70–71
Fragmentation events, 44
  catalog discovery, 223–229
  performance evaluation, 230–231

Galaxy IV, 308–309
Gaposhkin, Edward Michael, 40, 351,
  354, 359
GEO, 179
GEODSS, 120, 122, 154–155, 163, 168,
  187, 190, 213, 223, 257, 306,
  332–334
Geostationary orbit, 299–303, 307–308
Geosynchronous belt, 191–194,
  323
Geosynchronous orbit, 18–19, 130
Globus-2, 17
GLONASS, 187, 189, 206
GMWS, 308–326
Goldstone, 261
Gorizont, 220–221
GPS, 36, 206, 246–247
Guernsey and Slade, 20

Haystack, 3, 6–7, 10, 12, 16, 19, 22, 34,
  43–44, 47–50, 77, 255, 262–263,
  266, 273, 295
HAX, 43, 47, 68, 79–80, 295
Helin, Eleanor, 331–332
High eccentricity orbit, 18
Hubble Space Telescope, 346
HUSIR, 78

I and Q, 50
I-Ebsicon, 121–122, 136–137, 144, 146,
  153, 157, 163, 166
Image
  high-resolution, 100–101

linear, 86
  stroboscopic, 101
  three-dimensional, 102–103
  ultrawideband, 109
IMP-6, 42
IRAS, 329, 351–359
ISS, 261
ITU, 242, 257, 297, 306

Jodrell Bank Observatory, 4
JSpOC, 257

Keck Observatory, 346
Kessler, Don J., 271
KDS, 79
Kitt Peak, 130, 132
KOSMOS-520, 1–4, 8–10, 12
KOSMOS-606, 8, 10–11
KOSMOS-775, 149
Kuiper, Gerard, 331
Kwajalein, 13, 17, 41, 65, 75, 265

LAGEOS, 36–38, 40–41
Launch activity, 204–207
LCS-1, 34–35
LEO, 15, 179
LES-6, 7, 19–22
LINEAR, 329, 332, 334–350
LONEOS, 332
Lowell Observatory, 121
LRIR, 43–46, 48–49, 51–52, 57–60,
  68, 77–78, 80, 85–86, 101, 107,
  295
LSSC, 41, 215
LSSAC, 257

Measurements
  photometric, 274
  polarimetric, 275–276
Midori, 90–91

MIDYS, 32
Mie region, 46
Millstone Hill Radar, 1, 3–6, 10–13,
  17–33, 35–36, 38, 179, 210,
  218–220, 222, 238–243, 245,
  250–252, 255, 264, 266–267, 273,
  316
MIR, 91
MMW, 46, 77–79
Molniya, 9, 246, 248–249
Monopulse, 28, 59–64
Moon, 6–7, 180
Mount Palomar, 130, 132
Moving-target indication, 149–151
MPC, 331, 333, 337–338, 341,
  345
MRCS, 38, 40
MSX, 181, 186–187, 354
MTI, 148–150
MWIR, 124

NaK, 271, 283–289
Narrowband radar, 15–17, 239
NASA, 6–7, 16, 37, 41–44, 47, 51, 55,
  57–58, 79, 261–263, 295, 300, 329,
  333, 338, 350
NAVSPASUR, 207–208, 223
NEA, 329–332, 341, 349
NEO, 337–338, 345
NORAD, 3, 45, 161
NSSCC, 4

ODAS, 51, 56, 62
OPAL, 210–211
Operation Moonwatch, 4
Orbit
  control, 303–306
  estimation, 306–307
Orbital Debris Program Office, 16, 68,
  71, 204, 223, 295–296

PACS, 50–51
Palomar Planet-Crossing Asteroid Survey, 331
PANSKY, 124, 155
Paschal, Lee, 12
PAVE PAWS, 207–208
Pegasus, 225, 228
Phase
  angle, 180
  curve, 199–200
Photometer, 152–153, 168
Pinch points, 194, 196–198
Polarization ratio, 269
Processing, polarimetric, 26–28
Prognoz, 222

Quad camera, 166–168, 171

Radar calibration, space debris, 58–59
Raduga-18, 249
Range
  Doppler imaging, 84–87
  profiles, 81
  resolution cell, 86
  time-intensity plot, 81–83
  walk, 88, 99, 101
Raup, Richard, 25
RCA-A, 243–244
RCS, 15, 44, 55, 66, 266, 276
RCS_TOOLBOX, 211
Reagan Test Site, 75–76, 78, 316
Reinmuth, Karl, 331
RLOS, 81, 87
RORSAT, 261, 263, 271
RSO, 4–8, 15–20, 22–23, 25, 26, 28, 31–33, 126, 128, 130, 179, 190–191, 203–205, 208–209, 212, 237–238, 241, 260–262, 295–296, 298, 329, 351–359

SAGE, 3
Salyut, 44
Salyut-6, 161
SAMSO, 7, 48
SAR, 47
SATCIT, 11, 16, 27, 32, 34
Satellite
  acquisition and tracking, 161–162
  brightness from reflected sunlight, 124–130
  brightness in visual magnitudes, 139
  detection, 157–161
  DSCS, 127
  DSP, 127
  IDCSP, 125, 127
  motion determination, 96–97
  newly detected, 154–155
  orbital maneuvers, 155–157
  unknown, 155
SBSS, 186
SBV, 179–200
  enhancement of productivity, 194–198
  focal plane arrays, 183–184
  photometry, 198–200
  sensor, 181–184
  signal processor, 184–185
  space surveillance demonstration, 185–186
  surveillance of geosynchronous belt, 191–194
  visible metric and photometric processing, 186–189
  wide area search capability, 189–191
SCC, 12–13
SDSS, 343–344
SEM, 65
Sensor
  low-light-level television, 133–134
  implementation, 146–157

intensified-image Ebsicon, 136–137
  model, 137–143
Shoemaker, Eugene, 331
SIGINT, 2
Sky
  background, 139–140
  brightness, 130
SkyDOT, 345
Skylab, 44
SNR, 15, 20, 25, 27, 140–143, 146, 150, 238
SOI, 8
SOP, 7–9
Space debris
  data collection, 51–54
  detection and measurement, 55–56
Space Shuttle, 92, 321
Spacetrack, 2, 6
Spacewatch, 331–332
Specular, 87–89, 242, 245–246
SPIRIT III, 181
SPOCC, 186
Sputnik-1, 1, 4, 18, 43, 179, 237
Sputnik-2, 1
SRM, 58
SRP, 303
SSC, 53
SSN, 297
STARS, 33
Stokes, Grant, 350
Straddle factor, 138
Sum beam, 60
Swerling model, 55

TDAR, 119
Television sensor, 133–134
Telstar 401, 241–243, 308–311, 320, 326
TIRA, 80–81, 88, 90–93
TRADEX, 41, 76–77, 263, 265, 316

Tsiolkovsky, Konstantin, 299
Tukey, John, 19

UCT, 33, 147, 223–224, 226–227, 229–230, 233
Ultrawideband coherent processing, 109–114

Vega, 126, 128
Visual magnitude, 126, 128–130, 139

Wideband radar imaging, 75–116

XIPS, 320

Yerkes-MacDonald Survey, 331
YORP effect, 348

Printed in the United States
by Baker & Taylor Publisher Services